Chemistry of the Cell Interface

Contributors

Harry Darrow Brown

Swaraj K. Chattopadhyay

Julius W. Dieckert

Laylin K. James, Jr.

K. J. Laidler

P. V. Sundaram

H. T. Tien

John T. Van Bruggen

CHEMISTRY OF
THE CELL INTERFACE

Part A

Edited by

HARRY DARROW BROWN

CANCER RESEARCH CENTER
AND
THE UNIVERSITY OF MISSOURI
COLUMBIA, MISSOURI

1971

ACADEMIC PRESS   New York and London

CHEMISTRY

ACADEMIC PRESS, INC.
111 Fifth Avenue, New York, New York 10003

United Kingdom Edition published by
ACADEMIC PRESS, INC. (LONDON) LTD.
Berkeley Square House, London W1X 6BA

LIBRARY OF CONGRESS CATALOG CARD NUMBER: 76-162934

PRINTED IN THE UNITED STATES OF AMERICA

Contents

List of Contributors

Numbers in parentheses indicate the pages on which the authors' contribution begins.

HARRY DARROW BROWN (73), Cancer Research Center and University of Missouri, Columbia, Missouri

SWARAJ K. CHATTOPADHYAY (73), Cancer Research Center, Columbia, Missouri

JULIUS W. DIECKERT (33), Texas A & M University, College Station, Texas

LAYLIN K. JAMES, JR. (205), Lafayette College, Easton, Pennsylvania

K. J. LAIDLER (255), Department of Chemistry, University of Ottawa, Ottawa, Canada

P. V. SUNDARAM (255), Department of Chemistry, University of Ottawa, Ottawa, Canada

H. T. TIEN (205), Department of Biophysics, Michigan State University, East Lansing, Michigan

JOHN T. VAN BRUGGEN (1), University of Oregon School of Medicine, Portland, Oregon

Preface

Biochemistry has developed from the *fiction* that components of a biological cell, separated from each other as "pure" entities, can interact in a way that is illustrative of the chemistry *in vivo*. Despite *prima facie* improbability, this approach has been successful. Many studies, however, have drawn attention to the fact that structure contributes to the nature of the chemical events. Because reactions cannot proceed independently in the same place at the same time, there is an inherent requirement for the existence of compartmentation. Morphologists, using electron optics, have vividly defined subcellular structure. It is now clear that many (and perhaps *most*) of the reactions occurring in cells are not solution reactions, and *none* can be equated with those which occur in solution under "ideal conditions." A present view is that biological reactions generally are interfacial phenomena in which one or all of the reactants are spatially restricted.

The contributions to this two-volume work are involved with the relationship of structure to biochemical reactions. They approach in several ways the nature of the generalities which have grown out of the scientific literature. Too, the authors bring understandings of interface chemistry to bear upon conditions within the biological cell. Hence, "Chemistry of the Cell Interface" is a consideration of reactions involving the cell's structured elements and of interfacial reaction systems which are extrapolations from the conventional methodology of solution biochemistry.

This treatise is comprised of two volumes: Part A (Chapters I–V) and Part B (Chapters VI–VIII). Chapters I through III deal with components of complex subcellular systems. In Van Bruggen's Chemistry of the Membrane (Chapter I), the various models have been related to the source data, and the implication of the acceptance of one model over another has been made overt. Thus, while recognizing that biomembranes are chemically as well as functionally diverse, the discussion does much to unify the cell membrane literature. Dieckert, in Chapter II, describes cell particles and the concepts which have developed from their study, thereby further elaborating on interrelationships inherent in subcellular organization. Consideration of enzymes associated with the membranous organelles (Brown and Chattopadhyay, Chapter III) focuses upon catalytic activity as a function of structural ties.

An *in vitro* interface relationships model, that in which the lipid membrane provides the structured environment, has been considered by Tien and James (Chapter IV) as an aspect of lipid–lipid, lipid–protein interactions. Laidler and Sundaram (Chapter V) deal with the interpretations of chemical phenomena in systems having restricted degrees of freedom, with emphasis on the reaction model.

In Part B, water's contribution to the reaction systems is elucidated (Drost-Hansen, Chapter VI). Perhaps in this literature, more strongly than elsewhere, it is apparent that the biological cell is not the solution system of convention. Hence it is essential that the nature of the cell's aqueous phases be understood if we are to correctly interpret the chemistry. The two concluding chapters deal with modified proteins as model reactants. Matrix-supported enzymes, the technology of the model, and the properties of enzymes bound to polymeric matrices have been contrasted with solution and membrane particle systems (Brown and Hasselberger, Chapter VII). The final words are about aspects of protein chemistry pertinent to the design of such interface experimental systems (Habeeb, Chapter VIII).

I wish to express my appreciation for the clerical, editorial, bibliographic, and graphic arts support which has been provided by associates at the Cancer Research Center, and to acknowledge, in particular, the help of Helen Estes, Cynthia Cunningham, Florence Brown, Libby Forbis, Mary Dorward, Robert Hahn, and Yvonne Chapman.

H. D. BROWN

Contents of Part B

CHAPTER I

Chemistry of the Membrane

JOHN T. VAN BRUGGEN

I. Introduction

Few fields of biological endeavor are currently receiving the attention that is being given to the determination of the structure and function of biological membranes. Workers from practically every discipline, armed with a formidable array of exotic machinery, are simultaneously approaching the basic problem from many angles. It would be pleasant to be able to report that this massive front has resolved the problem of the structure and function of biological membranes. Not only has the problem not been resolved, but the controversies that have originated from the review of accumulating data have divided the field into a variety of camps, each representing its viewpoint with model systems that in many cases are the exact reverse or inverse of the model systems of a rival camp. Surely the battles that are raging in this field will not only yield a developing picture of the structure and function of biological membranes, but also will likely provide valuable spin-off in areas of biological activity previously not considered relevant to the problem of membrane structure. Subsequent chapters of this monograph will reveal the potential application of many of these findings to specific biological or physiological phenomena. This first chapter will not champion a single model system but will try to point out some of the chemical data on which the various models have been built and will attempt to show the implications of the acceptance of one model or another. Adjudication of the question of membrane structure and function must await further presentations by the proponents of the various models.

A word must be said about the problem of scientific advance versus the state of the technology of the field. In some areas of research, a single new instrument or technique sufficed to cause research in the area to advance rapidly. In the case of the chemistry and structure of biomembranes, multiple approaches have been brought to bear on the problem. The field of irreversible thermodynamics has contributed much to our knowledge of membrane function and has introduced its own precise terminology, but it cannot in itself create a model. Electron microscopy (EM) has come close to enabling us to "see" biomembranes but falls short of revealing aspects of molecular interaction that are a key to structure. X-Ray diffraction studies coupled with EM analysis have given further insight, particularly into molecular aggregates thought to reflect principles of membrane structure. The chromatographic analysis of complex lipids and their component, fatty acids, has revealed much about the lipid "building blocks" of membranes, just as has amino acid sequencing about membrane proteins. The application of optical rotatory dispersion and circular dichroism methods to the membrane protein research has begun to give a vision of the structure of proteins as they exist *in situ*.

To the reader who would be fully informed goes the task of interpretation of the many languages of biological research. Because of the excessively large literature now associated with "membranology," the chief literature citations will be review or summary articles, each of which is replete with relevant original citations.

II. Historical Background

The chemistry of the cell membrane is not an old subject. Some 40 years ago the structure, now known as the cell membrane, was demonstrated to be a finite component of the interface between the interior and the exterior of the cell. Experiments on the composition of this structural interface originated from a number of fields. It was apparent that the structure of nervous tissue required a special organization of molecular material on the surface of the nerve. It was also apparent that the solubility of certain dyes could be partially explained on the basis of a cell interface containing lipid. An early chemical fractionation of a cell interface came with the work of Gortner and Grendel (1925), who extracted red blood cell membranes and determined the ratio of the extracted lipid to the cell area. From these measurements they concluded that the cell membrane might well be a bimolecular leaflet of lipid. This work seemed to fit in well with the pioneering work of Langmuir (1927), who had described the molecular layering of lipids. Upon inspection and evaluation of the bilipid layer concept, Danielli and Davson (1935) found it necessary to introduce another chemical component, namely protein, to the red blood cell membrane structure. According to these authors, the cell membrane was pictured as having a lipid core, the interior of the core being the hydrophobic portion of the lipid and the exterior being the polar ends of the lipid coated with a layer of protein. Such a model seemed to answer the permeability and electrical requirements for some membranes. The paucimolecular theory of membrane structure by Danielli and Davson served as a starting point for many of the concepts of contemporary biology. Most biologists would concede that biological membranes are composed of protein and lipid, but beyond this point the agreement between investigators begins to disappear. It is important to state that at the present time no one knows the full details or the final structure of any biological membrane. The information regarding the composition and interaction of the components for a number of biological membranes will be given to set the stage for some of the functional considerations of the following chapters. Our choice of subject material follows both personal taste and the relative abundance of literature in the particular area.

The development of a model for a cell membrane is contingent upon the restrictions one places upon the functional nature of the membrane. If the membrane, as was originally conceived, is to be considered as a wall or barrier for the isolation of the cell contents and is, thus, to be some structure for the packaging of cellular functions, then the lipid and protein components and their lipid–protein interactions require one form of structure. However, if, in addition to the concept of the inert, restrictive barrier, there is added to the membrane structure a functional requirement such as control of the "traffic" across the membrane, cell adhesion, cell recognition, differentiation, and/or cell functional organization, then the model of the cell membrane becomes very complex. Now, as the model becomes complex it becomes more and more specific for the particular cell or membrane structure that is being described. The concept of the universality of membrane lipoprotein structure is a useful one but one that must be tempered with a consideration of the functional requirements of the membrane being described. The cell membrane may also be an expression of the genome of the cell, as stated by Salton (1967), "The evolution of the surface membrane possessing functions which are under specific gene control would have great survival value in nature and would undoubtedly account for the development and individuality of cell types in both unicellular micro-organisms and within structually complex animal and plant species."

III. Isolation of the Membrane

To describe the composition and eventually the function of the cell membrane, it is necessary that pure membrane preparations be at hand for the isolation and identification of their components. Ideally, the cell membrane to be studied in detail should begin as a preparation containing all of the natural components of the cell membrane uncontaminated with components of either interior or exterior phases of the cell. As is pointed out by Maddy (1966), there is great difficulty in determination of the components of the membrane because not only may the membrane have components not previously detected by histochemical or other type of study, but also some of the components of the cell may be only transiently located on the cell membrane. To isolate pure cell membranes, as difficult as it may be, three considerations must be met. First, there must be a technique for the disruption of the cell and the release and/or removal of components not related to the cell membrane. Second, there must be some demonstration that the material isolated is only that of the cell membrane. Third, it is necessary to prove that the isolated components of the cell membrane are pure. A variety of methods are available for disintegration of biological material: osmotic shock, sonic disintergration, homogenization, mincing, and/or differential extraction and centrifugation in a variety of hypo- or hypertonic solutions. Criteria applied to the purity of the membrane preparation depend upon the particular tissue being studied. For the red cell membrane, a popular membrane for study, the criterion of freedom from hemoglobin is used. For other membranes the absence from the preparations of morphologically or histochemically defined components is used. In the case of the red blood cell, it is not known for certain that the architecture of the cell is independent of hemoglobin and its possible binding with the cell membrane. For the attempted isolation of each particular membrane, the worker will need to establish his criteria of purity. Surely, from such an ill-defined set of ground rules there will arise an infinite number of membrane preparations of which the biochemical and functional activities will be stoutly defended by adherents of the particular approach. An excellent discussion of criteria for membrane isolation and characterization is given by Wallach (1967).

IV. Composition of Biomembranes

A few general comments about the chemical composition of membranes are in order. The lipid content of biological membranes can vary from 25 to 80% by weight with protein making up the balance of weight. In the case of some membranes, only a single lipid represents the larger part of the total lipid present. For example, in myelin, cholesterol makes up some 40% of the lipid weight, but in mitochondria, cholesterol is only present to about 5% and is not found at all in chloroplasts. In myelin, the sphingolipids are major components of the total lipid, but in red blood cells and mitochondria the sphingolipids are less prominent. The glycerol phosphatides make up the bulk of lipid in mitochondria, in myelin and red blood cells they represent about one-third of the total lipids, and in chloroplasts they are only minor components. In microsomes the chief lipids are glycerol phosphatides, and in bacterial membranes the phosphatides reach the ultimate in specialization of components, for in some species a single phosphatide, such as phosphatidylserine, alone is present. From the extremes cited for the lipid composition of biological membranes, it is clear that there exists in nature a broad spectrum of differences in the composition of membranes. Next, membrane structures and their detailed chemical analyses will be presented as examples of classes of biological membranes.

A. MEMBRANE LIPIDS

That lipid and protein make up the major components of animal and bacterial membranes is generally acknowledged. As shown in Table I, membrane may contain as much as 75% lipid, as in myelin. The difference between the total lipid cited and the total weight is assumed to be protein. This gross description of composition serves to set the stage for other data to be presented which suggest that, although lipid is ubiquitous in biomembranes, its content and composition varies greatly in different membrane structures. As is shown in Table I, some membranes are rich in cholesterol, and others are low. Of all the lipid classes, the glycerol phosphatides appear to be the chief component. Within this class of lipids, however, membranes show an amazing diversity. Whereas *Escherichia coli* has phosphatidylethanolamine as the sole lipid component, another microorganism has two classes of phospholipids. Bacterial membranes in general do not contain phosphatidylcholine. Animal membranes generally have a variety of phosphatides, but species structural specificity is still

TABLE I

MAJOR LIPIDS OF BIOMEMBRANES[a]

Lipid compound	% Total lipid					
	Myelin	Red blood cells	Mito-chon-drion	Micro-some	Es-cherichia coli	Bacillus mega-terium
Cholesterol	25	25	5	6	0	0
Phosphatidyl						
-ethanolamine	14	20	28	17	100	45
-serine	7	11	0	0	0	0
-choline	11	23	48	64	0	0
-inositol	0	2	8	11	0	0
-glycerol	0	0	1	2	0	45
Sphingomyelin	6	18	0	0	0	0
Cerebroside	21	0	0	0	0	0
Cardiolipin	0	0	11	0	0	0
Total lipid (% dry wt)	75	50	30	50	Low	18

[a] Taken in part from Korn (1966), O'Brien (1967), and Maddy (1966). Analytical data have been rounded to whole numbers.

present. The data reported in Table I are illustrative of the wide variations seen in lipid constitutents. It must also be remembered that lipid components, present in seemingly small amounts, cannot be excluded from the possibility of being biologically significant, particularly when the matter of nonhomogeneity of the membrane surface is considered.

Special attention has been focused on the lipids of certain membranes. The original papers of O'Brien and Sampson (1965; O'Brien, 1967) should be consulted for information on myelin lipids in health and disease. Although myelin has been extensively studied and is often used as a model for all membranes, its unique composition and structure put it in a class by itself. In the following tables some description will be given of the lipids of plasma and organelle membranes.

Table II describes the lipids of red blood cell membranes from six species of animals. Lipids make up from one-fourth to one-third of the weight of ghosts. This total lipid is about one-half phosphatides and one-fourth each neutral lipid and cholesterol. The wide range of values for lecithin is to be noted. The relationship of these values to red blood cell hemolysis will be described below.

Table III presents data taken from Ashworth and Green (1966) for rat membrane lipids. It is clear that there is almost an order of magnitude difference in the phospholipid or sterol content of the various membranes. Table IV, taken largely from Salton (1967), lists both lipid and protein contents for microorganisms.

TABLE II

LIPIDS OF RED BLOOD CELL GHOSTS[a]

Species	Total[b]	Total lipid (%)			Total phospholipid (%)[c]		
		Neutral	Cholesterol	Total	Lec	Ceph	Other
Sheep	30	28	27	60	1	36	63
Ox	28	33	32	56	9	32	61
Pig	31	27	22	53	29	35	36
Human	25	29	23	58	39	24	37
Rabbit	28	33	21	58	44	27	29
Rat	22	28	28	61	56	18	26

[a] Adapted from de Gier and van Deenan (1961) and van Deenan (1965).
[b] Percent lipid from lyophilized ghosts.
[c] Lec = lecithin, Ceph = cephalin, Other = sphingomyelin + lysophosphatides.

To this point we have seen not only that membranes of different species differ in lipid contents but also that the different membranes of any one species have a wide diversity of composition. The chemical and physical properties of many of the lipids shown in the tables are a reflection of their fatty acid components. Figure 1 characterizes the fatty acids of a number of membranes, and Table V presents some details of the fatty acid makeup of red blood cell lipids. As summarized by O'Brien (1967), myelin has predominantly saturated acids and red blood cells are rich in poly-unsaturated and medium chain acids. Both mitochondria and chloroplasts are rich in unsaturated fatty acids, and these latter organelles have special branched and cyclic fatty acids. Species variation is represented by a threefold difference in red blood cell acids such as palmitic, oleic, and arachidonic.

B. Membrane Proteins

1. Classification

For many years, the only membrane protein fraction to be isolated was that of red blood cells. A variety of lipid–carbohydrate–protein materials identified by names such as stromatin, stromine, and elinin have been recognized. Only within recent years has protein isolation seen material progress.

TABLE III

Lipids of Rat Cell Membranes[a]

Membrane	Phospholipid (mg/100 mg protein)	Sterol	
		Total phospholipid (mg/100 mg)	Moles per mole phospholipid
Red blood cell ghosts	16	45	0.89
Brain myelin	92	67	1.32
Intestinal microvilli	11	23	0.46
Liver plasma	42	13	0.26
Muscle sarcolemma	28	12	0.24
Mitochondrial			
brain	46	26	0.51
intestine	14	30	0.60
liver	28	6	0.11
muscle	11	8	0.15
Microsomal—liver			
fraction I	30	14	0.27
fraction II	80	6	0.13
fraction III	35	6	0.11

[a] Adapted from Ashworth and Green (1966).

It is now recognized that many membrane structures have associated with them two general kinds of proteins. On the one hand, a complex, partially soluble class of proteins is associated with the backbone or structure of the membrane itself. The term "structural protein" (SP) has been applied to this in the case of some membranes. This protein may have no functional responsibilities other than that of organization. On the other hand, associated with the membrane are many proteins with functional roles; but again, this is not to say that these latter proteins are not involved in the structure of the membrane. These functional proteins are, however, characterized by being more easily removed from the membrane and by possessing enzyme or enzymelike activity.

At this time at least three key kinds of proteins must be considered.

a. *Structural Protein.* This material is characterized by limited solubility in aqueous and organic solvents and by its high affinity for lipids. Structural protein has been isolated from membranes of mitochondria, microsomes, ascites cells, red blood cells, *Neurospora*, etc. The ubiquitous nature of SP implies a common role for these proteins, but little is known

TABLE IV

COMPOSITION OF MEMBRANES FROM GRAM-POSITIVE BACTERIA AND FROM MYCOPLASMAS[a]

Organism	Percent of				
	Protein	Total lipid	Hexose	RNA[b]	
Bacillus licheniformis	75	28	—	0.8	
Bacillus megaterium KM	75	23	0.2–1	—	
Bacillus megaterium M	67	18.5	5	1.3	
Bacillus megaterium	70–85	6–9	1.5	10–15	
Micrococcus lysodeikticus	53	28	16–19	—	
Sarcina lutea	57	23	—	5.4	
Staphylococcus aureus	41	22.5	1.7	2.4	(Small particle fraction)
Streptococcus faecalis	46	32	—	2.7	
Streptococci Group A	68	25	2.1	2.0	(Protoplast membranes)
	67	19	2.2	0.6	(Membrane particles)
Streptococcus pyogenes	68	15.3	1.7	—	
Mycoplasma laidlawii	47.3	36.0	6.0	2.8	
Mycoplasma bovigenitalium	59.2	37.3	6.8	2.0	

[a] Adapted from Salton (1967).
[b] Ribonucleic acid.

of their structure. These proteins may also have a functional component to their structural role, as will be discussed below.

b. *Metabolic Proteins.* (1) Enzymes of intermediary metabolism, e.g., enzymes of fatty acid biosynthesis and citric acid cycle; (2) components of mitochondrial electron transport systems of terminal oxidation and/or of chloroplast function.

c. *Transport Proteins.* Particularly in microbial systems, certain transport processes are closely linked to isolatable proteins that are present only in species having the specific transport processes.

This classification of membrane proteins may be useful for textual organizational purposes, but it is not suggested as a limiting or restrictive construction.

2. Nature of Membrane Proteins

a. *Structural Protein.* The history of structural protein is closely linked to the major developments regarding mitochondrial structure and function. In the initial isolation of mitochondrial particles and their separa-

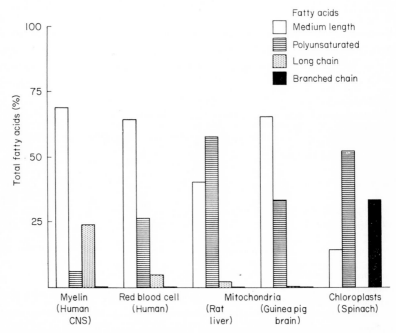

FIG. 1. Fatty acids of several membranes. The type of fatty acid is shown by a representative bar, and the proportion by the height of the bar; (CNS) central nervous system. (Modified from O'Brien, 1967.)

tion into discrete components, the oxidation–reduction proteins were easily and completely separated (Richardson *et al.*, 1963), but there remained a colorless, water-insoluble protein. This protein, SP, constitutes nearly half of the total protein of mitochondria. It was noted rather early that SP had an unusual ability to combine with phospholipid. This characteristic has been used as a criterion for the degree of "naturalness" of isolated SP. The combination of SP and phospholipids to form a water-insoluble complex occurs under conditions that simulate the physiological environment. The essentially hydrophobic nature of the lipid protein interaction has opened new vistas for the larger problem of membrane structure. Restoration of certain mitochondrial functions by recombination of redox components, phospholipid, and SP indicates that current research is coming close to an understanding of mitochondrial organization and function. Although the major effort on SP has come from studies on mitochondrial SP, many other interesting reports are appearing. For example, Woodward and Munkres (1966a,b) have shown that a respiratory-deficient mutant of *Neurospora* differs in a single amino acid replacement of its structural protein. The studies of Racker (1967) also indicate that SP may have a functional–structural role for a factor, F_4, similar to SP, which was shown to be required for coupling oxidation to respiration. Future studies on the structure and function of SP from a wide variety of membranes are anticipated to add greatly to our understanding of cell and organism function.

TABLE V

MAJOR FATTY ACIDS OF ERYTHROCYTES OF SEVERAL MAMMALIAN SPECIES[a,b]

Fatty acid	Rat	Man	Rabbit	Pig	Horse	Ox	Sheep
Palmitic	44	37	36	30	21	13	12
Stearic	22	15	11	14	14	14	7
Oleic	18	26	25	35	30	52	61
Linoleic	14	17	23	17	29	15	10
Linolenic	—	2	1	1	2	2	5
Arachidonic	17	8	2.5	2	—	0.5	0

[a] Values expressed in mole percent.
[b] Adapted from Maddy (1966).

b. *Metabolic Proteins.* Many enzymes associated with metabolic processes are known to be associated with membranes. Table VI lists a few of these for rat liver and red blood cells. The enzymes of the citric acid cycle and fatty acid biosynthesis are associated with mitochondrial membranes. Enzymes of protein and fatty acid biosynthesis are well-known components of microsomes. In bacterial systems, a variety of enzymes are associated with the mesosome, a structure closely related to the bacterial cell membrane but also somewhat analogous to mitochondrial structures. [See Salton (1967) for a description of other bacterial membrane-associated enzymes.]

Since the earliest identification and characterization of a set of monomeric proteins containing oxidation–reduction prosthetic groups, there has been much speculation about their physical arrangement in the mitochondrion. As suggested by Green and Perdue (1966a) and Green and MacLennan (1969), the active unit of electron transfer is not the monomeric protein but a series of at least four complexes each containing enzyme, phospholipid, and protein. The intergrated functioning of multienzyme complexes requires a physical organization, and such organization

TABLE VI

ENZYMES OF CELL PLASMA MEMBRANE ACTIVITY[a,b]

Enzyme	Rat liver plasma	Red blood cell
Adenosine triphosphatase	46.4	—
Na- and K-activated	11.7	1.6
5'-Nucleotidase	51.0	—
Glycerophosphatase		
alkaline	0	—
acid	0.4	—
p-Nitrophenylphosphatase		
alkaline	1.5	1.5
alkaline, K-activated	1.5	—
acid	5.8	2.5
Acetyl phosphatase	11.4	—
K-activated	10.8	—
Phosphodiesterase		
alkaline	3.6	—
acid	0.7	—
Ribonuclease		
alkaline	1.1	—
acid	0.27	—
Glucose-6-phosphatase	1.4	0.7
Esterase		
α-naphthyl laurate	0.75	—
α-naphthyl caprylate	34.0	—
Triose-3-P dehydrogenase	2.04	6.5
NADH$_2$–cyt. c reductase[c]	7.68	1.5
Adenosine diphosphatase	0–0.4	—
Inosine diphosphatase	30	—
NAD[d]–pyrophosphatase	5.7	—
NAD[d]–nucleosidase	0.5	—
Leucyl-β-naphthylamidase	3.7	—

[a] Data given in micromoles product per milligram protein per hour.
[b] Adapted from Dowben (1969) and Benedetti and Emmelot (1968).
[c] Reduced nicotinamide adenine dinucleotide–cytochrome c reductase.
[d] Nicotinamide adenine dinucleotide.

is accomplished through binding at membrane surfaces. The extensive literature available from the enormous research endeavors of a variety of workers must be considered for an appreciation of this field of study. For one of the pivotal areas of life, that of the release, storage, and transfer of metabolic energy, membrane support and organization of components are required.

c. Transport Proteins. Electrolytes and nonelectrolytes can be transported across membranes even in the face of electrical and chemical gradients. A variety of "black box" mechanisms have been devised to explain these transport processes. Although no one transport system is fully characterized or understood, some of the proposed models seem to fit the bulk of the data. One of the mechanisms proposed for the transport of solutes, such as SO_4^{2-}, leucine, galactose, in bacterial systems involves the participation of a protein in the transport process. Table VII lists characteristics

TABLE VII

PROPERTIES OF MEMBRANE TRANSPORT PROTEINS[a]

Organism	Substrate	Protein characteristics[b]		
		Mol wt ($\times 10^{-4}$)	f/f_0	K (mmole/ liter)
Salmonella typhimurium	SO_4^{2-}	3.2	1.3	0.03
Escherichia coli	β-Galactosides	3.1	—	—
Escherichia coli	Leucine	3.6	—	0.001
Escherichia coli	Leucine	3.6	1.28	0.002
Escherichia coli	Galactose	3.5	1.25	0.001
Escherichia coli	PEP	0.9	—	Covalent
Chick duodenum	Ca^{2+}	2.8	—	0.004
Beef brain	Na^+, K^+	67	—	—

[a] Adapted from Pardee (1968).

[b] The f/f_0 is the ratio of the frictional coefficient of the molecule to the frictional coefficient of a sphere of the same mass; K is the dissociation constant.

of some of the transport proteins that have been identified. Generally, these are small proteins of some 3×10^4 molecular weight showing a high substrate specificity and a dependence upon cellular energy. These proteins are considered to be a part of the cell membrane or to be intimately associated with it. Translocation of substrate is presumed to take place through a conformational change of the transport protein (Pardee, 1968; Kennedy, 1968).

C. OTHER MEMBRANE COMPONENTS

A wide variety of substances other than protein and lipid are associated with membranes. Some of these are peculiar to specific membranes and many are present in only small amounts. For example, red blood cell

ghosts contain glycolipids (globosides) with lactose, *N*-acetylgalactosa-
mine, and glucosylacylsphingosine. The glycolipids of human, sheep, goat,
rabbit, and pig are of the globoside type, whereas dog, horse, and cat
have sialic acid and little hexoseamine (Maddy, 1966). In bacteria, a
large number of special components are present (Salton, 1967). Those
materials that give to cells their special immunological or adhesive
properties are also found in or on the cell membrane.

V. Models of Membrane Architecture

Knowledge of the lipid and protein composition of biomembranes
should be useful for the construction of membrane models—just as a sup-

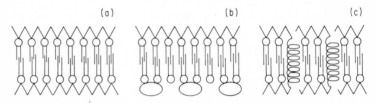

Fig. 2. Artist's concept of some membrane models based on the lipid bilayer concept. The center portion represents an area of hydrophilic interaction of the hydrocarbon components of the phosphatides. The outer layers represent interaction of
polar protein with polar head groups of the phosphatides.

plies inventory of lumber, bricks, etc. would be used by a building con-
tractor. However, the current plethora of membrane models attests to the
individuality and versatility of membrane model builders. These many
models will, however, serve their purpose, for progress in any field is
often catalyzed by the opportunity to criticize models submitted by
others. This truism is well substantiated by the state of the art of membrane building.

The earliest concept, that the red blood cell lipid might arrange itself in
a bilayer, seemed compatible with existing data on lipid content and the
behavior of lipid films. This concept was quickly followed by the models of Danielli and Davson (1935) who suggested that polar protein
might interact with the outer oriented polar lipid components to form
a protein, lipid-bilayer, protein sandwich (Fig. 2). This sandwich has
provided "soul food" for many researchers from a wide variety of
disciplines. The concept of a lamellar structure was advanced at about
the time two reasonably new research techniques for biological materials

became available, i.e., electron microscopy and X-ray diffraction analysis. These new techniques were applied to the complex problem of membrane ultrastructure. It is not at all surprising that multiple interpretations of EM and X-ray pictures have been advanced to the extent that rival camps champion widely differing concepts of membrane structure.

The principal controversy, however, revolves around the unit membrane concept proposed by Robertson (1959) which is in reality an extension of the paucimolecular model of Danielli and Davson. As pointed out by Korn (1966, 1968), the dogma of the unit membrane concept is that there is one basic structure to which most portions of all membranes of all species must conform. This structure consists of a bilayer of lipid, chiefly phospholipid, of which fatty acyl chains are oriented perpendicular to the plane of the membrane. The mutual solubility or van der Waals interaction of the hydrocarbon acyls leads to orientation of the polar ends of the phosphatides to the outside of the bilayer. Polar protein, generally considered to be in the β configuration, is considered to cover both sides of the lipid bilayer, but the type of protein on the two sides need not be identical. This model allows for a diversity in phosphatide composition and for the presence of other lipids, such as cholesterol, in the lipid bilayer.

The unit membrane concept was developed primarily from studies on myelin and largely from the interpretation of EM and X-ray diffraction of this tissue. From what is now known of the unique structure and the metabolic inertness of myelin, it would appear to be a poor prototype for all members.

Many are familiar with the EM representation of the unit membrane as a pair of dark lines separated by a clear space. Indeed, Elbers (1964) presents more than 50 different OsO_4 fixed preparations of cell membranes all showing the dark–light–dark conformation. To the proponents of the unit membrane theory, these representations are *prima facie* evidence for the unit bilayer.

For many, the unit membrane theory serves well to describe a diffusion-limiting barrier offering good electrical insulating properties. For the scientist concerned with the great diversity of membrane composition and function, the unit membrane concept is unnecessarily restrictive.

As pointed out by Sjöstrand (1968), EM study of cellular membranes shows two main types of structural organization. The first is illustrated with the plasma membrane, which appears to be triple-layered having a light middle layer between two opaque layers. The opaque layer on the cytoplasmic side appears somewhat thicker and stains more intensely than the outer layer. The total thickness is around 90 Å; the light layer is around 30 Å. Mitochondrial membrane elements are some 50–60 Å, and smooth-surfaced cytomembranes are some 60 to 70 Å in total thickness. The appearance of the plasma membrane in intestine columnar epithelium may change appreciably during fat absorption so that in descriptions of structure one must consider the functional aspects of the system. According to Sjöstrand (1968) the relatively simple geometrical pattern of the plasma membrane is not seen consistently in other membranes. Electron micrographs of cytoplasmic rough- and smooth-surfaced membranes, as well as of mitochondrial elements, show a much more complex pattern. These latter membranes under certain conditions show a two-dimensional array of globular structures. Globular, tubular, and also hexagonal structures are consistently seen by a number of workers.*

Electron-microscopic studies with lipid–water preparations reveal that a wide variety of "figures" may be seen in these nonbiological systems. The size, shape, and/or conformation of oil-in-water or water-in-oil micelles vary with temperature, ion environment, etc., suggesting that the organization of lipid and protein in biological membranes may assume a wide variety of three-dimensional configurations. These studies also make it necessary to interpret carefully EM records of biological membranes that have been subjected to a variety of physical and chemical manipulations. Many workers are aware of the circumstantial evidence available from EM records. Korn (1966, 1968) has been active in calling attention to the inadequacies of our interpretative skills. Electron microscopists are also aware of the problem and the imperfections or limitations of membrane EM studies.

The development of a concept of the true three-dimensional structure or architecture of cell membranes hinges upon many factors other than the problem of interpretation of EM pictures. A few such matters will be reviewed here.

VI. Roles for Structural Components

A. LIPIDS AND LIPID–LIPID INTERACTION

The idea that the central core of membranes is a hydrophobic array of hydrocarbon structures has long been of concern for those involved in the study of membrane permeability. Such permeability may be simple and related only to such matters as concentration gradients, diffusion

---

* The reader is referred to the many excellent published photographs and the accompanying commentary by EM specialists (see Glauert and Lucy, 1968; Benedetti and Emmelot, 1968; Sjöstrand, 1968; Fernández-Morán, 1962).

coefficients, and/or partition coefficients. For some membranes, it is clear that "lipid solubility" is a key determinant for permeation. Solubility in the lipid phase of the membrane serves to create a required gradient. Just how the hydrophobic interior of the cell would receive the lipid-soluble materials is another matter. Of great concern to many workers has been the problem of the rapid and selective movement of nonlipid-soluble materials across the membrane. To account for the traffic of metabolites through the membrane, a variety of phenomenological explanations have been offered. These can be separated into two classes. The first involves the postulation of a "pore" through the membrane. In its simplest form is the concept that tunnels or channels connect outer bathing solutions with the interior of the cell. Through these pores solutes may travel in or out of the cell. The rate of transmembrane movement of solutes is considered to be influenced by concentration gradients, pore size, bulk solvent flow, solute–pore interaction, and, as recently described (Franz et al., 1968), solute–solute interaction. The pore concept has been a useful one; in fact, it has been calculated that for the red blood cells the effective pore radii is 7 to 8 Å and that pores occupy some 0.06% of the surface (Solomon, 1960). This pore through the lipid phase of the unit membrane is somewhat easier to work with than to visualize. The second class involves a diphasic system. Instead of the membrane lipid being a solid construction, as it would appear to be in artistic two-dimensional presentations, it is more likely in a liquid–liquid or liquid–crystal state. Nuclear magnetic resonance studies reveal that the lipids have a considerable freedom of movement, with the polar groups being susceptible to attack by phospholipases (Lehninger, 1968). The presence of a pore in such a fluid structure is difficult to conceive of, and the behavior of transmembrane movement is subject to gross uncertainty. That an intermediate physical state of such a liquid–crystal system has dynamic implications for the biological system is clear. The possible transfer of ions, the interactions of light with matter, and changes in cell shape may relate to a liquid–crystal system (Fergasen and Brown, 1968). In spite of the uncertainties of the state of membrane core lipids, elaborate but highly speculative concepts of molecular placements have been advanced by Vandenheuval (1963, 1965). His papers, however, do contain a body of useful information regarding interatomic and intermolecular spacings which serves to illustrate the mechanism of the van der Waals interaction.

Another aspect of lipid composition that relates to membrane function is the matter of the effect of the physical interaction of membrane lipids. O'Brien (1967) has nicely summarized some of these interactions particularly as they relate to myelin lipid. In brief, the area oc-

cupied by lipid in film balance studies is determined by the nature of the
fatty acids present and by the influence of one lipid upon another. Long-
chain fatty acids form more compact films, and unsaturated fatty acids
form less compact films; the degree of contraction or expansion is related to chain length and the amount of unsaturation. Since many
membranes have their chief lipid phosphatides with both long-chain
and unsaturated fatty acids, the interplay of surface forces must be
considered. Cholesterol influences the area of monolayers, causing a compacting of the layer proportional to the amount of sterol present. The
effects of chain length, unsaturation, and cholesterol are illustrated in
Table VIII.

TABLE VIII

EFFECT OF LIPID COMPOSITION ON AREA OF LECITHIN MONOLAYERS[a,b]

Fatty acid composition			Lecithin ($Å^2$)	Lecithin with 1:1 cholesterol ($Å^2$)
Chain length	Double bonds			
C_{10}	0	Didecanoyl-	80	80
C_{14}	0	Dimyristoyl-	72	61
C_{18}	0	Disteroyl-	46	46
C_{18}	1	1-Stearoyl-a oleoyl-	75	64
C_{18}	2	Dioleoyl-	83	77
C_{18}	4	Dilinoleoyl-	99	99
		Egg lecithin	96	56

[a] Adapted from O'Brien (1967).
[b] At 12 dynes/cm, except egg lecithin which was 33 dynes/cm.

Note how the surface area decreases from 80 to 46 $Å^2$ in the progression
of C_{10} to C_{18} saturated fatty acids and then increases from 46 to 99
$Å^2$ in the series 0 to 4 double bonds. The compacting effect of cholesterol
can be seen particularly in the case of oleoyl lecithins and egg lecithin.
These measurements on model lipid systems can at best only approximate the summation of molecular effects in the complex milieu of biological membrane structure; for, in addition to lipid–lipid interaction,
protein is now known to play a determinate role in membrane organization.

B. PROTEIN–LIPID INTERACTION

The earliest models of the lipid bilayer–protein structure proposed by
Danielli showed the covering of enveloping layer of protein in the β or

extended configuration. This association was assumed to be the result of electrostatic interaction between polar protein groups and the polar "head groups" of phosphatides in the lipid bilayer. Such interaction of polar components is a well-known phenomenon with biological polyelectrolytes. The early models did recognize that protein configurations other than the β form might exist in membranes. Since globular proteins were possible components and since helical proteins might answer a requirement for pore structures, a variety of models have been advanced in an attempt to cover all the possibilities. A number of postulated structures are shown in Fig. 2. Each diagram is seen retaining the basic lipid bilayer configuration of the so-called unit membrane theory.

As techniques for lipid and protein isolation improved, it became obvious that protein and lipid could be separated under mild conditions (Wallach and Gordon, 1968; Green and Tzagoloff, 1966), and it is now popular to consider that hydrophobic bonding has a major role. Although the classic unit membrane theory allows for hydrophobic interaction of the nonpolar lipid residues, the present concept of interaction adds the hydrophobic components of protein structure to this environment. Evidence from the field of hemoglobin structure is cited by Wallach and Gordon (1968) to show how the tertiary structure of these compounds is determined to a large extent by hydrophobic interaction. The concept of hydrophobic interaction of protein and lipid, however, does not preclude a functional role for ionic sites on both of these structural materials. Micelles of phosphatides that have a net charge do form ionic complexes with charged peptides, such as protamines, lysozyme, and hemoglobin, some of which are stable in organic solvent. Indeed, electrostatic interaction of cytochrome c with mitochondrial protein is still an important concept favored by the research groups supporting the hydrophobic bonding theory (Green and Tzagoloff, 1966). In the red blood cell membrane, three degrees of "tightness of bonding" are recognized. In the final analysis, the structural form chosen by the combination of polar and apolar portions of lipids and proteins will depend upon the interaction of multiple forces (Salem, 1962; Maddy, 1966; Wallach and Gordon, 1968; Scheraga, 1966). The total "organizing forces" in a membrane will reflect the composition of the structure and the tendency of the components to attain the lowest free energy of the system. Among these the electrostatic forces of attraction or repulsion are rather long range, some 4 kcal/mole existing at a distance of 5 Å. The van der Waals forces are only 0.1 kcal/mole at 5 Å, but, as pointed out by Maddy (1966), this force between two stearyl residues totals some 8.4 kcal/mole at 5 Å. When one considers that membranes differ not only in their content of protein and lipid, but also in the types of protein and lipid present, it would appear unwise to attempt to settle upon final details of membrane structure at this time.

Since the concept of enveloping, β-conformation protein covering does not permit a high degree of hydrophobic interaction due to spatial separation of pairable forces, it is necessary to consider alternative structures.

The early concepts proposed for membrane structure were used to show the lipid core as the unifying component, but as infrared (IR), optical rotatory dispersion (ORD), and circular dichroism (CD) were applied to membrane proteins, a new role for protein appeared. It seemed that protein, rather than lipid, could express cellular individuality through genetically controlled mechanisms so that protein might then be the unifying principle. Knowledge of the amino acid composition and sequence of the proteins and clarification of their conformation became critical. This field of research is new, and much is yet to be learned about membrane proteins. It is recognized that interpretations of data on proteins suffer from a lack of precise meaning—just as is the case for X-ray and EM approaches. However, considerable progress is being made (Wallach and Gordon, 1968; Scheraga, 1966). Information on proteins of red blood cells, ascites cells, and some bacterial membranes is at hand. Apparently, the β conformation is essentially absent; some 20 to 50% of the protein is in the α-helix configuration, and the remainder in a random coil form (Wallach and Gordon, 1968; Korn, 1968). Estimates of helicity are not precise.

This knowledge of protein structure coupled with the necessity for hydrophobic lipoprotein interaction permitted Wallach and Gordon (1968) to draw certain tentative conclusions:

a. Protein is considered to be on the membrane surfaces as well as in the apolar portions.

b. The surface peptide is coiled irregularly, and the interior or penetrating protein is oriented normally to the surface and is likely to be α-helical.

c. The α-helical proteins tend to make up subunit assemblies, the outer surfaces of which are hydrophobic and considered to be the sites of interaction with apolar lipid.

d. The polar groups of the subunits are oriented centrally, possibly ending in the formation of aqueous channels.

e. The various subunits are assembled in membrane formation to form a protein lattice penetrated by cylinders of lipid.

These suggestions about membrane structure lead to the model proposed by Wallach (Fig. 3). It is recognized that this model is only one way of depicting present information. Seen in two dimensions, a so-called protein, lipid-bilayer, protein construction is present; the top view, however, is one of a nonhomogenous membrane surface. Wallach (1967) fractionated Ehrlich ascites cell membranes into diverse vesicle types that were distinguishable by their content of different complements of the functions of the intact cell surface. He, thus, suggested that the ascites cell surface is organized into a mosaic of large and functionally discrete

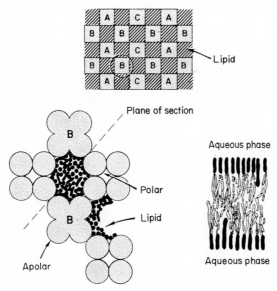

FIG. 3. Upper—surface view of membrane lipoprotein lattice with three types of interacting protein and penetrating lipid cylinders. Left—transverse section through apolar region of membrane showing five associated protein units, each consisting of four apposed helices. The external aspects of the proteins are hydrophobic and are shown in association with hydrocarbon chains of phosphatides (black circles). Cholesterol (black rods) is not in direct contact with protein. Four protein units are shown to form the walls of lipid cylinder. Two types of protein are shown: those with and those without a central aqueous channel. Right—sagittal section through lipid cylinder. The hydrocarbon chains are depicted in an irregular array, a consequence of hydrophobic interaction with membrane protein. (Used by permission of Wallach and Gordon as well as by Elsevier Publishing Co., New York.)

macromolecular units. A concept of mosaic lipoprotein is not incompatible with the suggestions of Benson (1964, 1966, 1968), who greatly expanded our ideas of membrane structure by suggesting that lipoproteins and phosphatides can associate, as shown in Fig. 4 (a and b). These suggested molecular aggregates for the chloroplast membrane may be useful concepts for other membrane structures. Using the model proposed by Benson, it is not difficult to conceive of specific areas of the membrane surface that could contain components essential for a particular membrane function. Benson's description of the membrane construction as a

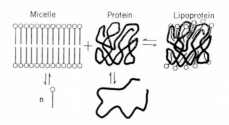

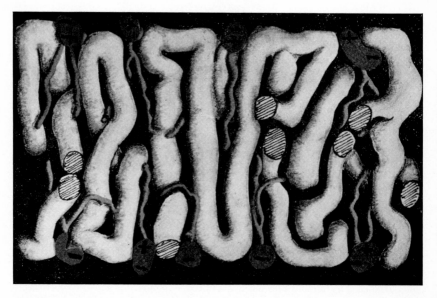

Fig. 4. (A) Equilibria involved in hydrophobic association of surfactant lipids and membrane protein. Vertical equilibria strongly favor aggregation of lipid into micellar structure and hydrated protein into globular protein. Horizontal equilibrium describes association of lipid with hydrophobic interior of membrane lipoprotein subunit. (From Benson, 1966.) (B) Molecular arrangement of lipids within a section of membrane lipoprotein. (From Benson, 1968.)

two-dimensional aggregate of lipoprotein subunits seems particularly appropriate.

VII. Membrane Structure versus Function

Although a case can be made for the unity of lipoprotein membrane composition, a strong case can also be made for the diversity of membrane structure. Such a diversity of composition in fact appears to be a requirement when the matter of membrane function is considered. A brief description of representative membrane structures will illustrate this point. The summary of protein–lipid and cholesterol–phospholipid compositions cited by Korn (1968), Table IX, serves well to illustrate the order of magnitude of variations in the basic structural components.

A. MYELIN

Extensive EM photographs and artists' conceptions have widely publicized the myelin sheath as an example of a biological membrane. These representations show a multilayered structure as a series of light and dark lines separated by spacings often characteristic of the method of

TABLE IX
PROTEIN AND LIPID OF PLASMA MEMBRANES[a]

Membrane	Protein–lipid (w/w)	Cholesterol (%)	Cholesterol–phospholipid (mole/mole)
Myelin	0.25	25	0.9
Red blood cell	2.5	25	0.9
Liver cell	1.5	14	0.4
Gram-positive bacteria	3	0	0
Halophile bacteria	2	0	0

[a] Adapted from Korn (1968).

cytological preparation. This particular structure, the myelin sheath, has been an important point of reference for proponents of the unit membrane (Robertson, 1966), but this concept has been severely criticized by others (see Korn, 1966, 1968; Green and MacLennan, 1969; Racker and Bruni, 1968). It appears that serious consideration must be given to the interpretation of stained and fixed preparations. If the lipid phase of the membrane is a liquid–crystal, micellar arrangement of which the forms will vary with pH, temperature, degree of hydration, and presence of critical cations, then it appears hazardous to attempt too precise a localization of individual structural components. Visualizations of potential molecular spacings such as those of O'Brien (1967), Vandenheuval (1963, 1965), and Finean (1953), have many shortcomings. However, the thought that myelin is made up of multiple layers of protein and lipid is in keeping with the concept that a likely function of these multiple layers is a supportive structural and insulative one. The myelin layers would appear to have a low permeability and lasting structural integrity.

As was stated earlier, the preponderance of long-chain fatty acyls together with the compacting effect of cholesterol leads to "tight" films with
low ionic conductions. These fatty acyls are also of low metabolic activity. O'Brien (1967) points out that the glycerolphosphatides of myelin
have a turnover time of 1 to 7 months, whereas sphingomyelin and
cerebrosides turnover times extend from 10 to 13 months. These are in
contrast to the fast turnover of red blood cells and mitochondrial lipids.
The myelin structure then appears to represent a compact, reasonably
metabolically inert structure. It may be that for this structure, the simple
bilayer configuration fairly well describes the membrane architecture.

So as not to oversimplify the concept of myelin structure and stable
function, it must be stated that even this structure is subject to biological
modification. O'Brien (1967) points out that in Refsum's disease, a de-
myelinating disease, there is an accumulation of a branched fatty acid,
phytanic acid, and a lack of ability to degrade this acid. Branched chain
acids, just as polyunsaturated acids, cause less compact lipid structures.

B. ERYTHROCYTES

Washed red blood cell ghosts have served many workers as model
membrane structures. Early in these studies it became apparent that the
nonnucleated red blood cell with its high content of hemoglobin and its
rather specialized metabolic systems was in many ways a unique source
material. Table X shows the great variation of red blood cell composition
in different species, e.g.: K^+, 18 to 104 meq/liter of cells; Na^+, 10 to 84
meq/liter of cells; diphosphoglycerate, 0.7 to 45.0 mg/liter of cells. The
K^+ and Na^+ contents of the two kinds of sheep red blood cells are to be
particularly noted. As shown in Tables I, II, and V, and in Fig. 5, the
phosphatide and fatty acyl content of red blood cell ghosts show wide
variation. That the lipid content of the membrane may be related to red
blood cell composition is likely. Figure 6 shows a relationship between the
lecithin content and hemolysis in glycol, whereas Fig. 5 shows hemolysis
times in relation to the amounts of specific fatty acid classes. The rat
red blood cells with short hemolysis times have a high content of polyunsaturated fatty acids but a low content of sphingomyelin. The ox red
blood cells with their prolonged hemolysis times have lesser amounts of
polyunsaturated fatty acids and much sphingomyelin. The rat red blood
cells have a K^+/Na^+ ratio of about 9, whereas this ratio for the ox red
blood cells is only 0.25. Certainly, at this time, it is not possible to go
much beyond the present comments on structure versus function. The
point to be made is that function is related to structure. It may be that

the architecture of the red blood cell membrane corresponds to a more open type of lipoprotein complex so that the interdigitation of lipid and protein, as suggested by Wallach and Gordon (1968), may best describe

TABLE X

COMPOSITION AND PROPERTIES OF ERYTHROCYTES[a]

Species	Potassium[b]	Sodium[b]	Diphosphoglycerate[c]	Hemolysis (sec)		Glucose permeability
				Glycol[d]	Glycerol[d]	
Rat	100	12	34	4	18	250
Rabbit	99	16	45	5	32	100
Man	104	10	38	4	51	—
Pig	100	11	45	16	132	10
Ox	22	79	<0.7	27	2169	0
Sheep, hi K+ lo K+	64	16	<0.8	25	1397	0
	18	84				

[a] Adapted from van Deenan and de Gier (1964).
[b] Expressed as milliequivalents per liter cells.
[c] Expressed as milligrams per liter cells.
[d] Isotonic solutions.

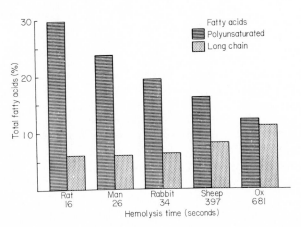

FIG. 5. Histogram showing relationship between the fatty acids of erythrocytes and their hemolysis time in 0.3 M glycerol. (Modified from O'Brien, 1967.)

red blood cell membranes. This kind of structure would be compatible with the relative ease of equilibration of red blood cell lipid with the lipid of plasma.

C. MITOCHONDRION

The membranes of the mitochondrion illustrate still another type of lipoprotein system, one which differs in many respects from myelin and red blood cell ghosts. The mitochondrial structure is unique not only because of its key position in energy transduction but also because it appears to be a coordinated assembly of two membrane systems.

The mitochondrion is pictured as a closed tubular structure less than 1 μ wide and some 3 to 10 μ long. It may appear granular or filamentous depending upon its tissue source and the functional state of the tissue. Microscopy reveals that the mitochondrion is made up of two membranes and two compartments. The outer membrane is reasonably smooth or

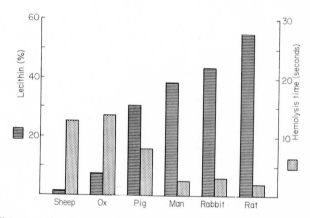

FIG. 6. Histogram showing relationship between the lecithin content of erythrocytes and their hemolysis time in glycol. (Modified from de Gier and van Deenan, 1961.)

orderly and is some 60 to 70 Å thick. The inner membrane is highly invaginated toward the center, the infoldings being called cristae. The outer compartment is the space between the outer membrane and the inner membrane, including the space of the infoldings of the cristae. The lumen of the inner membrane system seems to be continuous and is filled with a semiorganized fluid material (Lehninger, 1965a*).

The outer membrane is made up of lipoprotein units which, in the minds of some, indicates that the unit bilayer concept holds for this

* The vast literature on mitochondrial structure and structure versus function cannot be properly covered here. See also the following representative articles: Green and MacLennan (1969), De Robertis et al. (1965), Racker (1967, 1968), Lehninger (1965b), Green and Perdue (1966a), Brierley and Green (1965), Bulos and Racker (1968), Reynafarje and Lehninger (1969), Crane et al. (1968), Parsons et al. (1966), and Fleischer et al. (1962, 1967).

structure. Others see, microscopically, discrete units of structure which would indicate that the outer membrane was made up of units or aggregates having distinct structural and functional roles. It has been claimed (Green and Perdue, 1966b) that enzymes of citric acid cycle oxidations, fatty acid oxidations, amino acid oxidation, and even fatty acid elongation are localized in the outer membrane. These metabolic systems are suggested as being partitioned between the two sectors of repeating units envisioned by Green and associates. It is possible that certain of these enzymes are actually to be found in the intramitochondrial space or matrix.

The outer membrane is said to be freely permeable to solutes of small size but not to dextran of 60 to 90 $\times$ 10^3 mol wt, and the inner membrane is stated to be a perfect osmometer (Brierly and Green, 1965). It is to be hoped that techniques will soon be adequate to describe the transmembrane movement of solutes in terms of contemporary parameters such as permeability coefficients, reflection coefficients, and coefficients of hydraulic conductivity.

If the metabolic enzymes of biosynthesis and catabolism were functional components of the outer membrane, then it would seem reasonable to picture the membrane structure as a mosaic of enzyme complexes possibly united by "structural protein" and cooperating phosphatides. There is not, however, universal agreement regarding the location of these metabolic enzymes cited above. As stated earlier (see Wallach, 1967), it is most difficult to know beyond a reasonable doubt that isolated enzyme activities are or are not naturally present in a single phase such as the outer membrane or the mitochondrial matrix.

Discussion of the structure and function of the inner mitochondrial membrane cannot be dealt with adequately here because of the controversy involved. It is well know that many components of the electron transport system of biological oxidation exist in macromolecular configurations associated with the inner membrane. To some, EM techniques reveal this inner membrane as a highly organized structure characterized by three components. Of these, aggregates of globular repeating units are suggested to be the inner membrane itself. Stalks are pictured as attached to these base pieces, and knobs or headpieces are attached to the stalks. The unit of three components has been described by the Wisconsin group, Fernández-Morán et al. (1964) and others, as a tripartate repeating unit. The summary comment of Green and MacLennan (1969) may be consulted for a review of this field. Many workers have been active in the attempted elucidation of the structure and function of the inner membrane but not all agree with the concept of the tripartate unit (Racker and Bruni, 1968). Again we are faced with the problem of the interpretation of microscopic preparations and the correlation of this uncertain evidence with enzyme characterization after a variety of disruptive procedures. Most workers agree that special space orientations of electron transfer components are required, and it is not difficult to envision the membrane as packages of these units. The concept of stalk and knobs, however, remains a matter of controversy.

Mitochondrial membrane structure studies bear the unique distinction of showing that lipid–protein interaction and its associated hydrophobic bonding play a significant role in biology.

The basic structural role of protein was shown by Fleischer et al. (1962, 1967). It was also shown that most catalytic activity of the outer membrane does not have a lipid requirement, but that there is an absolute lipid requirement for inner membrane function. As Green and Tzagoloff (1966) point out, lipid-free complexes of the electron transport chain exist as highly polymerized, water-insoluble particles. When lipid is reintroduced into such material, membrane formation and biological activity are restored. We see then that the building blocks of protein and lipid cited in the preceding tables describe not only structural units but potentially functional units as well.

VIII. Conclusions

It is easy to conclude that biological membranes are composed of lipids, proteins, and other special components. The proportions of these building blocks vary greatly in different membranes. Wide qualitative differences are also seen in the lipid classes and in the constituent fatty acids of these lipids. The fact that qualitative and quantitative differences in composition have profound functional influences is also apparent.

Biomembranes are characterized by their chemical diversity as well as their functional diversity. At the one extreme, stable, inert myelin seems hardly a true member of the family of membranes. At another extreme are dynamic membranes engaged in metabolic functions including synthesis and degradation of cell components, and the active transport of materials in and out of the cell. Processes, such as these, seem certainly to require the catalytic participation of protein structures and possibly associated conformational changes in these units. That the chief lipid in many membranes is phospholipid is not coincidental. These amphipathic structures serve well as units of structure for the membrane that must face on both sides a generally dilute, aqueous environment. In this environment the membrane must serve as a limiting barrier for the purpose of isolation, concentration, and organization of cellular functions. The phosphatides aid in carrying out these functions, for they can provide, with polar protein, areas of high charge density as well as potential hydrophilic spaces. However, the phosphatidyl acyls can, through hydrophobic interaction, provide structural strength and regions of a milieu less aqueous in nature.

Still another aspect of phosphatide participation must be considered, for membranes are usually rather fluid stuctures capable of changing shape as in ameboid movement, pinocytosis, and red blood cell hemodynamic deformation. The ability of the phospholipids to exist in a variety of interchangeable micellar forms is a needed requirement for the lipid–crystal systems of dynamic membranes.

To this amazing diversity of composition and shape, the requirement of diverse function must be added. Membrane systems differ greatly in their metabolic activity and in the nature of the transmembrane transport processes. With all this diversity it is natural for scholars to seek out some unifying concept, particularly regarding structure. At the present time opinions on membrane structure are represented by two extreme views which bracket a large number of intermediate concepts.

The concept that the protein, lipid-bilayer, protein sandwich in cell membranes is universally present seems unnecessarily restrictive; the other extreme, namely, that practically all membranes are made of tripartate units is unacceptable to many workers and appears to be unnecessarily highly structured.

Other concepts of cell membranes are intermediate between these extreme views. To this writer the concept of a lipoprotein monolayer, as stated by Benson (1968), comes close to describing the dynamic interplay of molecular forces that characterize the cell interface.

It is possible then that there is no restrictive, underlying unit, architectural pattern for all biomembranes, and that membranes of one system or phase differ from each other as is dictated by the requirements of the mebrane's function. Some pattern of similarity may develop here, for plasma membranes, organelle membranes, and supportive or restrictive membranes may show common characteristics within these classes. Just so, within these classes of membranes, the common denominators of structure, namely, lipid and protein, may assume micellar forms and undergo electrostatic and hydrophobic interaction compatible with their functional needs. The idealized or model union of protein and lipid components as a rigid form seems only to limit the vision of those desiring to understand membrane function. The concept of a nonhomogenous, dynamically active, metabolically functional array of interdigitated molecules seems a pleasant working concept. To this basic picture there are easily added greater or lesser degrees of organization of individual membrane components to provide the membrane surface or core with a structure compatible with certain functions.

It is acknowledged that the role of special molecular species, such as sialic acid and glycoproteins, and of special chemical groups, such as —SH and ionic forms on or in the membrane, have been superficially mentioned. This is both by intent and through ignorance of their importance. With the variable and fluid membrane concept, discussed above, it is a simple matter to envision points of attachment for particular molecules dear to the hearts of particular investigators.

Acknowledgments

The author is grateful for partial support by the U. S. Atomic Energy Commission under research contract AT (45-1) 1754. The patient assistance of Dr. Marie Tichá, Mrs. Jean C. Scott, Mr. Barry Trowbridge, and Miss Priscilla Burk is gratefully acknowledged. Gratitude is also expressed to Drs. Green, Benson, Lehninger, Wallach, O'Brien, Racker, and others for their supplies of relevant reprints.

REFERENCES

Ashworth, L. A. E., and Green, C. (1966). *Science* **151,** 210.
Benedetti, E. L., and Emmelot, P. (1968). *In* "The Membranes" (A. J. Dalton and F. Haguenau, eds.), pp. 33–120. Academic Press, New York.
Benson, A. A. (1964). *Annu. Rev. Plant Physiol.* **15,** 1.
Benson, A. A. (1966). *J. Amer. Oil Chem. Soc.* **43,** 265.
Benson, A. A. (1968). *In* "Membrane Models and the Formation of Biological Membranes" (L. Boles and B. A. Pethica, eds.), pp. 190–202. North-Holland Publ., Amsterdam.
Brierley, G., and Green, D. E. (1965). *Proc. Nat. Acad. Sci. U.S.* **53,** 73.
Bulos, B., and Racker, E. (1968). *J. Biol. Chem.* **243,** 3901.
Crane, F. L., Stiles, J. W., Prezbindowski, K. S., Ruzicka, F. J., and Sun, F. F. (1968). *In* "Regulatory Functions of Biological Membranes" (J. Jarnefelt, ed.), pp. 21–56. Elsevier, Amsterdam.
Danielli, J. F., and Davson, H. A. (1935). *J. Cell. Comp. Physiol.* **5,** 495.
de Gier, J., and van Deenen, L. L. M. (1961). *Biochim. Biophys. Acta* **49,** 286.
De Robertis, E. D. P., Nowinski, W. W., and Saez, F. A. (1965). "Cell Biology." Saunders, Philadelphia, Pennsylvania.
Dowben, M. (1969). *In* "Biological Membranes" (M. Dowben, ed.), pp. 1–38. Little, Brown, Boston, Massachusetts.
Elbers, P. F. (1964). *Recent Progr. Surface Sci.* **2,** 443–503.
Fergason, J. L., and Brown, G. L. (1968). *Symp. Amer. Oil Chemists' Soc.* pp. 120–127.
Fernández-Morán, H. (1962). *In* "Ultrastructure and Metabolism of the Nervous

System" (S. R. Korey, A. Pope, and E. Robins, eds.), Vol. 40. pp. 235–267. Williams & Wilkins, Baltimore, Maryland.

Fernández-Morán, H., Oda, T., Blair, P. V., and Green, D. E. (1964). *J. Cell Biol.* **22**, 63.

Finean, J. B. (1953). *Int. Conf. Biochem. Prob. Lipids [Proc.], 1953*, pp. 82–91.

Fleischer, S., Brierly, G., Klouwen, H., and Slautterback, D. B. (1962). *J. Biol. Chem.* **237**, 3264.

Fleischer, S., Fleischer, B., and Stoeckenius, W. (1967). *J. Cell Biol.* **32**, 193.

Franz, T. J., Galey, W. R., and Van Bruggen, J. T. (1968). *J. Gen. Physiol.* **51**, 1.

Glauert, A. M., and Lucy, J. A. (1968). *In* "The Membranes" (A. J. Dalton and F. Haguenau, eds.), pp. 1–32. Academic Press, New York.

Gortner, E., and Grendel, F. (1925). *J. Exp. Med.* **41**, 439.

Green, D. E., and MacLennan, D. H. (1969). *BioScience* **19**, 213.

Green, D. E., and Perdue, J. F. (1966a). *Proc. Nat. Acad. Sci. U.S.* **55**, 1295.

Green, D. E., and Perdue, J. F. (1966b). *Ann. N.Y. Acad. Sci.* **137**, 667.

Green, D. E., and Tzagoloff, A. (1966). *J. Lipid Res.* **7**, 587.

Kennedy, E. P. (1968). *7th Int. Cong. Biochem.*, 1967, p. 5.

Korn, E. D. (1966). *Science* **153**, 1491.

Korn, E. D. (1968). *Proc. Symp., N.Y. Heart Ass., 1968*, p. 257.

Langmuir, I. (1927). *J. Phys. Chem.* **31**, 1719.

Lehninger, A. L. (1965a). "The Mitochondrion." Benjamin, New York.

Lehninger, A. L. (1965b). "Bioenergetics." Benjamin, New York.

Lehninger, A. L. (1968). *Proc. Nat. Acad. Sci. U.S.* **60**, 1049.

Maddy, A. H. (1966). *Int. Rev. Cytol.* **20**, 1.

O'Brien, J. S. (1967). *J. Theor. Biol.* **15**, 307.

O'Brien, J. S., and Sampson, E. L. (1965). *J. Lipid Res.* **6**, 537.

Pardee, A. B. (1968). *Science* **162**, 632.

Parsons, D. F., Williams, G. R., and Chance, B. (1966). *Ann. N.Y. Acad. Sci.* **137**, 643.

Racker, E. (1967). *Fed. Proc., Fed. Amer. Soc. Exp. Biol.* **26**, 1335.

Racker, E. (1968). *Sci. Amer.* **218**, 32.

Racker, E., and Bruni, A. (1968). *In* "Membrane Models and the Formation of Biological Membranes" (L. Boles and B. A. Pethica, eds.), pp. 138–148. North-Holland Publ., Amsterdam.

Reynafarje, B., and Lehninger, A. L. (1969). *J. Biol. Chem.* **244**, 584.

Richardson, S. H., Hulton, H. O., and Green, D. E. (1963). *Proc. Nat. Acad. Sci. U.S.* **50**, 821–827.

Robertson, J. D. (1959). *Biochem. Soc. Symp.* **16**, 3–43.

Robertson, J. D. (1966). *Symp. Int. Soc. Cell Biol.* **5**, 1.

Salem, L. (1962). *Can. J. Physiol. Biochem.* **40**, 1287.

Salton, M. R. J. (1967). *Annu. Rev. Microbiol.* **21**, 417.

Scheraga, H. A. (1966). *In* "Molecular Architecture in Cell Physiology" (T. Hayashi and A. G. Szent-Györgyi, eds.), pp. 39–61. Prentice-Hall, Englewood Cliffs, New Jersey.

Sjöstrand, F. S. (1968). *In* "The Membranes" (A. J. Dalton and F. Haguenau, eds.), pp. 151–210. Academic Press, New York.

Solomon, A. K. (1960). *Sci. Amer.* **203**, 146.

van Deenen, L. L. M. (1965). *In* "The Chemistry of Fats and Other Lipids" (R. T. Holman, ed.), Vol. 8, Part 1, pp. 1–127. Pergamon Press, Oxford.

van Deenen, L. L. M., and de Gier, J. (1964). *In* "The Red Blood Cell" (C. Bishop and D. M. Surgenor, eds.), pp. 243–307. Academic Press, New York.

Vandenheuval, F. A. (1963). *J. Amer. Oil Chem. Soc.* **40,** 455.
Vandenheuval, F. A. (1965). *J. Amer. Oil Chem. Soc.* **42,** 481.
Wallach,, D. F. H. (1967). *In* "The Specificity of Cell Surfaces" (B. D. Davis and L. Warren, eds.), pp. 129–161. Prentice-Hall, Englewood Cliffs, New Jersey.
Wallach, D. F. H., and Gordon, A. (1968). *Fed. Proc. Amer. Soc. Exp. Biol.* **27,** 1263.
Woodward, D. O., and Munkres, K. D. (1966a). *Proc. Nat. Acad. Sci. U.S.* **55,** 872.
Woodward, D. O., and Munkres, K. D. (1966b). *Proc. Nat. Acad. Sci. U.S.* **55,** 1217.

CHAPTER II

Cell Particles

JULIUS W. DIECKERT

I. Introduction

The cells of plants and animals contain a complex system of membrane-
bound compartments, each having a characteristic morphology, complement of macromolecules, and function. Some subcellular compartments
are considered to be cell organelles, e.g., the nucleus, mitochondria, endoplasmic reticulum, Golgi bodies (dictyosomes), lysosomes, peroxisomes,
and glyoxysomes. The nucleus contains the genetic material, deoxyribonucleic acid, and functions in the replication of the genome and in the
transmission of genetic information to the cytoplasm (Mirsky and Osawa,
1961). Mitochondria contain the enzyme systems for the oxidation of fats
and carbohydrates and act as energy transducers in the production of
adenosine triphosphate (ATP) (Novikoff, 1961a). Chloroplasts also function as energy transducers with light photons as the source of energy
(Granick, 1961). The endoplasmic reticulum and Golgi apparatus take
part in the biosynthesis, sequestration, and secretion of a variety of
macromolecules, including digestive enzymes and hormones (Porter,

33

1961). Lysosomes, peroxisomes, and glyoxysomes are membrane-bound vesicles of the cytoplasm which may qualify as cell organelles. Lysosomes contain hydrolytic enzymes and are said to function in digestive, auto-lytic, and necrotic processes (de Duve and Wattiaux, 1966). The func-tion of peroxisomes seems in doubt (de Duve, 1969), but the related glyoxysomes take part in the conversion of fat to carbohydrate in ger-minating oil seeds (Breidenbach et al., 1968).

The cell organelles are often highly organized structures morphologi-cally and biochemically. They are certainly not morphologically simple inert bags containing a solution of randomly arranged enzymes and sub-strates. In mitochondria, for example, there is an outer and an inner cyto-membrane (Novikoff, 1961a). The inner membrane is highly invaginated into the inner compartment and is studded with numerous stalked pro-trusions on its inner surface (Fernández-Morán, 1962b; Parsons, 1963). Among other proteins the inner membrane contains the complex array of enzymes of the electron transport system (Novikoff, 1961a). Chloroplasts, likewise, exhibit an intricate array of membranes (Granick, 1961). The chloroplast is bounded by a double membrane and contains stacks of thy-lakoids forming the grana and single thylakoids penetrating the stroma of the chloroplast. The membranes of the thylakoids contain chlorophyll, the enzymes for photophosphorylation and certain other photosynthetic proc-esses. The rough endoplasmic reticulum can be used to illustrate the same point. The polysomes that coat the outer surface of the rough endoplasmic reticulum are active components in the synthesis of proteins, and the mem-brane itself contains a system of oxidative enzymes (Dallner and Ernster, 1968; Porter, 1961).

The cellular organelles are not generally static systems. Instead, their structural and catalytic components are in a continual short-term state of flux. A recent study by Swick et al. (1968) showed that the turnover rates of six protein fractions of rat liver mitochondria were rapid, with half-lives ranging from 4 to 6 days. Moreover, not all mitochondrial pro-teins were found to turn over at the same rate. Alanine and ornithine amino transferase, two inducible mitochondrial enzymes, exhibited half-lives of 0.73 and 1.03 days. A similar result was noted for the membrane of the endoplasmic reticulum of rat hepatocytes (Omura et al., 1967). The total proteins of the rough endoplasmic reticulum and the smooth endoplasmic reticulum had similar half-lives of 75 to 113 hours. The half-life of re-duced nicotinamide adenine dinucleotide phosphate (NADPH)–cyto-chrome c reductase was almost exactly that of the total membrane pro-teins, whereas that of cytochrome b_5 was about 50% longer. The half-lives of the proteins of the endoplasmic reticulum are short compared to the half-life of 160 to 400 days for the hepatocytes, as reported by Rotman and Spiegelman (1954), Swick *et al.* (1956), and MacDonald (1961), implying that the protein component of the cytomembranes is replaced many times during the life of a cell.

The study of the origins of the proteins of the cell organelles is in its infancy; but, already there is evidence that cytoplasmic and organelle polyribosomes contribute to the protein complement of chloroplasts and mitochondria. These bodies (Kirk, 1966; Kislev *et al.*, 1966; Moustacchi and Williamson, 1966) contain deoxyribonucleic acid that is distinct from that of the nucleus; and, therefore, each of these organelles presumably contains a unique genome. The deoxyribonucleic acid of the chloroplast is transcribed into ribonucleic acid in the chloroplast (Kirk, 1966). Also, the chloroplasts (Ohad *et al.*, 1967; Loening, 1968) and the mitochondria (Kroon *et al.*, 1967; Linnane *et al.*, 1967) contain ribosomes that are distinct from the cytoplasmic ribosomes outside the organelles. Presumably, the mitochondrial and chloroplastic ribosomes are engaged in the synthesis of some of the proteins of the corresponding organelle. Recent studies (Hoober *et al.*, 1969) on the effects of antibiotic inhibitors on protein synthesis and the formation of chloroplast membranes in the mutant unicellular green alga, *Chlamydomonas reinhardi y-l*, indicate that products of both the cytoplasmic and the chloroplastic protein-synthesizing systems are required for disc membrane and disc enzyme production in this alga. A somewhat similar conclusion has been reached concerning the origins of mitochondrial proteins (Henson *et al.*, 1968a,b). The process by which the large-scale transport of macromolecules across double membranes occurs is not known. The cooperation between the nuclear and chloroplastic (mitochondrial) genomes opens up interesting possibilities for studying nuclear–cytoplasmic interactions and control.

The ultrastructural analysis of cells by the technique of electron microscopy shows clearly that the structural characteristics of the cell organelles vary considerably from tissue to tissue. These structural variations may often be related to differences in the metabolic activity of the cells. An interesting illustration of this is presented by Fawcett (1966) for mitochondria. The complexity of the infoldings of the inner membrane of mitochondria usually increases with the energy requirements of the cells in which they are found. For example, there are relatively few cristae in mitochondria from hepatic cells of a salamander emerging from dormancy. However, the mitochondria from bat liver cells contain numerous thin, platelike cristae of varying length that are usually short with respect to the width of the mitochondrion. And in the case of very active skeletal muscle mitochondria, the cristae are numerous, long, and packed in close parallel array. More recently, an effect of oxygen on the structure and enzymic composition of mitochondria in the yeast *Saccharomyces cerevisiae* was reported (Criddle and Schatz, 1969; Paltauf and Schatz, 1969; Plattner and Schatz, 1969).

In the preceding survey the cell is depicted as a system of dynamic, highly integrated, and variable cell organelles. This theory is developing rapidly as a consequence of the interplay between modern biochemistry, molecular biology, and structural cell biology. The development of high-resolution biological electron microscopy and the methodology for preparing large quantities of native, highly purified subcellular particles have made feasible the study of relationships among the ultrastructure, chemical composition, and function of the cell organelles. The use of antibiotics as inhibitors of specific processes and selected genetic mutants, coupled with the above technologies, have already advanced our knowledge of the relationships among cell organelles and should be valuable in working out the structural relations among the macromolecules composing the membranes of the organelles.

In this chapter, cell fractionation methodology and biological electron microscopy are discussed. Case histories are given which illustrate the interplay of these disciplines and biochemistry in the solution of important problems in cell biology. No attempt will be made to summarize all of the information on the structure, chemistry, and function of the principal subcellular particles. The references cited, although not exhaustive, include enough recent reviews, books, and original work to give the reader access to the literature. It is hoped that the discussion here will stimulate students of diverse scientific disciplines to contribute new ideas and original research to this fascinating field.

II. Methods

It seems impractical to assign a definite place in a linear hierarchy of credit to each method that has contributed to our knowledge of the cell organelles. However, there is wide agreement that biological electron microscopy and cell fractionation methodology would be high on such a list. The desirability of large-scale fractionation procedures is made clear from the following considerations. Characterization of most natural products requires at least a few milligrams and, more often, grams of material; but, very little matter is present in most cells. The dry weight of an average cell weighing 1 ng is 100 to 200 pg. Of this, at most, only a few picograms of dry matter are associated with a particular compound in a specific cell organelle. Even if procedures were available for conveniently isolating individual organelles in a native condition from a living cell, the quantity of material would seriously limit the work of the biochemist. It might be argued that methods could be devised for fixing cell components so that cytochemical tests could be applied and the result examined by light or electron microscopy. Cytochemistry of this type is actively being developed; but, in view of the complexity and lability of the enzymes and other macromolecules of the cell, it seems doubtful that such an approach will be sufficient to discover the location and function of all of these molecules in the cell.

With the availability of methods for the large-scale preparation of highly purified subcellular particles, the composition, function, and structural arrangements of the macromolecules of subcellular particles can be determined by the highly productive procedures of biochemistry and physical chemistry. Furthermore, cell fractionation procedures based on the biochemical properties of unique cell fractions can lead and have led to the discovery of new cell organelles (e.g., lysosomes and glyoxysomes) before the organelles were recognized as such by the more conventional methods of cytology.

Much of the value of biological electron microscopy depends on the high resolution (now about 1 Å) attainable with modern transmission electron microscopes, on the availability of procedures for the faithful preservation of ultrastructure and the preparation of sections for electron-microscopic examination, and on the realization that ultrastructure and function are closely related. In view of their Protean nature, it is interesting that the ultrastructural features of an organelle observed in fixed intact cells are often conserved in isolation schemes which successfully preserve certain functional properties of the organelle. Also, recent developments in cytochemistry (Silverman and Glick, 1969a,b; Silverman et al., 1969), high-resolution radioautography, and stereology (Weibel et al., 1966, 1969; Staubli et al., 1969; Loud, 1968) demonstrate that electron microscopy can be used to describe quantitatively dynamic processes taking place in a cell. The introduction of negative staining procedures (Brenner and Horne, 1959) into biological electron microscopy has already advanced the study of the structure of cytomembranes (Parsons, 1963; Fernández-Morán, 1962b) and their subunits.

The joint application of cell fractionation techniques and biological electron microscopy has proven to be a powerful approach to many problems in cell biology. In the next two sections the methods and principles of cell fractionation and biological electron microscopy will be summarized. Later in the chapter three case histories will be discussed which show how these fields have been combined with biochemistry to clarify important problems in cell biology.

A. Cell Fractionation

The principles of cell fractionation were reviewed recently by de Duve (1964); and earlier by Anderson (1956), Hogeboom and Schneider (1955), and Allfrey (1959). In his brief introduction to the history of the field, Allfrey credits the systematic isolation procedures introduced by Behrens (1932), the isolation of mitochondria by Bensley and Hoerr (1934), and the preparation of chloroplasts by Hill (1937) and Granick (1938) as inaugurating an era in which cell fractionation techniques were explicitly used to study the biochemistry, function, and morphology of subcellular particles.

In general, cell fractionation procedures are designed for analytical or preparative purposes. In the former, quantitative estimation is given more importance than cytological purity, whereas in the latter, cytological purity is paramount. Ideally, a fractionation scheme should satisfy both the purity and quantitative conditions; but, so far, this goal has been elusive. The cell fractionation procedure for the analysis of cell structure is based on separation and analysis of particulates obtained from tissue brei. At present, the most successful procedures for fractionating tissue brei are centrifugal methods. The two basic principles of particle separation by centrifugation are differential centrifugation and isopycnic separation. In the former, advantage is taken of differences in density, size, or shape between populations of subcellular particles which result in a differential sedimentation of particle populations in a centrifugal field. Isopycnic separation results solely from density differences. In the simplest experimental arrangements for differential centrifugation, the field strength or the time of centrifugation of a brei is increased stepwise, and fractions are collected at the bottom of the centrifuge tube as a pellet after each step. In the classic four-step fractionation of a liver brei in sucrose solution (Hogeboom and Schneider, 1955) a low field strength of 700 g is used to sediment a "nuclear fraction"; a field strength of 5000 g is then applied to sediment a "mitochondrial" fraction; and finally, at 57,000 g, a "microsome" fraction sediments. A series of "washing" steps for cleaning up the precipitate were included in the four-step scheme that is described here. The fraction that fails to sediment is the "cell sap" fraction. None of these fractions represent populations of a single cell organelle; therefore, the cytological connotation of the fraction names may be misleading. In the above procedure the suspending medium is constant. In many cases, further fractionation can be attained by modifying the density of the medium through which the subcellular particles pass. This method is called "gradient differential centrifugation." The density changes can be made discontinuous by layering techniques (Allfrey,

1959) or continuous with gradient-forming devices (Anderson, 1956). After a suitable time in a centrifugal field, samples for analysis can be taken from top to bottom at various points along the gradient.

In isopycnic separations the particle suspension, in a medium with the density of the particle being purified, is centrifuged until all the lighter particles have collected at the top and all the denser particles at the bottom of the centrifuge tube. In isopycnic gradient centrifugation, a layer of particles is sedimented through a density gradient until each species finds its isopycnic level. Then, fractions are taken along the gradient for analysis. This is a type of zonal centrifugation. Zonal velocity or equilibrium centrifugation in density gradients generally yields the highest resolution of all of the centrifugation procedures. When used in conjunction with swinging-bucket rotors, the capacity is ordinarily low, because of the requirement for a thin zone of a dilute suspension of particles for good separation.

In order to appreciate more fully the revolutionary new procedures in zonal ultracentrifugation, it may be helpful to examine briefly some of the centrifuge artifacts that cause trouble. Since most centrifuge cells are not sector-shaped and since the sedimenting particles move outward along the radius of rotation, particles of all sizes collide with the wall and agglutinate. The agglutinated particles may stick to the wall or slide to the bottom of the cell. The result is a mixing of cell fractions. A second factor centers on temperature control in the tube and thermal convection. Thermal convection results from density and viscosity changes that occur regionally within the sample as a consequence of temperature changes in the medium. Mixture of components may take place when the medium is swirled, due to Coriolis forces associated with acceleration and deceleration. Hydrodynamic mixing occurs when a number of particles and the liquid suspension medium move as a unit. Hydrodynamic effects result in streaming of the unfractionated brei through a denser underlayer, even without centrifugation. With centrifugation the process is enhanced. Aerodynamic mixing is caused by windstream effects and can be corrected by closing the centrifuge tubes. A more complete description of centrifuge artifacts appears in Anderson's review (1956).

As stated earlier, the zonal velocity or equilibrium separation of particles in liquid density gradients generally offers the greatest resolution of all the centrifugation techniques. The amount of sample and the volume of the gradient is limited in these techniques by the size restrictions of the swinging-bucket rotor. Other technical difficulties, in addition to those described earlier as centrifuge artifacts, are related to handling time, temperature control, and mixing during fraction collection. Norman G. Anderson and his collaborators at the Biology Division of the Oak Ridge

National Laboratory are developing a series of zonal centrifuges under the MAN program which are the preparative counterparts of the analytical ultracentrifuge and which will cover a range of particle sizes from whole cells to protein molecules. The general principles of zonal centrifugation are presented in two papers (Anderson and Burger, 1962; Anderson, 1962a), and the design principles for and results obtained with several of the experimental versions of the ultracentrifuge are described in the literature (Anderson et al., 1964b, 1967; Anderson and Burger, 1962; Anderson, 1963; Baggiolini et al., 1969).

The capacity and speed limitations of swinging-bucket rotors, the acceleration of large-volume gradients without mixing, and the recovery of the gradient undisturbed at the end of the experiment are major practical problems in ultracentrifugation (Anderson et al., 1964a). Anderson et al. have developed solutions to these problems as follows. The high-speed zonal centrifugation is conducted inside the largest pressure vessel that can be rotated at a given speed in a refrigerated vacuum chamber. For ideal sedimentation, a hollow rotor with an axial core and septa that divide the internal space vertically into compartments having a sector shape is provided. A system of ports is provided for loading and unloading the rotor while it is spinning, so that a density gradient can be rapidly developed and stabilized in the chambers of the rotor; the sample can be applied in a stable thin band at a desired point in the gradient; the operating speed desired for the separation can be achieved quickly without stopping the rotor; and, the gradient can be recovered rapidly and continuously without losing the stabilizing effect of the spinning rotor.

In the process of developing zonal centrifuges for the mass separation of subcellular particles, viruses, and macromolecules, a family of zonal centrifuges was developed under the MAN program. As of 1967 (Anderson et al., 1967), 40 different rotor designs had been constructed. For convenience these have been assigned letter designations to indicate their speed range (A for low speed; B for intermediate speeds up to 60,000 rpm; C for speeds up to 160,000 rpm; and D for ultrahigh-speed rotors operating up to about 400,000 rpm) and a Roman numeral designation to indicate the order in which the rotors were developed. As of 1967, two of these rotors (B-XIV and B-XV) are available from commercial sources for use with unmodified, commercially available centrifuges (Anderson et al., 1967). A brief description of the design and operation of these rotors will be given here to demonstrate the power of the technique. A summary of the characteristics of the B-XIV and B-XV zonal rotors and a detailed description of the rotor, its operation, and performance are given by Anderson et al. (1967) and abstracted in the next paragraph.

The rotor (B-XIV and B-XV) consists of two semihemispherical halves held together by buttress threads and sealed with an O-ring. "The internal volume of the rotor is divided into four sector-shaped compartments by four septa which are integral with the core." A coaxial seal allows fluid to be pumped through the core to the rotor edge via a channel in each septum or to the inmost surface of the core by way of a core channel. "The seal is attached to the rotor during loading and unloading at low speed (about 3000 rpm)" and is removed and replaced with a small cap for high-speed operation. In the operation of the rotor (B-XIV and B-XV), the gradient is introduced light end first via the rotor edge channels with the rotor spinning at low speed. Relatively simple gradient-forming devices such as the one described by Anderson et al. (1962) can be used to produce simple or complex density gradients in a predictable manner for use in zonal centrifugation. The density gradient is followed by a dense layer which is added until the light end of the density gradient is displaced from the rotor via the core channel, thereby assuring that the centrifuge chambers are full. After the gradient and dense layer are in place, the sample is pumped to the top (low-density end) of the gradient as a thin band via the core channel. The equivalent volume of the dense layer leaves by way of the rotor edge channel. Finally, a layer of suspending medium lighter than the sample medium is introduced through the core channel to move the sample zone away from the core faces. The coaxial core seal is removed from the rotor and replaced by a cap. During all of the operations the rotor is spinning at a slow speed. The loaded rotor is now operated at the speed and time required to perform the separation. At the end of the run the rotor is decelerated to a low speed, the core cap is replaced with the coaxial core seal, and the density gradient is collected. This is done by funneling the gradient out through the core channel while pumping more of the dense underlay medium into the rotor through the rotor edge channel. The effluent from the core channel can be analyzed continuously by systems such as a UV monitor and collected in fractions with a conventional refrigerated fraction collector (Anderson, 1962b).

Anderson et al. (1967) evaluated the performance of the rotor and found that: (1) "little disturbance of the gradient occurs during passage through the rotor," and (2) there is some boundary widening when the rotor is accelerated from 3500 rpm, but the widening is not serious. Similarly, it was shown that sharp zones can be formed and recovered from the area close to the rotor wall. Also, the rotor has been shown to be useful for the separation of subcellular particles including nuclei, mitochondria, lysosomes, microsomes, glycogen, polysomes, and viruses (Anderson et al., 1967). More recently (Baggiolini et al., 1969), a B-XIV rotor was used to resolve the granules from rabbit heterophile leukocytes into distinct populations by zonal centrifugation.

From the above considerations, it seems clear that the preparative zonal
ultracentrifuges are going to be versatile and potent tools for the study
of cell structure and function by cell biologists, biochemists, and bio-
physicists.

Extensive experience with cell fractionation techniques has accumu-
lated since Bensley and Hoerr first isolated mitochondria from the cells
of guinea pig liver brei by differential centrifugation. From this experience some helpful principles have developed for the design of cell fractionation experiments and the interpretation of the results obtained from
them. Many of these principles have been formalized in a review by de
Duve (1964). The following discussion includes some of his views on the
design and interpretation of cell fractionation experiments. The design of
the experiment is determined largely by the working hypothesis that in-
spired it, but some features are probably common to most well-designed
experiments. In this connection there must be a thorough understanding of
the physical principles of centrifugal analysis, the basic requirements for
maintaining cell particulates in a reasonably native form, and a careful
control of variables, such as field strength, medium composition, and
temperature, which determine the behavior of cell particles in the experiment. Basically, a tissue fractionation experiment proceeds in three steps:
homogenization, fractionation, and analysis. The experimenter must work
backward from the results obtained through these three steps to interpret
his data in terms of properties of the original cells. In the first step in the
interpretation of the results, the quantitative biochemical properties are
assigned to the fractions, due regard being given to the accuracy and
specificity of the assays and the possibility of positive or negative inter-
ferences from substances or physical factors present (or absent) from the
fractions. For example, when enzymes are measured by kinetic experiments, it is necessary to be certain that optimum assay conditions are established and the proportionality between reaction velocity and enzyme
concentration is met. Needless to say, the enzymological research can be
complicated, e.g., by the presence of isoenzymes (Shannon, 1968), allosteric effectors (Koshland and Neet, 1968), or structure-linked latency
(de Duve, 1959). Next a "balance sheet" must be established to determine
net loss or gain of the component in processing the brei. When the data
are in order, distribution curves or diagrams are prepared relating the
quantity of enzyme or other component to the fraction. If the fractionation was in a density gradient, the distribution function can relate quantity of component as a function of the density of the medium. In isopycnic
density centrifugation, the quantity of component can be plotted as a function of the equilibrium density of the particle fraction. This operation is
performed for each component of the system. If it is assumed that the

normalized distribution for the biochemical constituent is the same as the normalized mass distribution of the particle with which it is associated, then the distribution curves for the components can be tentatively interpreted as a measure of the particle distribution. From the shape of the distribution and the position of its peaks, much can be learned about the physical properties of the particles. The correspondence, or lack of it, between the distributions of different components can be used to indicate whether or not the components belong to the same particle. It should be pointed out here that apparent association of components in one system does not necessarily mean that the components are present in the same particle. If the distributions of two or more components coincide in several different sets of experimental conditions known to differentially affect cell particles, this is important evidence in favor of the hypothesis that the components are in the same particle, but it is not absolute proof.

A final step in the analysis of the cell fraction is to identify the cell organelle in cytological terms. This can be a difficult task and the interpretation may be wrong. Evidence for the identity of cell particles may come from the ultrastructural analysis of the particles by biological electron microscopy. Another way is to show that the distribution of a marker enzyme coincides with the distribution of a component that is characteristic of the cell fraction under investigation. If no marker enzyme has a distribution similar to that of the unidentified particle, it may mean that a new particle is in hand.

From what has been said in the last two paragraphs, there is a striking similarity between the isolation and characterization of cell particles and the isolation and characterization of organic compounds from natural products. This point was stressed by de Duve (1964). It might be interesting and profitable to push this analogy further.

Since the cell organelles are highly labile systems, the possibility of artifacts must always be kept in mind and their possible effects on the interpretation of experimental results evaluated. Most populations of cell organelles are naturally polydisperse with respect to size, shape, and density. Cell and tissue comminution usually damages the cell organelles, more or less, and thereby contributes to the degree of polydispersity. Such changes result in a loss of resolution in fractionation schemes based on centrifugation, since the degree of separation is dependent on the effective differences in size, shape, and density of the classes of particles involved. Damage to cell organelles at some point in the execution of the experiment may also lead to artifactual populations of cell particles having no strict morphological counterpart in the parent cells. A case in point is that of the microsomes. The term "microsomes" or "microsome fraction" is defined operationally in terms of the conditions of ultracentrifugation used to prepare them, e.g., pellet obtained from a large-particle-free supernatant of specified composition by centrifugation at 54,000 g for 60 minutes (Allfrey, 1959). In the case of liver microsomes, Dallner and Ernster (1968) estimate that the complete removal of mitochondria from a 10% (w/v) brei in 0.25 M sucrose requires centrifugation at 10,000 g for 10 minutes and that about 50% of the microsomal material is lost with the large-particle fraction. In general the microsomes do not represent a single cell organelle. The fraction is derived from the rough endoplasmic reticulum, smooth endoplasmic reticulum, Golgi bodies, cell membranes, and even fragments of large particles such as lysosomes and mitochondria. Some simplification of the problem is derived from the fact that in some tissues one organelle will contribute most of the vesicles of the microsomal fraction. For example, with suitable homogenization procedures, most of the liver cells that are broken are hepatocytes. It so happens that most of the microsomal vesicles from this cell type arise from the rough and smooth endoplasmic reticulum. The particles of the microsome fraction differ greatly from the cell organelles from which they were derived. Due to breakage of the organelles by a type of pinching-off process during homogenization, the flattened sacules of the endoplasmic reticulum and Golgi bodies are converted into smaller, rounded vesicles. The size and density distribution for the microsomes varies widely. This polydispersity creates serious problems in the cytological identification of any fractions obtained, particularly of the smooth-surfaced vesicles. The problems and partial solutions in this area are reviewed by Dallner and Ernster (1968).

B. BIOLOGICAL ELECTRON MICROSCOPY

In the last section some of the procedures for the mass fractionation of subcellular particles and the general principles for the application of these procedures to problems in morphology, chemistry, and function of cell particles were discussed. Much of the value of the cell fractionation methodology stems from the fact that frequently the form and function of the cell organelles are preserved sufficiently well, and enough of the subcellular particles derived from the cell organelle can be obtained to permit the application of the principles of biochemistry and physical chemistry to the problem. A limitation to the method is that it is not yet possible to determine from cell fractionation studies alone just how closely the chemical and morphological properties of the isolated cell particles relate to the corresponding properties of a native cell organelle. The case of the microsomes considered earlier is apropos of this point.

1. Basic Techniques

High-resolution transmission electron microscopy has come of age. Several commercial electron microscopes are available which, under favorable conditions, are said to yield 1 to 2 Å resolution, and there is little doubt that 3 to 4 Å resolution can be obtained on a routine basis with these instruments. In principle, at least, this implies that the structure of individual macromolecules and certainly macromolecular complexes should be open to study by direct examination with the electron microscope. The following examples of the relationship between the molecular weight and dimensions of protein molecules were taken from the list compiled by Sjöstrand (1967): insulin, 6000 mol wt, 45×6-Å rod; sperm whale myoglobin, 18,000 mol wt, $43 \times 35 \times 23$-Å flat disc; beef liver catalase, 250,000 mol wt, $80 \times 64 \times 64$-Å block; β-galactosidase from *Escherichia coli*, 500,000 mol wt, $115 \times 115 \times 72$-Å pillow; and glutamic acid dehydrogenase from ox liver, 10^6 mol wt, 104 Å per side of a tetrahedron. Some of the proteins in the list and many that are not listed have been studied profitably by electron microscopy. Globular micelles of lipid molecules are also of sufficient size for study with the electron microscope.

The principles of transmission electron microscopy cannot be discussed adequately in the little space available, but it should be emphasized that experimenters must understand the working principles of the instrument in order to produce good results and to understand the strength and limitations of electron microscopy to evaluate its potential properly. There are several good books on electron microscopy. Hall's text (1966) provides a solid introduction to the theory of electron optics, magnetic lenses and their aberrations, image characteristics, and related topics. Somewhat simpler introductions to the electron microscope are available in the books by Sjöstrand (1967) and Pease (1964).

Of great importance to the study of cell ultrastructure was the development of techniques for looking at thin sections of well-preserved cells. The nature of electron microscopy requires that the subject be thin and able to survive high vacuum and bombardment by 25- to 125-kV electrons. The modern basic techniques for preparing tissues and cell particles for examination in thin sections with the electron microscope entail the following principal steps: fixation, dehydration, embedment in an epoxy resin, thin sectioning, section mounting, and staining. Two of the most widely used fixatives for plant and animal tissues and for isolated cell particles are buffered solutions of glutaraldehyde and OsO_4. Glutaraldehyde is an excellent primary fixative for preserving the ultrastructure and some enzyme activities of animal cells (Sabatini *et al.*, 1963). It cross-links proteins but does not react with lipids. The OsO_4 fixes both proteins and lipids and is often used as a postfixative to follow glutaraldehyde fixation. Since osmium has a high atomic number and, consequently, a dense electron cloud around its nucleus, the OsO_4 fixative also acts as an electron stain. More will be said about this point later. Excellent discussions of the virtues and faults of these two fixatives are given by Pease (1964) and Sjöstrand (1967). Precautions must be taken to ensure that the fixative reaches the cell promptly, otherwise serious artifacts may develop as the cell dies. Small cubes of tissue (at most a few millimeters on edge) or perfusion techniques seem to work best, especially with OsO_4 fixatives. After the tissue is fixed, the fixative is removed by extraction, and the tissue is dehydrated with a series of alcohol or acetone solutions of increasing strength. The reason for this step is that most of the monomers used to impregnate tissues prior to embedment are not miscible with water or react with it unfavorably. The dehydrated tissue cubes are infiltrated with the unpolymerized resin mixture in decreasing concentrations of a transition solvent. After infiltration is complete, the tissue cube is oriented as desired in a suitable capsule, and the plastic is polymerized by heat. The tissue, which is now embedded in hard plastic, is ready for sectioning.

As mentioned earlier, thin sections are required for electron microscopic viewing at high magnification. Usually only sections of 700 Å or less are suitably thin. Two types of microtomes are available for cutting sections in this thickness range. One type advances the specimen by mechanical means and the other by thermal expansion of the specimen-bearing rod. The design and operation of the mechanical advance microtome is described by Pease (1964); the thermal advance instrument is described by Sjöstrand (1967). So far, the best knives for ultrathin sectioning are made from glass or diamonds. The glass knives are cheaper to make but will not cut the harder plastics as the diamond knives will. The knife is mounted in a boat filled with water or dilute acetone to the knife edge. Sections float on the liquid surface as they are formed. Section thickness may be estimated by the interference color. The thin ones are picked up on grids of copper or other suitable metal. The sections come to rest over the pores on the grid. At this point the sections may be stained with one or more electron stains to differentially enhance the contrast of the cytomembranes and certain other cell components. The principles of electron staining are covered by Pease (1964) and Sjöstrand (1967).

After the sections are stained, they are ready for examination with the electron microscope. The field covered by the electron microscope becomes less as the magnification is increased until, at very high magnification, only a small region of the cell is in view. Even at low magnification usually only one or, at most, a few cells are in view at any one time. This fact means that sampling can be a problem unless a large number of fields are scanned for the structures of interest. Another approach is to consider the number of sections it takes to completely section a whole cell. If the diameter of a cell is 10 μ and 500-Å sections are taken, 200 sections will be needed to cut completely through one cell. More will be said about this when stereology is discussed. After the subject of interest is found and the electron microscope is focused properly, the image is recorded on plates or cut film. Photographic prints are made of the negatives to give the primary data for evaluation. Of course, the magnification of the primary image and the photographic enlargement factor must be determined by appropriate calibration procedures. Focus can be difficult. If proper focus cannot be achieved, false "structure" may be read where none exists. Sjöstrand (1967) and Pease (1964) discuss this problem in more detail.

It is clear from the above that the preparation and examination of thin sections by electron microscopy is a multistep process with many chances for the occurrence of artifacts. Even so, this approach has greatly extended our knowledge of the ultrastructure of cells, has led to the discovery of at least one new organelle, e.g., the endoplasmic reticulum (Porter, 1961), and has settled controversies concerning the existence of another, e.g., Golgi bodies (Beams and Kessel, 1968).

2. Quantitative Techniques

Considerable progress has been made with quantitating biological electron microscopy. So much progress has been made that a symposium was held on the subject in 1964 at the Armed Forces Institute of Pathology in Washington, D.C. The proceedings of the symposium were published as a book in 1964 (Bahr and Zeitler). In the introduction of the book the term "quantitative electron microscopy" was stated "to encompass procedures which render quantitation of properties of the object under study which can be obtained only by electron microscopy. These included counting particles, determining linear or cross-sectional extension, thickness, mass thickness and weight, and, finally, chemical composition of objects too small for light microscopy." Three examples of quantitative electron microscopy will be considered here. Each is included because it already has provided information concerning the function or form of subcellular organelles or seems likely to do so. The examples are taken from quantitative radioautography (a form of cytochemistry), stereological methods, and negative staining.

Progress in high-resolution radioautography (autoradiography in the older literature) is reviewed by Caro (1966). The field is relatively new, since the first report which clearly demonstrated a resolution greater than that previously achieved by conventional light microscopy techniques appeared in 1961 (Van Tubergen). Since that time many improvements have been made in the technique with respect to grain size of the nuclear emulsion, section thinness, the use of stains to improve image contrast, and the theoretical understanding of the conditions for high resolution. Tritium is the favorite radioactive label, primarily because it emits very weak β particles (circa 0.018 MeV) which, consequently, have a very short range in matter, and because hydrogen is a common element in most organic molecules of biological interest. The attempt to quantitate high-resolution radioautography is a recent development in electron microscopy, e.g., see the review by Bachman and Salpeter (1964). The problems of resolution in electron microscope radioautography are considered in detail by Salpeter et al. (1969). They showed, by analysis of grain distributions around a radioactive line of polystyrene-^{3}H, that the shape of the distribution is independent of factors that influence resolution, such as section and emulsion thickness, silver halide crystal, and developed grain size. However, they also demonstrated that these factors did affect the spread of the distribution and, thus, the resolution. Techniques were developed for interpreting radioautograms. These methods have been employed by Budd and Salpeter (1969) to determine the distribution of labeled norepinephrine within sympathetic nerve terminals. They concluded, on the basis of quantitative radioautography, that norepinephrine-^{3}H was bound inside vesicle-filled sympathetic nerve terminals but was not associated with either mitochondria or certain large granules with a dense core in these terminals. More recently, Salpeter (1969) has used the method to determine quantitatively the distribution of diisopropylfluorophosphate (DFP)-sensitive enzyme sites at the neuromuscular junction. Most of the sensitive sites were located in the subneural apparatus at a concentration of 90,000 sites/μ^3 of cleft tissue or 12,000 sites/μ^3 of postjunctional membrane surface area. Rogers et al. (1966) found that about one-third of the DFP-sensitive sites at the end plate can be reactivated by pyridine-2-aldoxime methiodide (2-PAM). This compound selectively reactivates phosphorylated acetylcholine esterase. The same result was obtained by quantitative radioautography (Rogers et al., 1966).

The work outlined in the above brief discussion on quantitative radioautography demonstrates clearly that it will be a very useful tool for studying cytochemistry at the ultrastructural level of organization. However, at the present time this interesting area of electron microscopy is only in its early infancy.

Since cells can be considered as aggregates of membrane-limited volumes with characteristic structure, it is of interest to introduce techniques for determining the relative volumes and volume-to-surface ratios of these cell components. The availability of reliable procedures for preparing large numbers of high-quality thin sections of cells and the small size of the subcellular compartments indicate that this could be a task for biological electron microscopy. New methods that show promise in the estimation of volumetric ratios, surface areas, surface-to-volume ratios, thickness of tissues or cell sheets, and the number of structures per unit volume are the stereological methods. The history of the method as applied to problems in geology, the quantitative analysis of cytoplasmic structures, and the general principles of the method are reviewed by Loud et al. (1964).

A brief description of the stereological method as practiced on tissues and cells by Weibel et al. (1966) follows. Sectioning preserves the integrity of the tissue in two of three dimensions, but the third is sacrificed for the benefit of resolution. If the section thickness is very thin with respect to the size of the structures being investigated, it can be considered as an infinitely thin, two-dimensional image of the internal structure. In sectioning cells the structures are cut at random. Compartments appear as areas, surfaces as lines, and lines as points in thin sections. There is a relationship between the area of a section occupied by sections of a given compartment and the fraction of the tissue volume occupied by that component, a relationship between the length of the contours of the boundary of a component compartment in a unit test area and the surface of the compartment per unit volume, and a relationship between the number of structures of a given component in a unit volume and the number of transections of these structures found on a unit area of a random section. These relationships can be exploited to get a statistical estimate of volumes, surface areas, and numbers of a given compartment per unit volume of tissue. How this may be accomplished and the errors involved are covered by Weibel et al. (1966). This type of analysis is called "morphometric analysis."

A quantitative stereological description of the ultrastructure of normal rat liver parenchymal cells has been carried out by Loud (1968). He determined the fractional volume of cytoplasm occupied by mitochondria, peroxisomes, and lysosomes and the surface densities of smooth- and rough-surfaced endoplasmic reticulum and mitochondrial envelopes and cristae. Weibel and his collaborators (Weibel et al., 1969; Staubli et al., 1969) made correlated morphometric and biochemical studies on liver cells. For the normal liver they found (Weibel et al., 1969) that 1 ml of liver tissue contains 169×10^6 hepatocyte nuclei, 90×10^6 nuclei of other cells, and 280×10^9 mitochondria. They also found that hepatocyte cytoplasm and mitochondria account for 77 and 18%, respectively, of the liver volume and that the surface area of the membranes of the endoplasmic reticulum in 1 ml liver tissue measures 11 m^2, of which two-thirds is in the form of the rough endoplasmic reticulum. The rough endoplasmic reticulum from 1 ml of liver tissue was estimated to carry 2×10^{13} ribosomes. The surface area of mitochondrial cristae per milliliter of liver was estimated at 6 m^2. The effects of a 5-day treatment of phenobarbital on rat hepatocytes were also determined (Staubli et al., 1969). Phenobarbital is known to cause a reversible adaptable hypertrophy of liver. It was found that the increase in cytoplasmic volume of the hepatocytes was responsible for most of the liver hypertrophy. The endoplasmic reticulum accounted for the bulk of the cytoplasmic change in volume. Up to the second day of treatment the smooth and the rough endoplasmic reticulum contributed to the increase in surface and volume of endoplasmic reticulum. After this time the volume and surface of the rough endoplasmic reticulum returned to normal values, whereas those of the smooth endoplasmic reticulum continued to increase. The specific volume of the mitochondria and peroxisomes remained constant during the experiment, while that of the dense bodies increased. The above observations on the rough and the smooth endoplasmic reticulum are in agreement with results obtained by the biochemical analysis of microsomes from liver taken from normal and phenobarbital-treated rats (Staubli et al., 1969; Omura et al., 1967; Palade, 1959).

The third example of quantitative biological electron microscopy is negative staining. In negative staining, an unstained biological specimen is embedded in a thin layer of solution containing an electron dense solute, and the preparation is allowed to dry. The relatively less dense specimen then appears, in the electron microscope, as an electron transparent object bounded by an electron dense border. Under favorable conditions, considerable information can be obtained about the ultrastructure of the specimen. Hall (1955) and Huxley (1957) first noticed the negative staining effect while working with viruses. The latter author used the method to demonstrate the central hole in tobacco mosaic virus. The technique was developed further by Brenner and Horne (1959), again demonstrating the structure of viruses. They stained the viruses by mixing them with 1% phosphotungstic acid adjusted to pH 7.5. The sample was sprayed on grids coated with carbon films. Horne (1964) reviewed the application of negative staining methods in quantitative electron microscopy, and Glauert (1965) reviewed the factors affecting the appearance of biological specimens in negatively stained preparations. In addition to the study of virus ultrastructure, the method has been used to study proteins (e.g., hemoglobin, myoglobin, and the erythrocruorins), subcellular structures (e.g., muscle filaments, bacterial flagella, and mitochondrial components), and to count virus and protein molecules. As pointed out by

Horne (1964), there is often good agreement between measurements of objects made on negatively stained material and those made on the same objects by other methods; however, sometimes considerable disagreement exists, and the reasons for the discrepancies are not clear. Glauert (1965) observed that the nature of the stain, the staining conditions, and the properties of the object being stained all had a bearing on the result. In 1962, Fernández-Morán (1962a,b,c) applied the negative staining technique (Brenner and Horne, 1959) to beef heart mitochondria and found the inner membrane coated with regularly spaced spheres connected by stalks to the unstained core. Later, Parsons (1963) confirmed the existence of the projecting subunits on the cristae of each of eleven types of mitochondria examined by electron microscopy. These consisted of a stem 30 to 35 Å wide and 45 to 50 Å long, and a round head 75 to 80 Å in diameter. Other workers confirmed these findings (Stoeckenius, 1963; Horne and Whittaker, 1962; Smith, 1963). In addition, the electron micrographs of Smith (1963) showed the negatively stained core of the inner membrane as an array of globular units connected to the stalk. Fernández-Morán et al. (1964) referred to these as the base plate and used the term "elementary particle" to include base plate–stem–headpiece complex. Sjöstrand suggested that the headpieces might be reproducible artifacts resulting from the damage of the inner membrane during the negative staining procedure (Sjöstrand et al., 1964). Most investigators feel that this is unlikely.

Much fruitful effort has been spent correlating the structures of the inner membrane with the biochemical components of the electron transport system and the ancillary systems of oxidative phosphorylation. Papers by Pullman and Schatz (1967) and Green and Silman (1967) review mitochondrial oxidations and include sections on the relationships between the observed structure and function. Without going into the detailed arguments in this controversial field, the current opinion seems to be that the headpiece is identified with mitochondrial adenosine triphosphatase (ATPase) (Racker et al., 1965; Kagawa and Racker, 1966) and that the electron transfer chain is localized in the base piece. The significance of the stalk remains in doubt.

III. Case Histories

Three case histories are given in this section to illustrate the successful joint application of cell fractionation methodology, biochemistry, and biological electron microscopy to problems of ultrastructure, composition, and function of cell organelles. The first example covers a sequence of studies by Beevers and his collaborators (Breidenbach *et al.*, 1968; Cooper and Beevers, 1969a,b; Breidenbach and Beevers, 1967) in which they demonstrated a new class of particles, the glyoxysomes, by the biochemical analysis of cell particulates isolated by density gradient centrifugation. The ultrastructure of the glyoxysome was evaluated by electron microscopy, and the probable natural existence of these cell organelles was confirmed by the same techniques. As a by-product, the work illustrates that mitochondria from different sources can have different metabolic capabilities. The second case history is concerned with the clarification of a controversy over the existence of mitochondria in yeast cells grown anaerobically. Anaerobically grown cells of *Saccharomyces cerevisiae* lack cyanidesensitive respiration and the classic cytochromes but adaptively regain these characteristics upon aeration (Lindenmayer and Smith, 1964). Wallace and Linnane (1964) concluded from studies with the electron microscope that mitochondria are absent from anaerobic yeast cells and that the organelles form *de novo* from low molecular weight precursors upon aeration of the cells. Schatz and co-workers (Criddle and Schatz, 1969; Paltauf and Schatz, 1969; Plattner and Schatz, 1969) resolved the problem by isolating promitochondria from anaerobic yeast cells and characterizing them biochemically and ultrastructurally.

The pancreatic exocrine cell of mammals synthesizes enzymes and zymogens, sequesters them temporarily as proteinaceous storage granules, and, then, under proper stimulation, discharges them into the lumen of the acinus. In the last case history the principal steps of the process are described in biochemical and ultrastructural terms. In addition to principles of cell fractionation and electron microscopy, the principles of high-resolution radioautography were employed. The last case history is based largely on the work of Palade and Siekevitz and their collaborators (Siekevitz and Palade, 1960a,b; Redman *et al.*, 1966; Caro and Palade, 1964; Jamieson and Palade, 1967a,b, 1968a,b). Data are included here suggesting that the aleurone grains of seeds develop by a kind of internal secretion and that the process is similar in many respects to zymogen granule formation in the acinar cells of the pancreas (Dieckert, 1969).

A. GLYOXYSOMES—CELL PARTICLES THAT CONVERT FAT TO CARBOHYDRATE

Germinating oil seeds, such as castor bean and peanut, synthesize sucrose from the acetyl-CoA from fatty acid oxidation by way of the glyoxylate cycle (Canvin and Beevers, 1961). When a crude suspension of particles from germinating castor bean was fractionated on sucrose density gradients, three major bands were obtained. The distributions of the glyoxylate enzymes were essentially the same, and the enzymes were present only in the band of densest particles (Breidenbach and Beevers, 1967). Consequently, on the assumption that the glyoxylate enzymes are all found in a single particle, these particles were named "glyoxysomes." In a later paper the three principal density classes of cell particles were characterized further by biochemistry and biological electron microscopy (Breidenbach et al., 1968). Electron microscopy of the particulate bands from a continuous density gradient fractionation was carried out as follows. From the distribution of protein in the fractions, two peak tubes from each of the three major bands were pooled to give samples of each component. These were fixed in glutaraldehyde and pelleted. The washed pellets were fixed in cold, buffered OsO_4. The pellets were dehydrated and embedded in epoxy resin. Thin sections were stained with lead citrate or uranyl magnesium acetate and lead citrate. The stained sections were examined by electron microscopy. The results were used to identify the cytological components of the three bands, which were mitochondria (density 1.19 gm/cm^3), proplastids (density 1.23 gm/cm^3), and glyoxysomes (density 1.25 gm/cm^3). The morphological homogeneity of the glyoxysomes was high, and the contamination with mitochondria or proplastids was exceedingly low. The isolated glyoxysomes are bounded by a single unit membrane and frequently contain transparent spaces surrounded by a single or, rarely, a double unit membrane. The stroma is finely granular in appearance. The distribution of a selected set of glyoxylate cycle enzymes and enzymes thought to be strictly mitochondrial enzymes were also determined. Essentially all of the particulate malate synthetase, isocitrate lyase, catalase, and glycolic oxidase were found in the glyoxysomes, whereas the characteristically mitochondrial enzymes— fumarase, reduced nicotinamide adenine dinucleotide (NADH) oxidase, and succinoxidase—were found only in the mitochondrial fraction. Malate dehydrogenase and citrate synthetase were present in the mitochondrial and the glyoxysome fractions. The morphological and enzyme distributional data confirmed the hypothesis that glyoxysomes are a distinct class of particles. Further evidence in favor of the natural occurrence of glyoxysomes was adduced from correlative experiments with germinating peanut and watermelon seedlings, which contain the glyoxylate cycle, and corn roots, which do not. Particles resembling castor bean endosperm glyoxysomes in enzyme content and density were obtained from peanut and watermelon, but no protein band of equivalent density was found in a corn root brei. Still further evidence for the natural occurrence of glyoxysomes was derived from the observation of glyoxysome-like organelles in electron micrographs of castor bean endosperm and squash cotyledon.

Next, the enzyme composition and catalytic capacity of the glyoxysomes

from germinating castor bean endosperm were investigated further, and
the results compared with similar data on mitochondria from the same
source (Cooper and Beevers, 1969a). The data from earlier studies suggested that in the overall conversion of fats to sucrose in the tissue, the
conversion of acetyl-CoA to succinate was accomplished in the glyoxysome and the subsequent oxidation of succinate to oxalacetate, in the mitochondria. To verify this hypothesis, it is necessary to show that the necessary enzymes are present in the glyoxysomes and that their activities are
sufficiently high to account for the rate of conversion of fat to carbohydrate in intact tissue. To this end, mitochondria and glyoxysomes from
germinated castor bean endosperm were analyzed for each of the enzymes
of the glyoxylate cycle and the citric acid cycle. The survey showed that
the glyoxysome fraction contained more than 85% of the total isocitrase
and malate synthetase activities and the mitochondria, less than 5%.
Aconitase was found in the glyoxysomes but appeared to be easily solubilized. All of the enzymes of the glyoxylate cycle except aconitase were
present in the glyoxysomes at levels sufficient to support rates of acetate
metabolism greater than those known to be occurring in the intact tissue.
However, the enzymes of the tricarboxylic acid cycle, which are not also
members of the glyoxylate cycle, are clearly associated with the mitochondria and are essentially absent from the glyoxysomes or proplastids.
Since succinic dehydrogenase and fumarase were not detected in glyoxysomes, the succinate produced from isocitrate cleavage must be transported to the mitochondria for oxidation to malate. This was confirmed
by labeling experiments in which it was shown that succinate is not oxidized by glyoxysomes but is oxidized to malate when mitochondria are
added. Glyoxylate, the other product of isocitrate cleavage, is apparently
completely utilized in malate synthesis, since malate synthetase is localized in the glyoxysome and since the potential competing reactions,
NADH–glyoxylate reductase or transaminase reactions with glyoxylate
as the acceptor, were not detected in this cell particle.

Two important problems remained to be clarified in order to strengthen
the hypothesis that the glyoxysome is the only site for succinate formation from acetyl–CoA. The first is the mechanism that channels essentially
all of the endogenous acetyl–CoA into the glyoxylate cycle, and, the second, is the problem of how NADH, which is produced in the glyoxysome,
is reoxidized. The answer to the first problem has been sought by Cooper
and Beevers (1969b).

In their experiments with germinated castor bean endosperm, they reported that most of the β-oxidation activity of the original brei is in the
particulate fraction. The glyoxysomes obtained by density gradient centrifugation contain more than 80% of the particulate β-oxidation activity,

but the mitochondria obtained by the same procedure contain very little. This result is unusual, since it is well known that mitochondria from mammalian liver contain the enzyme systems for β oxidation (Novikoff, 1961b). The presence of the enzymes for β oxidation in the glyoxysome could explain why acetate escapes oxidation by the tricarboxylic acid cycle in the mitochondria and is consumed, instead, by the glyoxylate cycle in the glyoxysomes. Glyoxysomes oxidize palmitoyl CoA with the consumption of O_2 and the production of acetyl–CoA and NADH in a stoichiometric ratio of 0.5:1:1. Cyanide does not affect NADH production but doubles the rate of O_2 uptake. These observations suggest that β oxidation in the glyoxysome is not coupled to a cytochrome-containing electron transport sequence. This is another interesting difference between β oxidation in the mitochondria of mammalian cells and β oxidation in the glyoxysomes. In the former, β oxidation of fatty acids is tightly coupled to a functioning cytochrome-containing electron transport system (Garland, 1968). The data are consistent with the O_2-requiring oxidation of reduced flavin adenine dinucleotide (FADH) of an acyl–CoA dehydrogenase to yield H_2O_2, which is broken down to H_2O plus $\frac{1}{2}$ O_2 by catalase in the glyoxysome.

The totality of data presented in the work of Beevers et al. provides strong evidence for the existence of a new cell organelle, the glyoxysome, with the net production of succinate. The reaction sequence for the conversion of succinate to carbohydrate remains to be determined. Other important questions concerning the biochemistry of the system are these: How is the NADH, which is produced in the β oxidation and glyoxylate cycle reactions, reoxidized to provide the needed NAD to maintain the process? How is the fatty acid substrate, which originally is present in the spherosomes, transferred to the glyoxysome? Important questions relative to the cytology of glyoxysomes need to be answered. For example, What are the cytological antecedents of these transient cell organelles? How are they related to the somewhat similar peroxisomes described by de Duve (1969) and others (Tolbert et al., 1968, 1969; Kisaki and Tolbert, 1969) and the microbodies of Jensen and Valdovinos (1968)? The transient nature and specialized distributions of the glyoxysomes raise a host of questions concerning factors that control their appearance and disappearance in the cell.

B. Promitochondria—A Case of Organelle Dedifferentiation

A vast literature exists concerning the biochemistry and ultrastructure of the mitochondrion. Since the present section is limited to the promitochondria of yeast cells, the reader who wants additional information on mitochondria is referred to the books by Lehninger (1964) and Roodyn and Wilkie (1968), the chapter by Novikoff (1961b), and recent reviews by Pullman and Schatz (1967) and by Green and Silman (1967). A clear statement of the problem of mitochondrial structure and function as understood before 1959 is given in an essay by Green (1959). The problem with promitochondria probably involves the control of gene action and cell differentiation. Reviews on the biochemistry of differentiation by Gross (1968) and Rosen (1968) are included to introduce the reader to this rapidly developing field.

The cells of the yeast *Saccharomyces cerevisiae* contain functionally and structurally competent mitochondria bearing a cyanide-sensitive cytochrome system when grown aerobically (Agar and Douglas, 1957; Vitols *et al.*, 1961). The anaerobically grown yeast cells do not contain cyanide-sensitive respiration or the classic cytochrome complement. Functionally competent mitochondria with these properties reappear when the cells are cultured aerobically again (Lindenmayer and Smith, 1964). These facts suggested that the yeast mitochondrial system might be useful in the study of the biogenesis of mitochondria and certainly could be a promising system for studying the biosynthesis of the respiratory enzymes (Criddle and Schatz, 1969). Until recently, however, the presence of mitochondria or their morphological antecedents in anaerobically grown yeast cells was in doubt. The uncertainty was mainly due to a failure to find distinct cytoplasmic precursor particles in sections of cells by electron microscopy which clearly transformed into typical yeast mitochondria (Wallace and Linnane, 1964). These authors concluded, then, that mitochondria arise *de novo* from low molecular weight precursors.

Schatz (1965) fractionated a particulate fraction from anaerobically grown nonrespiring yeast cells by density gradient centrifugation. A band of particles with a density of 1.15 gm/ml, the density of yeast mitochondria prepared the same way, was isolated. The fraction contained almost 100% of the succinate–phenazine methosulfate–oxidoreductase activity of the particulate fraction. The observation is important because this enzyme activity is almost completely confined to the mitochondria of aerobically grown yeast cells and mammalian cells. The result suggested that the antecedents of the mitochondria of aerobically grown yeast cells are cell particulates in the anaerobically grown yeast cells and that these differentiate, under the inductive effect of oxygen, to the characteristic, functionally competent mitochondria of the aerobically grown yeast cell. These particles, called "promitochondria," were explored in detail by Criddle and Schatz (1969), Paltauf and Schatz (1969), and Plattner and Schatz (1969).

In the first study (Criddle and Schatz, 1969), particles were isolated from cells of *Saccharomyces cerevisiae*, grown under strict anaerobiosis. First, the crude particulates of the yeast cell homogenates were collected by centrifugation. These were separated into the component particle classes by centrifugation on a linear, continuous sucrose gradient. The fraction that equilibrated at a density of 1.16 to 1.18 gm/cm^3, the density of yeast mitochondria, was studied. The particles lacked a respiratory chain but did contain the oligomycin-sensitive adenosine triphosphatase, F_1, and the "structural proteins" that are characteristic of aerobic yeast mitochondria. These particles corresponded to the promitochondria of Schatz (1965). The presence of the ATPase was verified by showing that the particles contain an oligomycin-sensitive ATPase which is inhibited by a specific antiserum against purified F_1 from aerobic yeast mitochondria. Other similarities in the ATPases from the two sources were noted. Also, other enzymes found in aerobic yeast mitochondria were found in the particles from anaerobic yeast cells. A number of similarities were noted between the structural proteins from the particles from anaerobic yeast cells and structural proteins from mitochondria from aerobically grown yeast cells. The most interesting observation was that an antiserum against structural protein from anaerobic yeast cells agglutinated structural proteins from the mitochondria-like particles from anaerobic yeast cells, whereas nonimmune serum did not. A deoxyribonucleic acid (DNA) was isolated from the promitochondrial preparation which had a buoyant density of 1.685. This DNA was interpreted to be mitochondrial DNA. These data confirm and extend the earlier work of Schatz (1965). Apparently, the mitochondria of aerobic yeast cells dedifferentiate under anaerobic conditions.

Second, Paltauf and Schatz (1969) examined the lipid composition of anaerobically grown yeast cells. They found that the lipids from promitochondria from anaerobic yeast cells differed from the lipid of mitochondria from aerobic yeast. The promitochondria contain the same major phospholipid classes found in yeast aerobic mitochondria. Some differences in relative proportions of the phospholipid classes were noted. The promitochondrial lipids had a low level of ergosterol and an unusual fatty acid composition. Unlike the lipids of the aerobic yeast mitochondria, the lipids of the promitochondria had an extremely low level of unsaturation and a high content of short-chained saturated fatty acids. The low level of unsaturated fatty acids in the promitochondrial lipids might explain why these bodies were not visible in the electron micrographs (Plattner and Schatz, 1969). The OsO_4, which is used to "fix" the lipids of the cytomembranes, reacts only with unsaturated fatty acids. The membranes may not be preserved well enough to withstand the extraction of organic solvents which accompanies the embedding procedure. This could lead to a

58 JULIUS W. DIECKERT

partial failure of the membranes to take up electron stains, upon which their visualization by electron microscopy depends.

Plattner and Schatz (1969) compared the ultrastructure of anaerobically and aerobically grown cells of *Saccharomyces cerevisiae* by the freeze-etching technique of Moor *et al.* (1961). Electron micrographs of freeze-etched aerobic yeast cells showed numerous mitochondria, 0.2 to 0.8 μ diameter, usually located close to the periphery of the cell. In cross section the mitochondria exhibit characteristic cristae. Frequently, only the surface view of the mitochondrion is exposed in the section. In these cases the presence of unique 100 $\times$ 1000-Å slits in the outer mitochondrial membrane is usually characteristic. Electron micrographs of freeze-etched anaerobically grown cells of the yeast contained mitochondria-like structures bearing cristae, a double-layered envelope, and a brittle outer membrane with slits measuring about 100 $\times$ 1000 Å. Therefore, the comparative study by electron microscopy confirms the biochemical experiments that the anaerobic antecedents of aerobic yeast mitochondria are mitochondria-like particles.

C. INTRACELLULAR SYNTHESIS, TRANSPORT, SEQUESTRATION, AND SECRETION OF PROTEINS

Gland cells in animals are specialized to produce a variety of secretory products including diverse substances such as hydrolytic enzymes, zymogen, mucus, milk, steroids, peptidic hormones, and hydrochloric acid. Some glandular cells secrete their product intermittently and sequentially synthesize and concentrate the product into storage granules before extruding it from the cell. In this class are the acinar cells of the pancreas, goblet cells of the intestinal mucosa, pancreatic islets, and the anterior pituitary gland cells. In cells secreting proteins, the process generally occurs in a definite pattern. The proteins are synthesized on the ribosomes attached to the endoplasmic reticulum and transferred into the lumina of the rough endoplasmic reticulum. The protein is transported through channels of the endoplasmic reticulum to Golgi bodies, where it is concentrated into membrane-bound secretory granules. Under an appropriate stimulus the secretory granules move to the plasma membrane. The membrane of the granule fuses with the cell membrane, and the secretory product is extruded from the cell without "breaking" the cell membrane. The following references are cited to indicate where more information can be obtained concerning gland cells (Gabe and Arvy, 1961), the ER (Porter, 1961; Loening, 1968), Golgi apparatus (Kuff and Dalton, 1959; Dalton,

1961; Mollenhauer and Morré, 1966; Beams and Kessel, 1968), and secretory granules (Dalton, 1961; Fawcett, 1966).

In the third case history the development of current concepts of zymogen granule formation and secretion is considered. It provides a classic example of the fruitful interplay of biochemistry and biological electron microscopy. Palade (1959) clearly recognized that modulations in cell structure reflect changes in the biochemistry and physiology of the cell. He and Siekevitz began a long series of studies correlating structural and biochemical changes occurring in the secretory cycle of the pancreatic exocrine cell. This work began with Heidenhain's hypothesis that the granules of the acinar cells are the precursors of the digestive enzymes in the pancreatic juice. It was already known that the basal part of the acinar cell contains relatively high concentrations of ribonucleic acid, that the ribonucleic acid exists as ribonucleoprotein particles on the outer surface of the microsomal membranes, and that cytoplasmic ribonucleic acid (hence the ribosomes) is probably involved in protein synthesis. Furthermore, the conjecture had already been put forth by Palade and his collaborators that the digestive enzymes are synthesized in the endoplasmic reticulum.

Palade (1959) showed by electron microscopy that in starved guinea pigs the endoplasmic reticulum of the pancreatic acinar cells is tightly packed with light content. Shortly after feeding, however, large dense granules appear in the extended cavities of the endoplasmic reticulum. These granules have a density and texture similar to that of zymogen granules. By the crude cell fractionation procedures of the day, it was shown that in starved animals the trypsin-activatable proteolytic activity is in the zymogen granule fraction and not in the microsome fraction. Shortly after feeding, both cell fractions contained about the same activatable proteolytic activity. By similar techniques the intracisternal granules of the microsomes were purified and found to be richer in activatable proteolytic enzymes than the original microsomes. The collective results strongly suggested that zymogen was synthesized in or on the endoplasmic reticulum.

Electron micrographs of guinea pig pancreas producing zymogen granules showed that vesicles in all stages of filling were in the neighborhood of the centrosphere (Palade, 1959), including mature zymogen granules. Numerous connections were observed between the rough-surfaced endoplasmic reticulum and the smooth-surfaced vesicles of the centrosphere region. The intracisternal granules were seldom seen in transit through the junctions. It was also noted that the membrane of the zymogen granule fused with the cell membrane, thereby resulting in the extrusion of the zymogen into the pancreatic acinus. It is clear that at this early stage the

four basic phases of the secretory cycle of the exocrine pancreatic cells
were correctly interpreted; but, the experimental support for the proposed
scheme was still weak.

Additional support for the basic, four-step secretory cycle of the acinar
cells comes from the deeper studies described below. A strong case could
be made for the first step, i.e., that the zymogen proteins are synthesized
on the ribosomes attached to the endoplasmic reticulum, if well-character-
ized zymogen proteins were found tightly bound to the ribosomes. Siekevitz and Palade (1960b) isolated and characterized ribosomes from the
microsomal fraction of guinea pig pancreas. They showed that the isolated
ribonucleoprotein particles contain amylase, ribonuclease, and trypsin-
activatable proteolytic activities which cannot be washed off or detached
by incubation in 0.44 M sucrose but can be released if the ribosome was
disintegrated by chelators. Evidence for the synthesis of a specific enzyme,
amylase, on the ribosomes of the endoplasmic reticulum and its transfer
to the lumen of the endoplasmic reticulum was obtained with pigeon pancreatic microsomes (Redman *et al.*, 1966). The properly supplemented
microsome fraction incorporated radioactive amino acids into amylase.
Tryptic digestion of the radioactive, highly purified enzyme showed that
the labeled amino acids were incorporated into the peptide chain. The labeled amylase first appears bound to the ribosomes and then, after a short
period of time, appears in the deoxycholate-soluble fraction of the microsomes. Caro and Palade (1964) made a radioautographic study of protein synthesis, storage, and discharge in the acinar cells of the guinea pig
pancreas. In this study advantage was taken of pulse-labeling and the
newly developed methods of high-resolution radioautography (Caro and
Van Tubergen, 1962; Caro, 1962) and DL-leucine-4,5-^{3}H as the label. The
radioautographs showed that, at about 5 minutes after injection of the
leucine-^{3}H into the guinea pig, most of the label is associated with the
rough endoplasmic reticulum. At about 20 minutes after injection, it is
found on the Golgi complex, and after about 1 hour, it is in the zymogen
granules. The data gathered by Palade and his associates suggested that
the zymogen proteins are synthesized in the rough endoplasmic reticulum,
transferred through the cisternal spaces to the Golgi apparatus, and then
concentrated into zymogen granules; however, other interpretations were
not excluded. For example, Morris and Dickman (1960) and others suggested that the secretory proteins leave the microsomes in solution and
condense directly into zymogen granules or go directly into the lumen of
the gland.

Several experimental difficulties contributed to the confusion (Jamieson
and Palade, 1967a). To begin with, the pulse-labeling experiments described above were not satisfactory. Pulses less than 15 minutes could

not be achieved *in vivo*. A second problem arose. The cell fractionation procedures did not yield pure fractions, and a number of subcellular components were not accounted for. Finally, the limitations in radioautographic resolution obtained did not permit the localization of the label in particles less than about 0.2 μ in diameter. To meet the first limitation, i.e., the inability to get well-defined short pulses *in vivo*, experiments were carried out with slices of pancreatic tissue (Jamieson and Palade, 1967a). It was established that pancreatic slices actively incorporated labeled amino acids into protein for as long as 3 hours *in vitro*, that the cells incurred minimal damage as determined by electron microscopy, and that short, stable pulse-labeling could be achieved. The latter provided some assurance that changes in the location of label in the cell is due only to intracellular transport and not continued synthesis. The problems with the cell fractionation procedure were partially met by making no attempt at a systematic recovery of cell fractions and, instead, attempting to prepare reasonably pure representative fractions of the structural components thought to be involved in the secretory process. For example, the large particle fractions were sequentially removed by differential centrifugation with a consequent loss of microsomes, and the microsome fraction was further fractionated by equilibrium density gradient centrifugation. At this point in the discussion, the characteristics of the microsome fraction must be reviewed.

As pointed out by Dallner and Ernster (1968), the cytological antecedents of the microsomes are numerous, and the relative contribution of each to the total microsome fraction is heavily dependent upon their relative proportions in the source tissue. From the ultrastructural analysis of the acinar cells, the principal contributors to the microsome fraction are probably the rough endoplasmic reticulum, smooth endoplasmic reticulum, and the Golgi bodies; but, contributions from the other cell organelles and the cell membrane cannot be completely ruled out. Jamieson and Palade (1967a,b) examined the role of the peripheral elements of the Golgi complex in the intracellular transport of secretory proteins to the condensing granules in the pancreatic exocrine cell. Advantage was taken of the principles and methodology discussed in the preceding paragraph. The zymogen proteins were labeled with radioactive amino acids in short, well-defined pulses. The kinetics of the transport of the secretory proteins were determined by pulse-labeling sets of slices for 3 minutes with leucine-^{14}C and incubating them for an additional 7, 17, and 57 minutes in a chase medium. At each time, the smooth microsomes and the rough microsomes were isolated from the microsomal fraction by density gradient centrifugation. Labeled proteins appeared first in the rough microsomes immediately after the pulse; they appeared maximally in the smooth microsomes after 7 minutes chase. At the end of the pulse, few radioactive proteins were found in the zymogen granule fraction. After 17 minutes in chaser, the zymogen granule fraction was labeled half-maximally. The zymogen granule fraction reached maximum labeling between 37 and 57 minutes in chaser. Finally, the specific activity of the proteins in the post-microsomal fraction showed a relatively constant specific activity throughout the experiment. These data were taken as confirming evidence for the original hypothesis. Although the conclusion is probably correct, there may still be a reasonable doubt as to the correctness of the identification of the antecedent to the smooth vesicles when only the cell fractionation data are considered. Parallel experiments were conducted on intact cells in slices pulse-labeled with leucine-^{3}H using electron microscopy to follow the pulse (Jamieson and Palade, 1967b). The results showed that the zymogen granules became labeled only after 57 minutes in the chaser. The apparent discrepancy between the cell fractionation work and the radioautography was thought to be due to the rapid labeling of the condensing vacuoles in the zymogen granule fraction. This point was confirmed by the results of high-resolution radioautography of zymogen granule pellets isolated from slices pulse-labeled with leucine-^{3}H and subsequently incubated in chaser. The interpretation placed on the data was that the secretory proteins pass through the small peripheral vesicles of the Golgi bodies from the site of synthesis and segregation in the rough endoplasmic reticulum to the site of temporary sequestration in the condensing vacuoles. No evidence was found from the radioautography experiments implicating the stacked Golgi cisternae directly in the role of transport or concentration of labeled secretory proteins. The authenticity of the basic steps in the secretory cycle of the pancreatic acinar cells seems to be assured. As pointed out by Jamieson and Palade (1967b), the forces, molecular events, and control mechanisms operating at each step remain for future research.

As already pointed out by Redman *et al.* (1966), proteins synthesized on the ribosomes of the rough endoplasmic reticulum of pigeon pancreatic microsomes are transferred to cisternae of these bodies. Later, Redman (1967) showed that in a puromycin-treated, rat liver, microsomal system the protein from the ribosomes of the rough endoplasmic reticulum was released into the lumen of the sacule. This process was not energy dependent. In a further analysis of the problem, Jamieson and Palade (1968a) initiated experiments to determine some of the factors that control secretory protein synthesis and transport. Conditions were sought for dissociating secretory protein synthesis from intracellular transport. The antibiotic cycloheximide is known to inhibit protein synthesis by cytoplasmic ribosomes (Siegel and Sisler, 1964; Ennis and Lubin, 1964; Colombo *et al.*, 1966). Jamieson and Palade (1968a) used cycloheximide to determine if the synthesis of the secretory proteins of pancreatic slices can be dissociated from intracellular transport of the proteins. Guinea pig pancreatic slices were pulse-labeled with leucine-^{3}H for 3 minutes and then incubated for 37 minutes in a chase medium containing enough cycloheximide to inhibit protein synthesis by 98%. Under these conditions the rate of transport was about 80% of the controls. In similar experiments, high-resolution radioautographs showed that each step from zymogen transport to zymogen discharge from the cell took place in the presence of cycloheximide. Consequently, each of these steps is independent of secretory protein synthesis. The data also indicated that, within the time limits of the test, the transport process is not dependent on the synthesis of "specific" nonsecretory proteins.

The demonstration that protein synthesis by the ribosomes of the rough endoplasmic reticulum can be dissociated from the transport process opened a way to study the energy requirements and source of the energy for the process. Various metabolic inhibitors were used by Jamieson and Palade (1968b) to poison specific energy-producing systems in pancreatic exocrine tissue slices in which protein synthesis was blocked by cycloheximide shortly after pulse-labeling. The pulse-labeled (3 minutes) slices were incubated in chase medium containing cycloheximide and inhibitors of glycolysis, respiration, or oxidative phosphorylation. The effect of the inhibitor was followed by measuring the accumulation of radioactive protein in the zymogen granule fraction. In principle, this measures the transport of secretory proteins from the sacules of the rough endoplasmic reticulum to the condensing vacuoles. Inhibitors of glycolysis (fluoride and iodoacetate) did not block transport, but respiratory inhibitors (cyanide, N_2, antimycin) and inhibitors of oxidative phosphorylation (dinitrophenol, oligomycin) did. The energy-dependent step appears to lie between the transitional elements of the rough endoplasmic reticulum and the condensing vacuoles. Since no open connections have been found between the rough endoplasmic reticulum and the condensing vacuoles, the connections must be intermittent. Jamieson and Palade (1968b) propose that the small peripheral vesicles of the Golgi complex participate in the connection. The details of the transport process remain to be discovered.

On the basis of high-resolution radioautographic evidence, Jamieson and Palade (1967b) suggested that the piled Golgi cisternae located at the periphery of the complex do not play a direct role in the transport and concentration of the secretory proteins of guinea pig acinar pancreas. However, in other systems the secretory product appears to fill completely one or more of the Golgi cisternae at the "maturing" face. The mature cisternae then seem to move away from the Golgi complex and become immature secretion granules. Under this model, new cisternae develop at the "forming" face. The transfer of the proteinaceous secretory product from the rough endoplasmic reticulum to the Golgi body might then be effected through the formation of vesicles of smooth endoplasmic reticu-

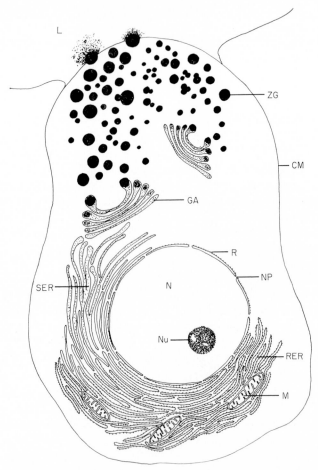

FIG. 1. Diagram of cell secreting protein. (L) Lumen of acinus; (ZG) zymogen granule; (CM) cell membrane; (GA) Golgi apparatus; (R) ribosomes; (SER) smooth endoplasmic reticulum; (N) nucleus; (NP) nuclear pore; (Nu) nucleolus; (RER) rough endoplasmic reticulum; (M) mitochondrion.

lum. These vesicles, which contain a dilute solution of the secretory product, then migrate to the vesicles of the forming face of the Golgi body and empty their contents into them by membrane fusion. A schematic diagram for this type of transfer is shown in Fig. 1. For reviews covering the literature on the transfer of material from the cisternae of the endoplasmic reticulum to the sacules of the Golgi body, see Mollenhauer and Morré (1966) and Beams and Kessel (1968).

Golgi bodies (dictyosomes) are found in plant cells, as well as in animal cells, and appear to function there in essentially the same way as in animal cells (Mollenhauer and Morré, 1966). Dieckert (1969) concluded, on the basis of ultrastructural analysis of the developing plant embryo, that the aleurone grains develop by essentially the same process as zymogen granules. Figure 2 shows an electron micrograph of a cell in a developing cotton embryo that is synthesizing aleurone grain proteins. The rough endoplasmic reticulum is highly developed, the sacules of the maturing face of the Golgi body, shown in the box, contain material staining exactly like the protein in the aleurone vacuole; and, the content of the Golgi sacule clearly stains differently from the content of the spherosomes. These relationships are seen more easily at higher magnification in Fig. 3. Dieckert proposed that the transfer of the aleurone grain proteins of the Golgi vesicles to the aleurone vacuole represents a kind of internal secretion, since the vacuole may be considered as topologically outside the cell.

IV. Conclusions

Two propositions seem to pervade the story of the subcellular particles. Cytomembranes occupy a dominant place as a determinant of the form and function of cell organelles. And, no single approach to the study of the structural and functional organization of the cell is immune from artifacts and misinterpretations resulting from them. In support of the first point, it may be said that cellular organization is based on a system of dissimilar cytomembranes with characteristic structures and functions. Many, and perhaps all, of the cytomembranes are in a dynamic state of flux over the short term. The turnover rate of individual components of the cytomembranes may vary with the component, indicating that the cytomembranes are pleomorphic at the molecular level. Environmental factors can influence the composition, form, and function of the cytomembranes. These changes may represent cytodifferentiation, in some instances. Some cell organelles contain a genome, and their genome may contribute to the formation of the cytomembrane of that organelle. In at least two important instances, however, part of the macromolecular complement of the cytomembrane is derived from cytoplasmic ribosomes and, therefore, not from the particle genome. This could contribute to an understanding of how the nucleus regulates the activities of these partially autonomous particles. The current structural theory of cytomembranes is sketchy in most areas.

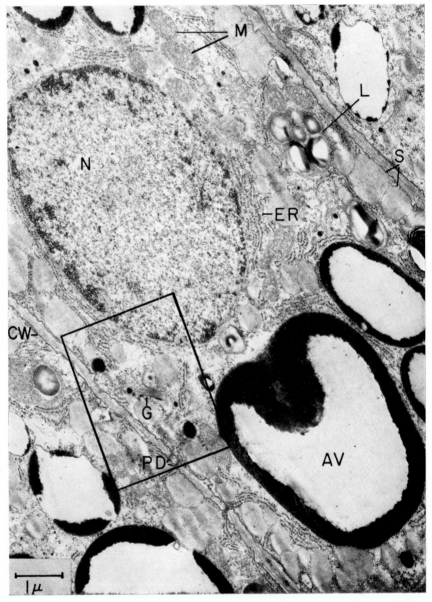

Fig. 2. An electron micrograph of part of a cotton cotyledon cell producing aleurone grain proteins. (N) Nucleus; (M) mitochondrion; (ER) rough endoplasmic reticulum; (L) leukoplast; (S) spherosome; (CW) cell wall; (G) Golgi body; (PD) plasmodesmata; (AV) aleurone vacuole.

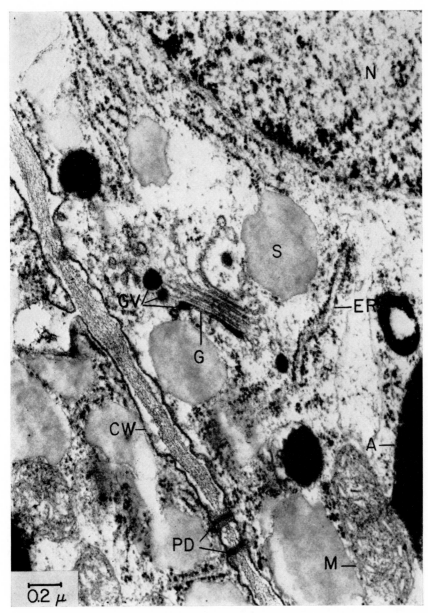

Fig. 3. Electron micrograph of the region in the rectangle in Fig. 2 at higher magnification. (N) Nucleus; (S) spherosome; (GV) Golgi vesicles; (G) maturing face of Golgi body; (ER) rough endoplasmic reticulum; (CW) cell wall; (PD) plasmodesmata; (M) mitochondrion; (A) aleurone vacuole.

68   JULIUS W. DIECKERT

General reviews on the structure of cytomembranes and membrane phenomena were written recently by Korn (1966) and Rothstein (1968).
The second point needs no elaboration. A partial remedy for the problem of artifacts is vigilance on the part of the investigator and the application of several approaches under a variety of conditions when exploring
new areas. The major approaches discussed here are based on biochemistry, cell fractionation procedures, and biological electron microscopy.
The examples and case histories given in the text may serve to demonstrate some of their advantages and disadvantages.

                                REFERENCES

Agar, H. D., and Douglas, H. C. (1957). *J. Bacteriol.* **73**, 365.
Allfrey, V. G. (1959). In "The Cell" (J. Brachet and A. E. Mirsky, eds.), Vol. 1, pp.
    193–290. Academic Press, New York.
Anderson, N. G. (1956). *Phys. Tech. Biol. Res.* **3**, 241–296.
Anderson, N. G. (1962a). *J. Phys. Chem.* **66**, 1984.
Anderson, N. G. (1962b). *Anal. Biochem.* **4**, 269.
Anderson, N. G. (1963). *Nature (London)* **199**, 1166.
Anderson, N. G., and Burger, C. L. (1962). *Science* **136**, 646.
Anderson, N. G., Bond, H. E., and Canning, R. E. (1962). *Anal. Biochem.* **3**, 472.
Anderson, N. G., Barringer, H. P., Babelay, E. F., and Fisher, W. D. (1964a). *Life
    Sci.* **3**, 667.
Anderson, N. G., Price, C. A., Fisher, W. D., Canning, R. E., and Burger, C. L.
    (1964b). *Anal. Biochem.* **7**, 1.
Anderson, N. G., Waters, D. A., Fisher, W. D., Cline, G. B., Nunley, C. E., Elrod,
    L. H., and Rankin, C. T., Jr. (1967). *Anal. Biochem.* **21**, 235.
Bachmann L., and Salpeter, M. M. (1964). In "Quantitative Electron Microscopy"
    (G. F. Bahr and E. H. Zeitler, eds.), pp. 303–315. Armed Forces Inst. Pathol.,
    Washington, D. C.
Baggiolini, M., Hirsch, J. G., and de Duve, C. (1969). *J. Cell Biol.* **40**, 529.
Bahr, G. F., and Zeitler, E. H., eds. (1964). "Quantitative Electron Microscopy."
    Armed Forces Inst. Pathol., Washington, D. C.
Beams, H. W., and Kessel, R. G. (1968). *Int. Rev. Cytol.* **23**, 209–276.
Behrens, M. (1932). *Hoppe-Seyler's Z. Physiol. Chem.* **209**, 59.
Bensley, R. R., and Hoerr, N. L. (1934). *Anat. Rec.* **60**, 449.
Breidenbach, R. W., and Beevers, H. (1967). *Biochem. Biophys. Res. Commun.* **27**,
    462.
Breidenbach, R. W., Kahn, A., and Beevers, H. (1968). *Plant Physiol.* **43**, 705.
Brenner, S., and Horne, R. W. (1959). *Biochim. Biophys. Acta* **34**, 103.
Budd, G. C., and Salpeter, M. M. (1969). *J. Cell Biol.* **41**, 21.
Canvin, D. T., and Beevers, H. (1961). *J. Biol. Chem.* **236**, 988.
Caro, L. G. (1962). *J. Cell Biol.* **15**: 189.
Caro, L. G. (1966). *Progr. Biophys. Mol. Biol.* **16**, 171–190.
Caro, L. G., and Palade, G. E. (1964). *J. Cell Biol.* **20**, 473.
Caro, L. G., and Van Tubergen, R. P. (1962). *J. Cell Biol.* **15**, 173.
Colombo, B., Felicetti, L., and Baglioni, C. (1966). *Biochim. Biophys. Acta* **119**, 109.

Cooper, T. G., and Beevers, H. (1969a). *J. Biol. Chem.* **244,** 3507.
Cooper, T. G., and Beevers, H. (1969b). *J. Biol. Chem.* **244,** 3514.
Criddle, R. S., and Schatz, G. (1969). *Biochemistry* **8,** 322.
Dallner, G., and Ernster, L. (1968). *J. Histochem. Cytochem.* **16,** 611.
Dalton, A. J. (1961). *In* "The Cell" (J. Brachet and A. E. Mirsky, eds.), Vol. 2, pp. 603–619. Academic Press, New York.
de Duve, C. (1959). *In* "Subcellular Particles" (T. Hayashi, ed.), pp. 128–159. Ronald Press, New York.
de Duve, C. (1964). *J. Theor. Biol.* **6,** 33.
de Duve, C. (1969). *Proc. Roy. Soc. London Ser. B* **173,** 71.
de Duve, C., and Wattiaux, R. (1966). *Ann. Rev. Physiol.* **28,** 435–492.
Dieckert, J. W. (1969). *Abstr. 11th Int. Bot. Congr.,* 1969 p. 47.
Ennis, H. L., and Lubin, M. (1964). *Science* **146,** 1474.
Fawcett, D. W. (1966). "An Atlas of Fine Structure." Saunders, Philadelphia, Pennsylvania.
Fernández-Morán, H. (1962a). *Res. Publ., Ass. Res. Nerv. Ment. Dis.* **40,** 235.
Fernández-Morán, H. (1962b). *Circulation* **26,** 1039.
Fernández-Morán, H. (1962c). *In* "Macromolecular Specificity and Biological Memory" (F. O. Schmitt, ed.), p. 39. M. I. T. Press, Cambridge, Massachusetts.
Fernández-Morán, H., Oda, T., Blair, P. V., and Green, D. E. (1964). *J. Cell Biol.* **22,** 63.
Gabe, M., and Arvy, L. (1961). *In* "The Cell" (J. Brachet and A. E. Mirsky, eds.), Vol. 5, pp. 1–82. Academic Press, New York.
Garland, P. B. (1968). *In* "Metabolic Roles of Citrate" (T. W. Goodwin, ed.), p. 41. Academic Press, New York.
Glauert, A. M. (1965). *In* "Quantitative Electron Microscopy" (G. F. Bahr and E. H. Zeitler, eds.), pp. 331–341. Armed Forces Inst. Pathol., Washington, D. C.
Granick, S. (1938). *Amer. J. Bot.* **25,** 558.
Granick, S. (1961). *In* "The Cell" (J. Brachet and A. E. Mirsky, eds.), Vol. 2, pp. 489–602. Academic Press, New York.
Green, D. E. (1959). *In* "Subcellular Particles" (T. Hayashi, ed.), pp. 84–103. Ronald Press, New York.
Green, D. E., and Silman, I. (1967). *Ann. Rev. Plant Physiol.* **18,** 147–178.
Gross, P. R. (1968). *Ann. Rev. Biochem.* **37,** 631–660.
Hall, C. E. (1955). *J. Biophys. Biochem. Cytol.* **1,** 1.
Hall, C. E. (1966). "Introduction to Electron Microscopy." McGraw-Hill, New York.
Henson, C. P., Weber, C. N., and Mahler, H. R. (1968a). *Biochemistry* **7,** 4431.
Henson, C. P., Perlman, P., Weber, C. N., and Mahler, H. R. (1968b). *Biochemistry* **7,** 4445.
Hill, R. (1937). *Nature (London)* **139,** 881.
Hogeboom, G. H., and Schneider, W. C. (1955). *In* "The Nucleic Acids" (E. Chargaff and J. N. Davidson, eds.:), Vol. 2, pp. 199–246. Academic Press, New York.
Hoober, J. K., Siekevitz, P., and Palade, G. E. (1969). *J. Biol. Chem.* **244,** 2621.
Horne, R. W. (1964). *In* "Quantitative Electron Microscopy" (G. F. Bahr and E. H. Zeitler, eds.), pp. 316–330. Armed Forces Inst. Pathol., Washington, D. C.
Horne, R .W., and Whittaker, V. P. (1962). *Histochemie* **58,** 1.
Huxley, H. E. (1957). *Proc. Stockholm Conf. Electron Microsc., 1956* p. 260 Almqvist & Wiksell, Stockholm.
Jamieson, J. D., and Palade, G. E. (1967a). *J. Cell Biol.* **34,** 577.
Jamieson, J. D., and Palade, G. E. (1967b). *J. Cell Biol.* **34,** 597.

Jamieson, J. D., and Palade, G. E. (1968a). *J. Cell Biol.* **39**, 580.

Jamieson, J. D., and Palade, G. E. (1968b). *J. Cell Biol.* **39**, 589.

Jensen, T. E., and Valdovinos, J. G. (1968). *Plant Physiol.* **43**, 2062.

Kagawa, Y., and Racker, E. (1966). *J. Biol. Chem.* **241**, 2475.

Kirk, J. T. O. (1966). *In* "The Biochemistry of Chloroplasts" (T. W. Goodwin, ed.), Vol. 1, pp. 319–340. Academic Press, New York.

Kisaki, T., and Tolbert, N. E. (1969). *Plant Physiol.* **44**, 242.

Kislev, N., Swift, H., and Bogorad, L. (1966). *In* "The Biochemistry of Chloroplasts" (T. W. Goodwin, ed.), Vol. 1, pp. 355–364. Academic Press, New York.

Korn, E. D. (1966). *Science* **153**, 1491.

Koshland, D. E., and Neet, K. E. (1968). *Annu. Rev. Biochem.* **37**, 359.

Kroon, A. M., Saccone, C., and Botman, M. J. (1967). *Biochim. Biophys. Acta* **142**, 552.

Kuff, E. L., and Dalton, A. J. (1959). *In* "Subcellular Particles" (T. Hayashi, ed.), pp. 114–127. Ronald Press, New York.

Lehninger, A. L. (1964). "The Mitochondrion." Benjamin, New York.

Lindenmayer, A., and Smith, L. (1964). *Biochim. Biophys. Acta* **93**, 445.

Linnane, A. W., Lamb, A. J., Christodoulou, C., and Lukins, H. B. (1967). *Proc. Nat. Acad. Sci. U.S.* **59**, 1288.

Loening, U. E. (1968). *Ann. Rev. Plant Physiol.* **19**, 37–70.

Loud, A. V. (1968). *J. Cell Biol.* **37**, 27.

Loud, A. V., Barany, W. C., and Pack, B. A. (1964). *In* "Quantitative Electron Microscopy" (G. F. Bahr and E. H. Zeitler, eds.), pp. 258–270. Armed Forces Inst. Pathol., Washington, D. C.

MacDonald, R. A. (1961). *Arch. Intern. Med.* **107**, 335.

Mirsky, A. E., and Osawa, S. (1961). *In* "The Cell" (J. Brachet and A. E. Mirsky, eds.), Vol. 2, pp. 677–770. Academic Press, New York.

Mollenhauer, H. H., and Morré, D. J. (1966). *Annu. Rev. Plant Physiol.* **17**, 27–46.

Moor, H., Mühlethaler, K., Waldner, H., and Frey-Wyssling, A. (1961). *J. Biophys. Biochem. Cytol.* **10**, 1.

Morris, A. J., and Dickman, S. R. (1960). *J. Biol. Chem.* **235**, 1404.

Moustacchi, E., and Williamson, D. H. (1966). *Biochem. Biophys. Res. Commun.* **23**, 56.

Novikoff, A. B. (1961a). *In* "The Cell" (J. Brachet and A. E. Mirsky, eds.), Vol. 2, pp. 299–421. Academic Press, New York.

Novikoff, A. B. (1961b). *In* "The Cell" (J. Brachet and A. E. Mirsky, eds.), Vol. 2, pp. 423–488. Academic Press, New York.

Ohad, I., Siekevitz, P., and Palade, G. E. (1967). *J. Cell Biol.* **35**, 521.

Omura, T., Siekevitz, P., and Palade, G. E. (1967). *J. Biol. Chem.* **242**, 2389.

Palade, G. E. (1959). *In* "Subcellular Particles" (T. Hayashi, ed.), pp. 64–83. Ronald Press, New York.

Paltauf, P., and Schatz, G. (1969). *Biochemistry* **8**, 335.

Parsons, D. F. (1963). *Science* **40**, 985.

Pease, D. C. (1964). "Histological Techniques for Electron Microscopy." Academic Press, New York.

Plattner, H., and Schatz, G. (1969). *Biochemistry* **8**, 339.

Porter, K. R. (1961). *In* "The Cell" (J. Brachet and A. E. Mirsky, eds.), Vol. 2, pp. 621–675. Academic Press, New York.

Pullman, M. E., and Schatz, G. (1967). *Ann. Rev. Biochem.* **36**, 539–610.

Racker, E., Tyler, D. D., Estabrook, R. W., Conover, T. E., Parsons, D. F., and Chance, B. (1965). *In* "Oxidases and Related Redox Systems" (T. E. King, H. S. Mason, and M. Morrison, eds.), Vol. 2, p. 1077. Wiley, New York.

Redman, C. M. (1967). *J. Biol. Chem.* **242**, 761.

Redman, C. M., Siekevitz, P., and Palade, G. E. (1966). *J. Biol. Chem.* **241**, 1150.

Rogers, A. W. Darzynkiewicz, Z., Barnard, E. A., and Salpeter, M. M. (1966). *Nature (London)* **210**, 1003.

Roodyn, D. B., and Wilkie, D. (1968). "The Biogenesis of Mitochondria." Methuen, London.

Rosen, R. (1968). *Int. Rev. Cytol.* **23**, 25–88.

Rothstein, A. (1968). *Ann. Rev. Physiol.* **30**, 15–72.

Rotman, B., and Spiegelman, S. (1954). *J. Bacteriol.* **68**, 419.

Sabatini, D. D., Bensch, K., and Barrnett, R. J. (1963). *J. Cell Biol.* **17**, 19.

Salpeter, M. M. (1969). *J. Cell Biol.* **42**, 122.

Salpeter, M. M., Bachmann, L., and Salpeter, E. E. (1969). *J. Cell Biol.* **41**, 1.

Schatz, G. (1965). *Biochim. Biophys. Acta* **96**, 342.

Shannon, L. M. (1968). *Annu. Rev. Plant Physiol.* **19**, 187–210.

Siegel, M. R., and Sisler, H. D. (1964). *Biochim. Biophys. Acta* **37**, 83.

Siekevitz, P., and Palade, G. E. (1960a). *J. Biophys. Biochem. Cytol.* **7**, 619.

Siekevitz, P., and Palade, G. E. (1960b). *J. Biophys. Biochem. Cytol.* **7**, 631.

Silverman, L., and Glick, D. (1969a). *J. Cell Biol.* **40**, 761.

Silverman, L., and Glick, D. (1969b). *J. Cell. Biol.* **40**, 773.

Silverman, L., Schreiner, B., and Glick, D. (1969). *J. Cell Biol.* **40**, 769.

Sjöstrand, F. S. (1967). "Electron Microscopy of Cells and Tissues," Vol. 1. Academic Press, New York.

Sjöstrand, F. S., Andersson-Cedergren, E., and Karlsson, U. (1964). *Nature (London)* **202**, 1075.

Smith, D. S. (1963). *J. Cell Biol.* **19**, 115.

Staubli, W., Hess, R., and Weibel, E. R. (1969). *J. Cell Biol.* **42**, 92.

Stoeckenius, W. (1963). *J. Cell Biol.* **17**, 443.

Swick, R. W., Koch, A. L., and Hamda, D. T. (1956). *Arch. Biochem. Biophys.* **63**, 226.

Swick, R. W., Rexroth, A. K., and Stange, J. L. (1968). *J. Biol. Chem.* **243**, 3581.

Tolbert, N. E., Oeser, A., Kisaki, T., Hageman, R. H., and Yamazaki, R. K. (1968). *J. Biol. Chem.* **243**, 5179.

Tolbert, N. E., Oeser, A., Yamazaki, R. K., Hageman, R. H., and Kisaki, T. (1969). *Plant Physiol.* **44**, 135.

Van Tubergen, R. P. (1961). *J. Biophys. Biochem. Cytol.* **9**, 219.

Vitols, E., North, R. J., and Linnane, A. W. (1961). *J. Biophys. Biochem. Cytol.* **9**, 689.

Wallace, P. G., and Linnane, A. W. (1964). *Nature (London)*, **201**, 1191.

Weibel, E. R., Kistler, G. S., and Scherle, W. F. (1966). *J. Cell. Biol.* **30**, 23.

Weibel, E. R., Staubli, W., Gnagi, H. R., and Hess, F. A. (1969). *J. Cell Biol.* **42**, 68.

CHAPTER III

Organelle-Bound Enzymes

HARRY DARROW BROWN AND SWARAJ K. CHATTOPADHYAY

I. Introduction

Progress in the development of enzymology has, importantly, been based upon a fiction which served to make possible an approach to the problem at hand. We have embraced the assumption that enzymes, whatever their structural affinities to the cell, might be isolated and studied in solution as chemical entities. Indeed much of the accomplishment of "biochemistry" has followed acceptance of this fiction although we know that, contrary to this assumption, much of the cell's population of protein catalysts exist bonded to subcellular elements. Only a portion of the enzyme pool is, by any definition, in solution. (Even these nonparticulate enzymes [*sic!* cytosol enzymes] are probably active under conditions which are quite different from the high dilution of the classic *in vitro* reaction mixture.)

73

There is now substantial effort in enzymology to consider conditions which are an attempt to put aside the convention of solution chemistry applied to biological material. Hence the literature of particulate enzymology is growing, and the numbers of investigators who find themselves confronted by the inappropriateness of solution conditions are certainly significant. Our attempt to introduce the subject in this chapter is necessarily truncated and it must perforce be selective.

It would appear that some enzymes are intrinsic to one subcell component; others are found in ubiquitous distribution. However, one cannot be sure, based upon the limits of the literature (extensive though it seems to be), whether proteins that catalyze the same reaction may not be quite different chemical entities. Differences in primary sequence, secondary, tertiary, and quaternary structure are known. A reaction at one subcellular location may be quite different in its kinetics and in its relationship to allosteric effectors than at another. Even where marked differences are shown, however, we cannot rule out the possibility that the enzyme forms may be in equilibrium with each other. Since the questions have only been touched upon, the literature has hardly defined the problem. Certainly, the many hundreds of individual catalytic activities have not been characterized as a function of their subcellular site. Despite this, the existing literature is imposing, and a review that was to provide exhaustive coverage of it under the subheadings used in this chapter would necessarily be too lengthy. Hence, our approach has been essentially didactic using the techniques of literature review to illustrate concepts the authors deem important. Tables and reference lists are more comprehensive but not exhaustive. Since selection is requisite, it would be ideal to use illustrations that will prove to be landmarks. No such omniference is to be hoped for, however, and we apologize in advance for mistakes of omission and particularly for the selection balance which results less from catholic judgment than from personal academic interest.

II. Endoplasmic Reticulum and Plasma Membrane Enzymes

A. CURRENT VIEW OF THE RETICULUM AS A SUBCELLULAR ORGANELLE

A much convoluted membrane system, the endoplasmic reticulum, together with discrete bodies often affixed to the membranes, the ribosomes, constitute the *microsomes* (Claude, 1943). (This operational term known from the centrifugist lexicon survives in the biochemical literature.) Microsomes are, in fact, bits and pieces which result from the time-honored if aesthetically unpleasing procedure of mechanical homogenization. Endoplasmic reticulum, where defined to include the plasma and nuclear membranes, Golgi apparatus, etc., contains 20% or so of the total cell protein. In a specially membranous tissue, such as liver, this fraction may contain 25 mg of protein and 5–8 mg of phospholipid per gram weight of tissue.

The endoplasmic reticulum provides a geographic separateness to the chemical events that occur within the cell. It should come as no surprise that before the electron microscope provided detailed pictures of cell organization, the necessity for a structured geography was postulated and considerable indirect methodology was brought to bear upon its nature. It is certainly a comfort, however, to have photographs (e.g., Figs. 1 and 2) which provide a useful base for further work and a corroboration of the conclusions which were based upon indirect evidence.

In most membranous cells, the network of interconnecting membranes is often surfaced by pebble-like ribosomes (ribonucleoproteins). These areas of reticulum are generally called *rough* reticulum; membrane devoid of the ribosomes is the *smooth* reticulum. The extensiveness of the membrane reticulum (Figs. 1 and 2) certainly varies from tissue to tissue; and in some pathologies departure from the norm in extent of endoplasmic reticulum compared to the tissue of origin (e.g., hepatoma) is dramatic. It appears that the nuclear envelope is part of the endoplasmic reticulum structure and may, with logical consistency, be considered together with the rough, smooth, and plasma membrane as an additional subform. Though interconnected, these specialized suborganelles are no doubt distinct in function and, therefore, in enzyme localization. Plasma membranes, for example, have been isolated from cellular tissue of a number of experimental animals. They represent a physiologically distinct entity in that the enzyme complement is different from that of cell solution, mitochondria, lysozymes, or, and of particular interest here, that of the microsome even though the two areas of membrane may be continuous and confluent. Benedetti and Emmelot (1968) reported that in plasma membranes isolated from liver the following enzyme activities are absent: hexokinase, 3′-mononucleotidase, monoamine oxidase, adenylate kinase, the succinate–cytochrome c reductase system, cytochrome c oxidase, cathepsin, phosphoprotein phosphatase, reduced nicotinamide adenine dinucleotide phosphate (NADPH) oxidase, and N-demethylase. Further illustrative of the distinctness of the plasma membrane, cytochrome P-450 characteristic of endoplasmic reticulum membrane and present there in large amount (Estabrook *et al.*, 1969) was also (Benedetti and Emmelot, 1968) absent from rat liver membrane.

Morphologists find that the endoplasmic reticulum appears in a variety of configurations associable with particular cell types and with particular

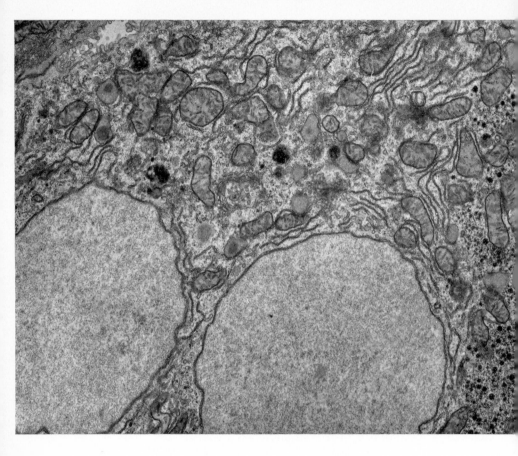

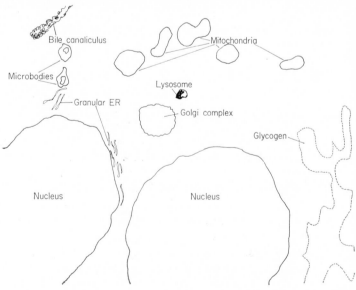

FIG. 1. Liver cell electron micrograph showing bile canaliculus, microbodies, lysozome, Golgi complex, nucleus, mitochondria, and other structures. (Photograph courtesy of R. Yates.)

Bile canaliculus

Microbodies

Granular ER

Nucleus

Mitochondria

Lysosome

Golgi complex

Glycogen

Nucleus

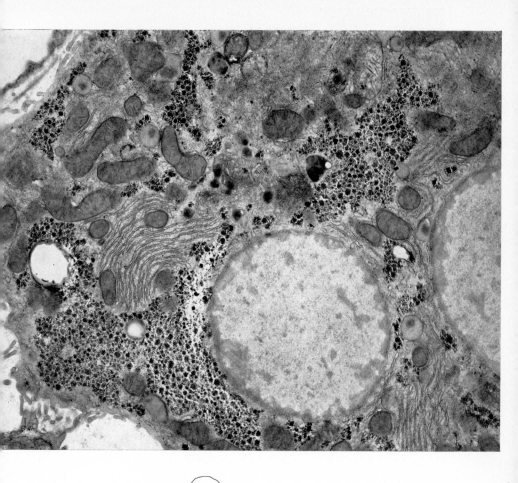
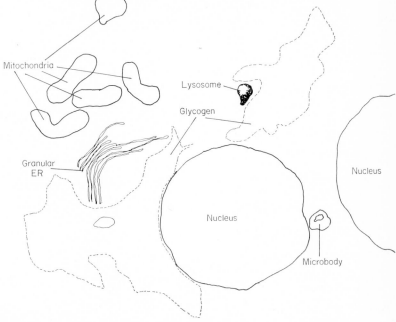

IG. 2a. Liver cell
tron micrograph
ing extensive gly-
n stores and other
ellular structures.
otograph courtesy
. Yates.)

Mitochondria

Lysosome

Glycogen

Granular
ER

Nucleus

Nucleus

Microbody

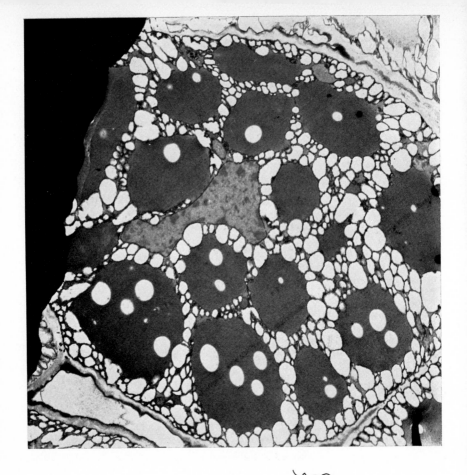

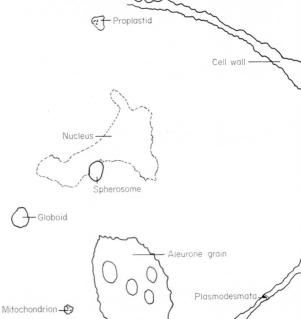

Fig. 2b. Cross section of cottonseed parenchyma cell. (Photograph courtesy of Lawrence Y. Yatsu.)

species. It is a generalization that cells having similar functions are recognizably similar in their endoplasmic reticulum morphology.

B. The Enzyme Systems

1. Introduction

Many biosynthetic reactions are catalyzed by enzymes bound to the endoplasmic reticulum. The central importance of the localized (insoluble) systems may relate to a role in the control of metabolism. There is inferential evidence that many parameters associated with the reactions catalyzed by these proteins are a function of their relationship to other components of the membrane. Indeed, membrane attachment may affect not only the kinetics and substrate specificities of the reactions but also impose vectorial character upon the system (Mitchell, 1963, 1968, 1969; H. D. Brown et al., 1966, 1967b, 1968c,d; D. E. Green and Baum, 1970). These key systems include those responsible for the electron transferring (energy flow) phenomena in which NADH or NADPH provide the reducing equivalents; biosynthesis of lipids including steroids and phospholipids, and the synthesis of mucopolysaccharides and other complex polysaccharide moieties. In addition, the localization of the $Na^+ + K^+$ stimulated adenosinetriphosphatases (ATPases) with membrane fractions seems to confirm that a transport phenomenon is localized on the membranes. Hence both synthesis and distribution systems are organized at or very near to the membrane structures. Table I outlines the several series of chemical pathways associated with the endoplasmic reticulum. The enzymes associated with the organelle are tabulated in Table II.

Complexities of enzymology were certainly not lessened by formalization of the concept of isozymes (different proteins of the organism having similar catalytic activities). A well-studied isozyme system is the lactic dehydrogenase (LDH) family of enzymes. The variability of LDH is a function of arrangement of basic residues which are put together in one of several ways. Given variants tend to predominate in given tissues of the organism. In a number of isozyme families, however, multiple components (even of crystalline preparations) have been obtained from a single tissue. This fact and subcell fractionation studies may indicate (e.g., H. D. Brown et al., 1969a,b) that a particular isozyme may be associated intrinsically with a given subcell organelle. Thus molecular populations, within a single cell, appear to vary with regard to arrangement of active residues.

Lists of isozymes have grown extensive. The symposium volume edited by Vesell (1968) illustrates the diversity of enzymic activities which exists

TABLE I

MAJOR CHEMICAL PATHWAYS ASSOCIATED WITH ENZYMES OF THE
ENDOPLASMIC RETICULUM

Summary notation	Review[a]
I. NADH $\rightarrow$ O_2	Estabrook et al. (1969)
	Sato et al. (1969)
	Estabrook and Cohen (1969)
II. NADPH $\rightarrow$ O_2	Brodie et al. (1958)
	Hayaishi (1969)
	Estabrook and Cohen (1969)
	Sato et al. (1969)
	Gillette and Gram (1969)
III. Triphosphate $\rightarrow$ cyclic monophosphate	Sutherland et al. (1962)
	Butcher and Sutherland (1962)
	G. A. Robinson et al. (1967)
	G. A. Robinson et al. (1968)
IV. ATP $\rightarrow \dfrac{ADP + P_i}{AMP + 2\,P_i} \rightarrow$ In part the source of the energy of osmotic work	Mitchell (1963)
	Albers (1967)
	Heinz (1967)
	Glynn (1968)
V. Mucopolysaccharide synthesis	Glaser and Brown (1957a, b)
	Ginsburg and Neufeld (1969)
VI. Sterol synthesis	Bloch (1965)
	Dean et al. (1967)
	Frantz and Schroepfer (1967)
	Mukherjee and Bhose (1968)
	Scallen et al. (1968)
	Fillios et al. (1969)
	Kandutsch and Saucier (1969)
	Etemadi et al. (1969)
	Akhtar et al. (1969a, b)
VII. Triglyceride synthesis	Y. Stein and Shapiro (1958)
	Siekevitz (1963)
	Hübscher et al. (1964)
	M. E. Smith et al. (1967)
	Bickerstaffe and Annison (1969)
VIII. Phospholipid synthesis	Wilgram and Kennedy (1963)
	Siekevitz (1963)
	Hübscher et al. (1964)
	Shapiro (1967)
	Carter (1968)
	Sánchez de Jiménez and Cleland (1969)
	D. W. Foster and McWhorter (1969)
	Barden and Cleland (1969)
IX. Protein synthesis	Schweet and Heintz (1966)
	H. J. Vogel and Vogel (1967)
	Campbell et al. (1967)
	Tsukada et al. (1968)
	Lengyel and Soll (1969)
	Massaro et al. (1970)
	Rudolph and Betheil (1970)

[a] Includes original papers with substantial literature introductions. Inclusion of references in this table does not imply that endoplasmic reticulum systems are dealt with exclusively.

in a variety of molecular forms. N. O. Kaplan (1968), in fact, believes that enzymes which exist in only one form are the exception. Despite the fact that in those systems in which isozymes have been characterized, variations of primary structure have been shown and the concept must logically be extended to variants in secondary or tertiary structure and even of conformation. As we have suggested earlier (H. D. Brown et al., 1966, 1967a,b, 1968a–d), membranes themselves affect characteristics of the enzyme. The mathematical treatment by Laidler and Sundaram (Chapter V of this volume) is both a discussion of this fact and an illustration to the point. Hence the parameters associated with enzymic activity are derived not only from the protein's structure but from its conformation and the environment imposed upon it by its attachment to a membrane. Whether this view is considered a simple extrapolation from the concept of isozymes or one distinct in definition, it may be pertinent to the function of control mechanisms of intermediary metabolism.

Purification of reticulum-bound enzymes has been described in a large and imposing literature which collectively reports the refractory nature of endoplasmic reticulum enzymes to solubilization procedures of any description. For the most part, the procedures used have been those conventional to the protein biochemist: mechanical or sonic disruption in consonance with "salting-out," extraction, or detergent treatment. Many of the reticulum's physiologically active proteins which have been dealt with were solubilized, if at all, only by very drastic treatments. Reports confirm the alteration of properties which were incident upon the particular technique used (P. D. Swanson et al., 1964; H. D. Brown et al., 1966, 1967b,d). Many authors, including Tashiro (1957), have recognized that the enzyme in membranes must be presumed to be bound to lipids (and probably to other membrane components) and, hence, resistant to solubilization unless those bonds (hydrogen bonds, van der Waals forces, —S—S— bridges, electrostatic interactions, possibly others) were broken. Results have often been unsatisfactory with small residual activity being the rule. The literature is further made difficult by the lack of uniformity in definition of "solubility" and in the variations of rigor for establishing the uniformity of particle size. It has frequently been suggested that these enzymes require the lipid component of the membrane for activity. Hence, it may be that the difficulties in purification have resided with the fact that the lipids which were either chemically or, more probably, physically (by providing a structured surface) essential were lost during the purification. Fleischer et al. (1962) in his studies of the NADH–electron transport system of mitochondria have shown the dependence of a complex enzyme system upon membrane lipids. Other workers (Rothfield and Finkelstein, 1968) have confirmed functional relation-

TABLE II

Enzymes Associated with the Endoplasmic Reticulum

E. C. number	Enzyme (trivial name)	Comments and references
1.1.1.19	Glucuronate reductase	Enzymes of the ascorbate-forming systems are associated with "heavy microsomes," with a range of activity from high to very little. "Light microsomes" have no activity. Kanfer et al. (1959), rat liver; Kar et al. (1962), rat and goat liver; N. C. Ghosh et al. (1963), rat and goat liver
1.1.1.20	Glucuronolactone reductase	Centrifugal fractions used to identify activity with microsomes (Chatterjee et al., 1960). Isherwood et al. (1960) report that the mitochondria possess activity to a lesser extent. Chatterjee et al. (1959), goat liver; Chatterjee et al. (1960), rat and goat liver, chick kidney; Isherwood et al. (1960), rat liver; Suzuki et al. (1960), rat liver
1.1.1.22	Uridine diphosphate glucose (UDPG) dehydrogenase	This complex catalytic activity requires supernatant fraction as well as microsomes. It is believed that supernatant which contains UDP glucoside dehydrogenase oxidizes UDPG and the microsomal enzyme catalyzes the transfer of the glucuronyl group to form glucuronides. Strominger et al. (1957), guinea pig liver
1.1.1.27	Lactate dehydrogenase	Although lactate dehydrogenase is predominantly in the supernatant fraction, activity has been observed in liver and hepatoma microsomes. Approximately half of total activity has been associated with cytoplasm, the remaining activity is particulate. Novikoff (1960), hepatoma; Amberson et al. (1965), rabbit skeletal muscle; Agostoni et al. (1966), rat liver; Güttler (1967), calf liver, heart, kidney, and skeletal muscle; Güttler and Clausen (1967), calf liver, heart, kidney, and skeletal muscle; H. D. Brown et al. (1969a), rat liver and Morris hepatoma; H. D. Brown et al. (1969b), rat breast carcinoma
1.1.1.34	Hydroxymethylglutaryl-CoA reductase	This enzyme, an important catalytic agent in cholesterol synthesis, is found only in heavy and light microsomes. Microsomal reductase activity is very low relative to other activities in this complex biosynthetic pathway. Bucher and McGarrahn (1956), rat liver; Bucher et al. (1960), rat liver; Linn (1967a,b), rat liver
1.1.1.37	Malate dehydrogenase	A small percentage of enzyme activity has been reported in the microsomal fraction. Shull (1959), horse liver microsomes; de Duve et al. (1962)
1.1.1.47	Hexose 6-phosphate dehydrogenase (glucose dehydrogenase)	Most of the enzyme is localized in microsomal fraction. Beutler and Morrison (1967), mouse liver; Srivastava and Beutler (1969), rat liver
1.1.1.51	β-Hydroxysteroid dehydrogenase	Diffuse distribution has been reported for β-hydroxysteroid dehydrogenase. Two distinct 17-β-hydroxysteroid dehydrogenases have been identified in guinea pig liver. One, nicotinamide adenine dinucleotide (NAD)-linked, is mainly localized in microsomes; the other is NADP-specific and found in the cytoplasmic fraction. 3-β-Hydroxysteroid dehydrogenase obtained from bovine adrenal cortex is found to be NAD-dependent and mainly associated with microsomes. In rat liver, NADP-linked 11-β-hydroxysteroid dehydrogenase has been found mainly in microsomes. However, both NAD- and NADP-linked 17-β-hydroxysteroid dehydrogenases in dog prostate gland have been found in microsomes. Mouse liver microsomes have β-hydroxysteroid dehydrogenase capable of catalyzing the oxidation and reduction of C_{27} and C_{30} steroid (NADP used preferentially). Beyer and Samuels (1956), ox adrenal cortex; Hurlock and Talalay (1959), rat liver; G. L. Endahl et al. (1960), guinea pig liver; Villee and Spencer (1960), guinea pig liver; Kandutsch (1967), mouse liver; Cheatum et al. (1967), bovine adrenal cortex; Hussein and Kochakian (1968), dog prostate gland; Neville et al. (1969), bovine adrenal cortex; Koerner (1969), rat liver; Sulimovici and Boyd (1969), rat ovary
1.1.1.53	Cortisone reductase	Reduced NADP-dependent enzyme is present in microsomes of rat kidney. Mahesh and Ulrich (1960), rat kidney

82

1.2.1.9	Triosephate dehydrogenase	Emmelot and Bos (1966), rat liver and hepatoma
1.3.1.-	Δ,7-Cholesterol reductase	Found in both supernatant and microsomal fractions. Frantz et al. (1959), rat liver
1.3.1.-	Desmosterol reductase	35% of activity in the microsomal fraction. Avignon and Steinberg (1961), rat liver
1.3.2.-	Gulonolactone dehydrogenase	This enzyme, which catalyzes the last step in the formation of ascorbic acid, appears to be essentially associated with microsomal fraction. Chatterjee et al. (1959), goat liver; Isherwood et al. (1960), rat liver
1.4.3.4	Monoamine oxidase	Although monoamine oxidase is generally considered to be a mitochondrial enzyme, variable amounts of monoamine oxidase activity have been observed in microsomal fraction after differential centrifugation of various tissue homogenates. Normally a small but significant (as high as 24%) part of the total activity of liver homogenate has been recovered from rat liver microsomes. Variable levels of activity have been reported in other tissues. Hawkins (1952), rat liver; de Duve et al. (1960), rat liver; Jarrott and Iversen (1968), rat liver and vas deferens; A. G. Fischer et al. (1968), thyroid gland; Youdim and Sandler (1968), rat liver, brain, and heart; Stjarne et al. (1968), rat liver and other tissue; Thines-Sempoux et al. (1969), rat liver
1.6.2.2	Cytochrome b$_5$ reductase	Associated with microsomes in liver, cytochrome b$_5$ serves as a carrier in the reaction catalyzed by the enzyme. P. Strittmatter and Velick (1956), rabbit liver; P. Strittmatter (1958); Estabrook and Cohen (1969); Okada and Okunuki (1969), house fly larvae
1.6.4.1	Cystine reductase	In rat liver and kidney this enzyme is microsomal. L. T. Myers and Worthen (1961), rat liver and kidney
1.6.4.3	Lipoamide dehydrogenase	There are two or more enzymes in this system with a variable distribution of the acceptor-specific forms. C. H. Williams et al. (1959), rat liver; T. F. Slater (1959), rat liver; Reid (1961a,b), rat liver; de Duve et al. (1962); Ernster et al. (1962), rat liver
1.6.6.7	Azobenzene reductase	This enzyme is entirely associated with microsomes, especially with the "heavy" microsomal fraction. It requires reduced NADP and anaerobic conditions for optimum activity. Mueller and Miller (1949), rat liver; Fouts et al. (1957), guinea pig, mouse, rabbit, rat, and chicken liver; Emanoil-Ravicovitch and Herisson-Cavet (1963); P. H. Hernandez et al. (1967a,b), rat liver; Burba (1968), guinea pig, rabbit, and rat liver
1.6.99.1	Reduced NADP dehydrogenase	64% of this activity sediments with microsomal fraction and the level is higher in hepatic preparations than in any other tissue studied. It is distributed equally between heavy and light microsomes. Hogeboom and Schneider (1950), mouse liver; de Duve et al. (1955), rat liver; Baillie and Morton (1958), cow mammary gland; Giuditta and Strecker (1959), rat brain; T. F. Slater (1959), rat liver; C. H. Williams et al. (1959), rat liver; Kikuchi et al. (1959), Ascaris muscle; Novikoff (1960), hepatoma ascites, Novikoff hepatoma; Spiro and Ball (1961a), ox adrenal medulla; Spiro and Ball (1961b), ox adrenal medulla and cortex; Phillips and Langdon (1962), rat liver; Kamin et al. (1965), rat liver; Masters et al. (1965a,b), rat liver; Gillette (1966); Kamin et al. (1966), liver; P. H. Hernandez et al. (1967a,b), rat liver; Kato et al. (1968), rat liver; Kato and Takanaka (1968), rat liver; H. D. Brown et al. (1969b), rat liver and Morris hepatoma; Kato et al. (1969), rat liver; Dallman et al. (1969), rat liver; Ichikawa and Yamano (1969), rabbit liver; Gillette and Gram (1969); Thines-Sempoux et al. (1969), rat liver
1.6.99.3	Reduced NAD dehydrogenase	Three distinct NADH-cytochrome c reductases have been described in rat liver. One is attached to microsomal fraction and is antimycin-A insensitive. Its activity presumably requires cytochrome b$_5$ as an intermediate carrier. Other tissues do not give a clear-cut picture of the enzyme's subcellular distribution. However, in rat and guinea pig spleen and in Novikoff hepatoma ascites, the microsomal activity, as in liver, is higher than is the mitochondrial activity. In rat brain and nerve, ox adrenal medulla and cortex, cow mammary gland and in chick embryo, the mitochondrial type is more abundant than is the microsomal. Hogeboom and Schneider (1950), rat and mouse liver; Brody et al. (1952), rat cerebral cortex; de Duve et al. (1955), rat liver; Palade and Siekevitz (1956), rat liver; Eichel (1957a), rat spleen; Eichel (1957b), rat and guinea pig spleen; Novikoff (1960), hepatoma ascites; Chauveau et al. (1962), rat liver; L. Ernster et al. (1962), rat liver; Dallner (1963), rat liver; P. D. Jones and Wakil (1967), hen liver; Dallman et al. (1969), rat liver; Thines-Sempoux et al. (1969), rat liver

TABLE II (Continued)

E. C. number	Enzyme (trivial name)	Comments and references
1.8.3.1	Sulfite oxidase	Earlier indications that this enzyme was diffuse in distribution (Baxter et al., 1958) now have been supplemented by evidence which indicates that it is mostly microsomal and nuclear. Baxter et al. (1958), rat liver; McLeod et al. (1961), ox liver and rat liver; Joshi et al. (1969), rat liver and rat embryo
1.9.3.1	Cytochrome oxidase	Like other mitochondrial enzymes, cytochrome oxidase is occasionally used as a chemical marker. However, cytochrome oxidase has been isolated from mitochondria-free mouse liver homogenates. This may result from mitochondrial damage (especially in glycerol) but the question remains unresolved and the use of the enzyme as an organelle marker is probably not justified. Brody et al. (1952), rat cerebral cortex; Wattiaux et al. (1956), rat liver; Carruthers et al. (1959), mouse liver
1.10.3.1	Tyrosinase	Menon and Haberman (1970), B16 and Harding–Passey melanoma of mice
1.11.1.2	NADP peroxidase	Gillette et al. (1957), rabbit liver
1.14.1.–	Hydroxylases and related systems	The hydroxylases and related enzymes which are responsible for drug metabolism are usually associated with microsomal fraction. Axelrod (1955), rabbit and mammalian liver; Brodie et al. (1955), rabbit liver; Cooper and Brodie (1955), rabbit liver; McLagan and Reid (1957), rat liver; Stanbury (1957), rat liver; Cremer (1958), rat liver; Stanbury and Morris (1958), rat liver; Tata (1958), rat liver; Seal and Gutmann (1959), rat liver; Yamazaki and Slingerland (1959), rat liver; H. S. Mason et al. (1965), mammalian liver; Orrenius (1965), rat liver; Orme-Johnson and Ziegler (1965), mammalian liver; Fouts et al. (1966), liver; Imai and Sato (1966b), rabbit, guinea pig, and rat liver; Ziegler and Pettit (1966), pork liver; Machinist et al. (1966), pork liver; J. T. Wilson and Fouts (1967), opossum; Lotlikar et al. (1967), rat, hamster, mouse, rabbit, and guinea pig liver; E. C. Miller and Miller (1967), rat liver; Gilbert and Golberg (1967), rat liver; Schenkman et al. (1967), rat liver; Lin et al. (1967), rat liver; Gram et al. (1968a,b), rat liver; Stevenson and Turnbull (1968), rat liver; Imai and Sato (1968b), rabbit, guinea pig, and rat liver; Hoffman et al. (1968), rat liver; Jaffe et al. (1968), mouse liver; Kratz (1968), rabbit liver; Gurtoo et al. (1968), rat liver; Gigon et al. (1968), rat liver; Kato and Takanaka (1968), rat liver; Kato and Takahashi (1968), rat liver; de Baun et al. (1968), rodent liver; Machinist et al. (1968), rat liver; Schonbrod et al. (1968), housefly muscle; Anders (1968), rat liver; Orrenius and Thor (1969), rat liver; Kato et al. (1969), rat liver; Lu et al. (1969), guinea pig, rat, and rabbit liver; C. F. Strittmatter and Umberger (1969), chicken liver; Agosin et al. (1969), Triatoma infestans; Prabhu (1969), mouse liver; Guarino et al. (1969), rat liver; Levin et al. (1969), rat liver; H. D. Brown et al. (1969c), rat liver and Morris hepatoma; H. D. Brown et al. (1969d), rat liver; H. D. Brown et al. (1969e), rat liver; Pennington et al. (1970a,b), rat liver; Jerina et al. (1970), rat liver
	[Heme oxygenase]	The enzyme is specifically localized in microsomes; requires NADPH and molecular oxygen. It is inhibited by carbon monoxide thus suggesting it is a mixed-function oxidase. Tenhunen et al. (1969), rat liver
1.14.1.1	Aryl-4-hydroxylase	This enzyme has been described as exclusively a microsomal entity. Mitoma et al. (1956), liver; Conney et al. (1957), liver; Hultin (1957), liver; Hultin (1959), liver; Nebert and Gelboin (1968a,b), hamster fetus; Robbins (1968), mammalian tissue; Daly et al. (1968), rabbit liver; Anders (1968), rat liver; Wada et al. (1968), rat and mouse liver; Nebert and Gelboin (1969), mammalian tissues; Nebert and Gelboin (1969), mammalian cells; Daly et al. (1969), mammalian NADPH; Nebert and Gelboin (1970), mammalian cell culture
1.14.1.3	Sterol cyclase	Transfers 2,3-oxidosequalene to lanosterol. This reaction does not require oxygen or NADPH. Dean et al. (1967), hog liver; Yamamoto et al. (1969), hog liver; Yamamoto and Bloch (1969), hog liver

84

EC No.	Enzyme	Notes and references
1.14.1.7	17-α-Hydroxylase	The enzyme is responsible for 17-α-hydroxylation functions in a complex biological oxidation system and requires reduced NADP and O₂. It appears to be solely in the microsomes. Ball and Kadis (1965), sow ovary; Shikita et al. (1965), rat testes
1.14.1.8	Steroid 21-hydroxylase	Narasimhulu et al. (1965), steer adrenal cortex; Inano et al. (1969), rat adrenal
2.1.1.-	Phosphatidylethanolamine methyltransferase	K. D. Gibson et al. (1961), rat liver
2.1.1.-	Δ24-Sterol methyltransferase	This enzyme is firmly bound to microsomes. J. T. Moore and Gaylor (1969), yeast
2.3.1.-	3-Acylglycerophosphoryl-choline acyltransferase (from acyl-CoA)	Lands and Merkl (1963), rat liver
2.3.1.-	Acyl-CoA–phospholipid acyltransferase	Okuyama and Lands (1969), rat liver
2.3.1.13	Glycine acyltransferase	The activity has been reported to be associated with both supernatant and microsomal fractions. Bremer (1955), rat liver
2.3.1.15	Glycerolphosphate acyltransferase	Associated largely with microsomal fraction in rat liver. Y. Stein and Shapiro (1958), rat liver; Barden and Cleland (1969), rat liver; Zahler and Cleland (1969), rat liver; Sánchez de Jiménez and Cleland (1969), rat brain; Bickerstaffe and Annison (1969), pig, sheep, and chicken intestine
2.3.1.20	Diglyceride acyltransferase	Approximately 84% of the total activity sediments with the microsomal fraction. Wilgram and Kennedy (1963), rat liver; Bickerstaffe and Annison (1969), pig, sheep, and chicken intestine
2.3.2.-	γ-Glutamyl transpeptidase	In rat and pig kidney, the enzyme is associated with the microsomal fraction. Slight activity in mitochondrial and supernatant fractions are assumed to result from technical limitations. Binkley et al. (1950), pig kidney; Avi-Dor (1960), rat kidney; Binkley (1961), pig and rat kidney
2.4.1.1	α-Glucan phosphorylase, glycogen phosphorylase	After sucrose density-gradient centrifugation, about 70 to 80% of this enzyme was recovered from microsomal sediment in fed rats. In starved animals, however, 80% of the enzyme was present in the 105,000 g supernatant. Tata (1964), rat liver
2.4.1.-	Lactose synthetase	Present both in microsomal and supernatant fraction. Brodbeck and Ebner (1966), bovine and rat mammary tissue
2.4.1.16	N-Acetylglucosaminyl-transferase	The enzyme requires N-acetylglucosamine as substrate. D. C. Collins et al. (1968), rabbit liver
2.4.1.17	UDP glucuronyltransferase	This enzyme is found in rough microsomes primarily or equally distributed in rough and smooth microsomes. Dutton and Storey (1954), mouse liver; Strominger et al. (1957), guinea pig liver; Leventer et al. (1965), guinea pig liver; Gram et al. (1968a, b), rabbit liver; Graham and Wood (1969)
2.4.1.23	UDP galactosesphingosine galactosyltransferase, psychosine UDP galactosyltransferase	The formation of psychosine from UDP galactose and sphyngosine is reported to be largely localized in the microsomes. Cleland and Kennedy (1960), rat and guinea pig brain
2.6.1.5	Tyrosine aminotransferase	Tyrosine aminotransferase is a diffusely distributed particulate enzyme in rat brain. Highest specific activity is in the mitochondrial fraction; some activity has been observed in microsomal fraction. J. E. Miller and Litwack (1969), rat brain
2.7.-.-	Adenyl cyclase	Sediments with plasma membrane, nuclear membrane, mitochondria, and microsomal fractions with considerable variations reported as a function of tissue studied. Sutherland and Rall (1960), dog liver; Murad et al. (1962); de Robertis et al. (1967), rat brain cortex; Castañeda and Tyler (1968), sea urchin egg; G. A. Robinson et al. (1968); Vaughan and Murad (1969), fat cells; Entman et al. (1969), canine heart; Cryer et al. (1969), fat cell; Marinetti et al. (1969), rat liver

TABLE II (*Continued*)

E. C. number	Enzyme (trivial name)	Comments and references
2.7.1.1	Hexokinase	Diffuse in distribution but mostly particulate in guinea pig cerebral cortex. Crane and Sols (1953), rat tissue; McComb and Yushok (1959), Krebs-2-ascites tumor; Emmelot and Bos (1966), rat hepatoma; Bachelard (1967), guinea pig cerebral cortex; Hanson and Fromm (1967), rat skeletal muscle; H. D. Brown *et al.* (1968a), Pekin duck muscle; Newsholme *et al.* (1968), guinea pig cerebral cortex
2.7.1.11	Phosphofructokinase	Although this enzyme was thought to be associated exclusively with the cytosol, it is found in sarcoplasmic reticulum of frog skeletal muscle. Margreth *et al.* (1967), frog skeletal muscle
2.7.3.2	Creatine kinase	Although primarily found in supernatant, substantial activity is associated with mitochondria, and slight activity in microsomal fraction can, in rat cerebral cortex, be released by mild treatment with NaCl or ethylenediaminetetraacetate. Sullivan *et al.* (1968), rat cerebral cortex; H. D. Brown *et al.* (1968a), Pekin duck muscle; R. J. Baskin and Deamer (1970)
2.7.7.8	Polyribonucleotide nucleotidyltransferase, polynucleotide phosphorylase	Most activity (80%) is present in the soluble fraction and the balance is equally distributed between cell membranes and washed ribosomes. Kimhi and Littauer (1967), *Escherichia coli* cells
2.7.7.16	Ribonuclease	The enzyme has a diffuse distribution pattern: Earlier studies indicated that alkaline ribonuclease is localized in ribosomes but recent observations indicate that the endoplasmic reticulum membranes contain ribonuclease activity. Bartos and Uziel (1961), guinea pig pancreas; de Duve *et al.* (1962); Morais and de Lamirande (1965), rat liver; de Lamirande *et al.* (1966), rat liver; de Lamirande *et al.* (1967), rat liver
2.7.8.-	Deoxycytidinediphosphatase choline-1,2-diglyceride cholinephosphotransferase	50% or more of the total catalytic activity sedimented with microsomes. Schneider (1963), rat liver; Brindley and Hübscher (1965); Benjamins and Agranoff (1969), guinea pig brain
2.7.8.2	Cholinephosphotransferase	97% of the total activity is in the microsomal fraction. Hübscher (1962), rat liver; Wilgram and Kennedy (1963), rat liver
2.8.1.2	3-Mercaptopyruvate sulfurtransferase	Half of the activity has been observed in nuclear fraction and the rest in other fractions. Activity was found in microsomes when fractionation was performed in KCl. Kun and Fanshier (1959), rat liver
3.1.1.-	Lipoprotein lipase	40% of the total enzyme activity sediments with microsomes. Shibko and Tappel (1964), rat liver
3.1.1.1	Carboxylesterase	About half of the catalytic activity is associated with the microsomes. Omachi *et al.* (1948), mouse liver; Heller and Bargoni (1950), rat and mouse liver; Aldridge and Johnson (1959), rat brain; Marker and Hunter (1959), mouse liver; Heymann *et al.* (1969), swine liver and kidney
3.1.1.2	Arylesterase	Approximately 58% of the arylesterase activity is associated with the microsomal fraction. Underhay *et al.* (1956), rat liver; Carruthers and Baumler (1962), mouse liver
3.1.1.4-5	Phospholipase A and B	61% of the activity is in the microsomes. Shibko and Tappel (1964), rat liver; Björnstad (1966), rat liver; Waite and Van Deenen (1967), rat liver
3.1.1.6	Acetylesterase	In rat liver, acetylesterase is found only in the microsomal fraction. The enzyme is particulate in other tissues studied also but its distribution is not well established. Ludewig and Chanutin (1950), rat liver; Novikoff *et al.* (1953b), rat liver; Underhay *et al.* (1956), rat liver; Novikoff (1957), Novikoff hepatoma; Fenwick (1958), fat body of the desert locust; Krishnamurthy *et al.* (1958), chicken and rat liver; Giuditta and Strecker (1959), rat brain; Seshardri Sastry *et al.* (1961), rat liver

86

3.1.1.7-8	Acetylcholinesterase and cholinesterase	The enzyme has a diffuse distribution in rat and rabbit brain, in insect nerve and in torpedo electric tissue. Activity is present in all subcellular fractions. In rat and rabbit liver and in human muscle the enzyme is predominantly (40–60%) in the microsomal fraction. Goutier and Goutier-Pirotte (1955a,b), rat, rabbit, and guinea pig liver; Nathan and Aprison (1955), rabbit brain; Hagen (1955), ox adrenal medulla; Smallman and Wolfe (1956), insect nerve; Underhay et al. (1956), rabbit liver; Frontali and Toschi (1958), torpedo electric tissue; Aldridge and Johnson (1959), rat brain; Toschi (1959), rat brain; J. C. Smith et al. (1960), human muscle; Trotter and Burton (1969), rat brain
3.1.1.9	Benzoylcholinesterase	Similar in distribution pattern to acetylcholinesterase in rat, guinea pig, and rabbit liver with very little activity in supernatant fraction. In dog pancreas, 40–50% of the total activity is associated with the supernatant fraction and the rest is distributed among particulate fractions. Goutier and Goutier-Pirotte (1955a), rat, guinea pig, and rabbit liver; Goutier and Goutier-Pirotte (1955b), dog pancreas
3.1.1.12	Vitamin A esterase	In rat liver this enzyme is found in the microsomal fraction predominantly, but in chicken liver the enzyme is also found at substantial levels in the nuclear fraction. Krause and Powell (1953), rat liver; Ganguly and Devel (1953), rat liver; Ganguly (1954), rat liver; Krishnamurthy et al. (1958), rat and chicken liver; Shibko and Tappel (1964), rat liver; Hayase and Tappel (1969), rat liver
3.1.1.13	Cholesterol esterase	Same distribution as vitamin A esterase. Schotz et al. (1954), rat liver; Krishnamurthy et al. (1958), rat and chicken liver; Fillios et al. (1969), rat liver
3.1.1.19	Uronolactonase	Associated with the microsomal fraction in 0.25 M sucrose homogenates. Winkelman and Lehninger (1958), rat liver
3.1.3.1	Alkaline phosphatase	This enzyme's localization has not been established; two distinct types are known. An Mg^{2+}-insensitive alkaline phosphatase has a diffuse distribution among particulates with greatest activity in nuclear and microsomal fractions in liver. In kidney and intestine, however, Mg^{2+}-insensitive enzyme has been found largely in the microsomal fraction. The Mg^{2+}-sensitive enzyme activity is found mostly in the supernatant fraction. Alkaline phosphatase is found to be present in other structures, such as Golgi apparatus, epididymis, and secretory granules. Kabat (1941), mouse kidney; Chantrenne (1947), mouse liver; Ludewig and Chanutin (1950), rat liver; Novikoff et al. (1950), rat liver; Hers et al. (1951), guinea pig kidney; Kallman (1951), rabbit kidney; Abood et al. (1952), rat brain; Rosenthal et al. (1952), rat liver; Tsuboi (1952), mouse liver; Novikoff et al. (1953b), rat liver; Abood and Gerard (1954), rat nerve; Dianzani (1954b), rat liver; Morton (1954a), calf intestinal mucosa; Morton (1954b), cow mammary gland; Rosenthal and Vars (1954), rat liver; Straus (1954), rat kidney; Allard et al. (1954), rat liver; Baccari and Guerritore (1955), rat kidney; Emery and Dounce (1955), rat liver; Wakid and Kerr (1955), dog brain; Allard et al. (1956), rat liver; Stark and Jung (1956), human placenta; Allard et al. (1957), rat azodye hepatoma and Novikoff hepatoma; Goodlad and Mills (1957), rat liver; Baillie and Morton (1958), cow mammary gland; Krishnamurthy et al. (1958), rat and chicken liver; La Bella and Brown (1958), pig anterior pituitary: Ahmed and King (1959), human placenta; de Vincentiis and Testa (1959), ox retina; Kasbekar et al. (1960), human placenta; T. F. Slater and Planterose (1960), Novikoff hepatoma; Binkley (1961), pig and rat kidney; Fernley and Walker (1967), calf intestine; Eaton and Moss (1967), human liver and intestine; Moss et al. (1967), human liver and small intestine; Melani et al. (1968), rat kidney
3.1.3.4	Phosphatidate phosphatase	In rat and chicken liver this enzyme seems to be associated with microsomal fraction. S. B. Weiss and Kennedy (1956), chicken liver; Sedgwick and Hübscher (1965); M. E. Smith et al. (1967), rat liver; Johnston et al. (1967); Bickerstaffe and Annison (1969), pig intestine

TABLE II (Continued)

E. C. number	Enzyme (trivial name)	Comments and references
3.1.3.5	5'-Nucleotidase	This enzyme is bound to membranes of the endoplasmic reticulum and the plasma membrane in rat liver. Abood and Gerard (1954), rat nerve; Wakid and Kerr (1955), dog brain; Goodlad and Mills (1957), rat liver; Novikoff (1957), Novikoff hepatoma; de Lamirande et al. (1958a), Novikoff hepatoma; de Lamirande et al. (1958b), rat liver; Ahmed and King (1959), human placenta; Giuditta and Strecker (1959), rat brain; Cerletti et al. (1960), human placenta; Emmelot and Bos (1965), rat liver; Song and Bodansky (1967), rat liver; Song et al. (1968), rat liver; Emmelot and Bos (1968), rat liver, plasma membrane; Thines-Sempoux et al. (1969), rat liver
3.1.3.9	Glucose-6-phosphatase	In various tissues from several animals the activity has been isolated with the microsomal fraction. This same enzyme has been reported to have pyrophosphatases and pyrophosphate glucose phosphotransferase activity. M. A. Swanson (1950b), rat liver; Hers et al. (1951), rat liver; Glock and McLean (1953), rat liver; Kuff and Schneider (1954), rat liver; Beaufay and de Duve (1954), rat liver; de Duve et al. (1955), rat liver; Gianetto and Viala (1955), rat liver; Birbeck and Reid (1956), rat liver; Reid et al. (1956), rat liver; Underhay et al. (1956), rat liver; Wattiaux et al. (1956), rat liver; Goodlad and Mills (1957), rat liver; Krishnamurthy et al. (1958), rat and chicken liver; Beaufay et al. (1959a,b), rat liver; Segal and Washko (1959), rat liver; Caravaglios and Franzini (1959), rat liver; Ginsburg and Hers (1960), guinea pig intestine; Leloir and Goldemberg (1960), rat liver; Sellinger et al. (1960), rat liver; Van Lancker (1960), rat liver; Brosemer and Rutter (1961), rat liver; C. J. Fisher and Stetten (1966), rat and kidney liver; Stetten and Burnett (1966), rat liver; Feuer et al. (1966), rat liver; Nordlie and Lygre (1966), rat liver; Baginski et al. (1967), rat liver; Nordlie and Snoke (1967), rat liver; Nordlie et al. (1967), rat liver; Stetten and Burnett (1967), rat and kidney liver; Soodsma et al. (1967), rat and kidney liver; Feuer and Golberg (1967), rat liver; Jakobsson and Dallner (1968), rat liver; Nordlie et al. (1968), rat liver; Nordlie and Johns (1968), rat liver; Feldman and Butler (1969), rat liver; Lygre and Nordlie (1969), rabbit intestine; Dyson et al. (1969), rat liver; Thines-Sempoux et al. (1969), rat liver; Hanson and Nordlie (1970), rat liver
3.1.4.1	Phosphodiesterase	Razzell (1961), rat liver and kidney, hog kidney; de Lamirande et al. (1966), rat liver; de Lamirande et al. (1967), rat liver; de Robertis et al. (1967), rat brain cortex; Breckenridge and Johnston (1969); Unemoto et al. (1969), Vibrio alginolyticus
3.1.4.2	Glyceorylphosphorylcholine diesterase	Rat kidney has highest activity of eight tissues studied. The enzyme is diffusely distributed among subcellular fractions, with highest activity in microsomes. Baldwin and Cornatzer (1968), rat kidney and other tissues
3.1.6.1	Arylsulfatase	There are three distinct arylsulfatases, two (A and B) are localized in lysozymes; arylsulfatase C is microsomal (smooth reticulum). Dodgson et al. (1954), rat and mouse liver; Dodgson et al. (1955), rabbit, rat, ox, and mouse liver; Gianetto and Viala (1955), ox liver; Roy (1958), chicken liver; French and Warren (1967), human placenta; Milsom et al. (1968), rat liver
3.1.6.2	Sterol sulfatase	Activity is exclusively microsomal; high in rat liver, low in mouse and guinea pig liver. Activity is very low in the fetus; it increases steadily to the eightieth day. Burstein (1967), rat liver; Burstein et al. (1967), rat fetal stage; Zuckerman and Hagerman (1969)
3.2.1.1-2	Amylase	Approximately 52% of the total enzyme sediments with the microsomes. The predominant latent form is activated by treatment with detergents. Hokin (1955), dog pancreas; Van Lancker and Holtzer (1959), mouse pancreas; Brosemer and Rutter (1961), rat liver; Redman et al. (1966), pigeon pancreas; Siekevitz and Palade (1966), guinea pig pancreas; Mordoh et al. (1968), rat liver

88

3.2.1.31	β-Glucuronidase	Though predominantly lysosomal, 40% of the total activity sediments with the microsomes. Walker (1952), mouse liver; de Duve et al. (1955), rat liver; Ide and Fishman (1969)
3.2.2.5	NAD nucleosidase	This enzyme is associated with microsomal fraction; exceptions are the Ehrlich ascites cell and carp liver cells in which activity is found in mitochondria. Activity in the nuclear fraction has also been reported. Banay-Schwartz et al. (1969), however, believe that NADase associated with the nucleus is a distinct enzyme from the microsomal form. Kun et al. (1951), Ehrlich ascites cells; Sung and Williams (1952), rabbit liver; Dianzani (1955), hamster liver; Bojarski and Wynne (1957), rat liver; Jacobson and Kaplan (1957a), pigeon liver; Jacobson and Kaplan (1957b), pigeon liver, rabbit kidney and brain; Raeznska-Bojanowski and Gasiorowska (1963), carp liver; Blake et al. (1967), rat liver; Banay-Schwartz et al. (1969), rat liver
3.4.-.-	Peptidase	Three distinct pH optima are found in peptidase activity: pH 3.5, 5.5, and 7.5. pH 3.5 peptidase shows some characteristics of lysosomal enzymes, pH 5.5 and pH 7.5 enzymes are microsomal. Wong-Leung et al. (1968), rat and rabbit kidney; Wong-Leung and Kenny (1968), rat kidney
3.4.3.1	Glycylglycine dipeptidase	In most reports this peptidase is found in the cytoplasmic fraction but an enzyme capable of attacking di- and tripeptides at pH 7.8 is found in bovine thyroid microsomes. B. Weiss (1953), bovine thyroid
3.4.3.5	Cysteinylglycine dipeptidase	In rat and pig kidney this peptidase activity is found largely in microsomal fraction. Binkley (1961), rat and pig kidney
3.5.1.-	Nicotinamide deamidase	Liver microsomes have a higher activity than any of the eight preparations studied. Petrack et al. (1963), rat liver; Kirchner et al. (1966), rabbit liver
3.5.1.-	N-Deacylase	Deacylation of certain carcinogen requires the microsomal fraction. Seal and Gutmann (1959), rat liver; Krisch (1963), hog liver
3.5.3.1	Arginase	Arginase activity of mouse liver was found in the microsomal fraction when the tissue homogenate was prepared in 0.25 M sucrose but half of the relative activity was lost when preparations were made in 0.9% NaCl solution. The enzyme of the rat liver sediments with the nuclear and microsomal fractions in nonionic media, and in supernatant from salt-containing homogenates. Ludewig and Chanutin (1950), rat liver; Schein and Young (1952), rat liver; Thomson and Mikuta (1954), rat liver; Rosenthal and Vars (1954), rat liver; Rosenthal et al. (1956), rat liver; Carruthers et al. (1959), mouse liver and Ehrlich ascites cells
3.5.4.6	Adenosine monophosphate deaminase	This enzyme is largely microsomal; entirely particulate. Abood and Gerard (1954), rat nerve
3.6.1.-	Thiamine pyrophosphatase	This enzyme, glucose-6-phosphatase and inorganic pyrophosphate-glucose phosphotransferase, in rat liver and kidney, occur together in the microsomal fraction. Kiessling and Tilander (1960), rat liver; Novikoff and Heus (1963), rat liver
3.6.1.1	Inorganic pyrophosphatase	A spectrum of catalytic activities are considered under this heading. Stetten and Burnett (1966), rat liver; C. J. Fisher and Stetten (1966), rat and kidney liver; Soodsma and Nordlie (1966), rat heart; Ogawa et al. (1966), rat liver; Stetten and Burnett (1967), rat liver; Feuer and Golberg (1967), rat liver; Eaton and Moss (1967), human liver and small intestine; Moss et al. (1967), human liver and small intestine
3.6.1.3-5	Adenosinetriphosphatase	Chantrenne (1947), mouse liver; Hers and de Duve (1950), rat heart; Schneider et al. (1950), mouse liver; Allard and Cantero (1952), rat liver; Hechtand Novikoff (1952), rat azodye hepatoma; Novikoff et al. (1952), rat liver; Perry (1952), rabbit skeletal muscle; Christie and Judah (1953), rat liver; Maxwell and Ashwell (1953), mouse spleen; Novikoff et al. (1953a,b), rat liver; Sacktor (1953), housefly flight muscle; Abood and Gerard (1954), rat nerve; Dianzani (1954b), rat liver; Thomson and Mikuta (1954), rat liver; Straus (1954), rat kidney; Nelson (1954), bull spermatozoa; Birbeck and Reid (1956), rat liver; Jordan and March (1956), rat brain; Masko et al. (1956), Walker carcinoma; Reid et al. (1956), rat liver; Allard et al. (1957), Novikoff hepatoma; de Lamirande and Allard (1957), rat liver; Novikoff (1957), Novikoff hepatoma; Scevola and de Barbieri (1957),

89

TABLE II (Continued)

E. C. number	Enzyme (trivial name)	Comments and references
3.6.1.3–5 (Continued)		chick embryos; Thomson and Klipfel (1957), rat liver; Novikoff et al. (1958), rat liver; Kasbekar et al. (1959), rat liver; Thorell and Yamada (1959), bone marrow from normal and leukemic hens; Blecher and White (1960), lymphosarcoma; Cerletti et al. (1960), human placenta; Schwartz (1962), rat heart; Rendi and Uhr (1964), calf kidney; Schwartz and Laseter (1964), guinea pig heart; Hokin and Yoda (1964), beef kidney; H. D. Brown and Altschul (1964), peanut; H. D. Brown et al. (1964a) peanut; H. D. Brown et al. (1964b), rat brain; Hosie (1965), guinea pig brain; H. D. Brown et al. (1965), peanut; Emmelot and Bos (1965), rat liver K. G. Kennedy and Nayler (1965), toad heart; Israel et al. (1965), rat and guinea pig brain and eel electroplaque tissue; Nagano et al. (1965), rabbit brain cortex; Nakao et al. (1965), rabbit tissue; Kurokawa et al. (1965), nerve ending particles; H. D. Brown (1966a,b), rabbit heart; H. D. Brown et al. (1966), rabbit heart; Bader and Sen (1966), guinea pig kidney cortex; Thomas and Aldridge (1966), rat brain; Tice and Engel (1966), psoas muscle; Stahl et al. (1966), ox cerebral cortex; H. D. Brown et al. (1967a), Pekin duck muscle; Brown et al. (1967b), barley root ATPase; H. D. Brown et al. (1967c), rabbit ciliary process; H. D. Brown et al. (1967d), rabbit heart; Israel and Salazar (1967), beef brian; Medzihradsky et al. (1967), guinea pig brain; Stahl (1967), brain tissue; J. D. Robinson (1967), rat brain; G. J. Siegel and Albers (1967), electrophorus electric organ; Rubin and Katz (1967), rabbit skeletal muscle; Nakamaru et al. (1967), pig brain; Martonosi (1967), skeletal muscle; Goz (1967), adrenal medulla; Toda et al. (1967), guinea pig kidney cortex; Carell and Kahn (1967), Euglena gracilis; Chattopadhyay et al. (1968), Pekin duck muscle; Ikemoto et al. (1968), mouse leg muscle; H. D. Brown et al. (1968b), rabbit muscle; Fujita et al. (1968), pig brain; S. K. Ghosh and Ghosh (1968), rat brain; Duggan (1968), frog muscle; K. Gibson and Harris (1968), human heart; Nakamaru (1968), pig brain; Yoshida et al. (1968), guinea pig brain; Martonosi (1968), rabbit and rat skeletal muscle; Bowler and Duncan (1968a), rat brain; Rawson and Pincus (1968), rat and guinea pig brain; Kepner and Macey (1968), guinea pig kidney cortex; Bowler and Duncan (1968b), frog brain; Martonosi et al. (1968); Sullivan et al. (1968), rat cerebral cortex; H. D. Brown et al. (1968a), Pekin duck muscle; Nakamaru and Konishi (1968), pig brain; Rigdon et al. (1968), chick pipping muscle; Emmelot and Bos (1968), rat liver; Chignell (1968), pig brain; Chignell and Titus (1968), beef brain; beef brain; J. D. Robinson et al. (1968), rat brain; Atkinson et al. (1968), pig brain; K. Gibson and Harris (1969), human heart; Harold and Baarda (1969), Streptococcus faecalis; Uesugi et al. (1969), beef brain; L. S. Baskin and Leslie (1968), dystrophic mouse skeletal muscle; Goldfarb and Quigley and Gotterer (1969), rat intestine; Chattopadhyay et al. (1969), dystrophic mouse skeletal muscle; Goldfarb and Rodnight (1969); Inturrisi (1969), beef brain; Kepner and Macey (1969); Holland and Perry (1969), rabbit muscle; Tobin and Sen (1970), guinea pig kidney; Towle and Copenhaver (1970), rabbit kidney
3.6.1.6	Nucleoside diphosphatase (guanosine diphosphate–inosine diphosphate–uridine diphosphatase)	Nucleoside diphosphatase has the same distribution pattern as ATPase (3.6.1.3–5). Abood and Gerard (1954) rat nerve; Cerletti et al. (1960), human placenta; L. Ernster and Jones (1962), rat liver; Novikoff and Heus (1963), rat liver; Feuer and Golberg (1967), rat liver; Penniall and Holbrook (1968), rat liver; Wattiaux-De Coninck and Wattiaux (1969), rat liver
3.6.1.–	Adenosinediphosphatase (nucleoside diphosphatase)	L. Ernster and Jones (1962), rat liver; Novikoff and Heus (1963), rat liver; Penniall and Holbrook (1968), rat liver; Wattiaux-De Coninck and Wattiaux (1969), rat liver

3.6.1.7	Acylphosphatase	It has been reported that microsomal fraction of kidney and brain contains a K^+ acetylphosphatase which, it is speculated, may be part of the transport ATPase complex. The acetylphosphatase is ouabain sensitive and Mg^{2+} activated, stimulated by K^+, but not by Na^+. Bader and Sen (1966), guinea pig kidney cortex; Yoshida et al. (1966), guinea pig brain; Israel and Titus (1967), beef brain; Sachs et al. (1967), rat brain; Yoshida et al. (1968), guinea pig brain; Emmelot and Bos (1968), rat liver
3.6.1.-	Nucleotide pyrophosphatase	Jacobson and Kaplan (1957a,b), rat, hamster, rabbit, and pigeon liver and rabbit kidney
3.6.1.9	Nucleotide pyrophosphatase (NADP pyrophosphatase)	Jacobson and Kaplan (1957a,b), rat, hamster, rabbit, and pigeon liver and rabbit kidney
4.1.2.7	Ketose-1-phosphate aldolase	Although aldolase, a glycolytic enzyme, is associated with the supernatant fraction of rat liver, occasionally it is found attached to nuclei by weak bonds which are broken in ionic solution. The distribution of this enzyme in myopathic duck is, however, diffuse. Amberson et al. (1965), rabbit heart and skeletal muscle; H. D. Brown et al. (1968a), myopathic duck muscle
4.1.3.5	Hydroxymethylglutaryl-CoA synthetase	This enzyme, which forms an important intermediate in cholesterol synthesis, is found in mitochondria; Rudney (1957) has reported it present in the microsomal fraction of rat liver. Rudney (1957), rat liver
5.3.3.1	Steroid Δ-isomerase (Δ5-3-ketosteroid isomerase)	Oleinick and Koritz (1966a,b), rat adrenal particles; Neville and Engel (1968), bovine adrenal gland
6.2.1.2	Acyl-CoA synthetase	This enzyme is associated with microsomes in rat intestine mucosa and probably found in microsomal fraction in rat liver. Y. Stein and Shapiro (1958), rat liver; Senior and Isselbacher (1961), rat intestinal mucosa; Creasey (1962), rat liver; Bar-Tana and Shapiro (1964), rat liver; Farstad et al. (1967), rat liver
6.2.1.-	Carnitine palmitoyltransferase	This enzyme has a dual (mitochondrial and microsomal) localization in rat liver. Van Tol and Hülsmann (1969), rat liver
6.2.1.-	Palmitoyl-CoA synthetase	This enzyme has been demonstrated in both mitochondria and microsomes. Van Tol and Hülsmann (1969), rat liver
6.3.1.2	Glutamine synthetase	Localized essentially in microsomal fraction. Wu (1961), rat liver
6.3.4.-	Serine ethanolamine phosphate synthetase	Found in the microsomal fraction. It is specific for serine and both stereoisomers are active. The enzyme requires Mg^{2+} for its optimum activity; however, Mn^{2+} and Co^{2+} are less effective and Ca^{2+} is inhibitory. A. K. Allen and Rosenberg (1968), chicken intestinal mucosa
6.4.1.2	Acetyl-CoA carboxylase	Dilution of homogenate and presence or absence of Mg^{2+} plays an important role in enzyme isolation. Although found in supernatant, it is believed to be essentially a membrane-borne enzyme. Margolis and Baum (1966), pigeon liver

ship of mitochondrial phospholipid to the several enzyme systems. It has proven possible, with relatively crude systems, to reconstruct active complexes by "adding back" lipid isolated from the membranes (Racker, 1963, 1965). Lipid dependence of membrane enzyme systems is by no means limited to mitochondria; bacterial systems also can be reactivated in this same way, i.e., adding back phospholipid after purification (Rothfield et al., 1966; Andreoli et al., 1967). In the endoplasmic reticulum–plasma membrane system, the $Na^+ + K^+ATPase$ similarly is dependent upon membrane phospholipids. It has been possible to reactivate the enzyme with synthetic phosphatidyl serine (Tanaka and Abood, 1964; Tanaka and Strickland, 1965; Tanaka and Sakamoto, 1969). H. D. Brown and associates (1966, 1967b, 1968c,d) have reported experiments using nonlipid polymeric matrices which resulted in isolated $Na^+ + K^+ATPase$ systems with characteristics that resembled those of the membrane–enzyme system.

Upon the rationale of this thesis that lipid support was requisite, a literature has developed which includes the classic studies of Langmuir and Schaefer (1938). Models of enzymes on synthetic lipid membranes were constructed to represent the native situation. This literature has been reviewed by James and Augenstein (1967) and by Tien and James in Chapter IV of this work.

2. Electron Flow and Microsomal Oxidation

Microsomes of many tissues almost universally throughout the plant and animal kingdoms contain a family of hydroxylating enzymes (Posner et al., 1961) which are physically and functionally closely tied to the NADPH–electron transport system. Although the electron transport system in which NADH furnishes reducing equivalents is associated primarily with the mitochondria, there have been frequent reports of NADH-requiring enzymes[*] and electron flow intermediates associated with microsomal fractions. The physiological role of this system has not yet been established, but it appears that the "microsomal NADH–electron flow system" is significant in the cell's economy.

Many chemically quite diverse compounds are metabolized by the NADPH–reduced microsomal system. An overall sequence has been described by H. S. Mason (1965) as mixed-function oxidation in which

---

[*] Reduced NAD–cytochrome c reductase (de Duve et al., 1955) exists in the microsomes and appears to have properties which are different from those of the mitochondrial equivalent. The same activity has been reported by Brody et al. (1952) and Hogeboom (1950) and has been further characterized by P. D. Jones and Wakil (1967) and by P. Strittmatter (1959).

NADPH supplies reducing equivalents, and atmospheric oxygen is incorporated into a substrate. The type reaction

$$H^+ + NADPH + AH + O_2 \rightarrow AOH + NADP^+ + H_2O$$

is a useful summarization, although it, of course, sheds no light upon the mechanisms of the true course of events.

Electron flow from NADPH leading finally to hydroxylation of a substrate probably serves, physiologically, to make "foreign" substances more water-soluble. The precise sequence of events remains conjectural, but most authors presume that the compound (substrate) forms a complex with P-450; this is reduced by NADPH–cytochrome c reductase. Reduced compound–P-450 complex then reacts with molecular oxygen to form an active oxygen–compound–P-450 complex which ultimately becomes oxidized compound and oxidized P-450. "Detoxication," a term associated with this system, is probably a poor word choice since the transformations are in many instances harmful to the organism. Reviews have been published by H. S. Mason et al. (1965), T. E. King et al. (1965), Gillette (1966, 1969), and a series of authors in the symposium volume edited by Gillette et al. (1969).

In higher forms mixed-function oxidases (MFO) (Siekevitz, 1963; de Duve et al., 1962) are associated with the endoplasmic reticulum although interaction with electron flow components apart from the reticulum is indicated by recent studies (Estabrook et al., 1969). An electron carrier intermediate, P-450, is firmly bonded to the endoplasmic reticulum and apparently is dependent upon structure for activity. It serves to activate the oxygen which is used ultimately in substrate (Cooper et al., 1965). The P-450 makes a convenient observation point for studying the system because it binds carbon monoxide specifically and can then be characterized by differential spectroscopy in the visible range.

Precise localization of the enzymes involved in electron flow and in the catalysis of transformation of the various intermediates is not completely known; it appears to be microsomal though not necessarily exclusively so. Hepatic endoplasmic reticulum has been best studied. Separation of rough and smooth microsomes (Fouts, 1961; Dallner, 1963) has indicated that the MFO complex is independent of the ribosomes (Gillette et al., 1957). In support of this, Gillette has reported that liver microsomes treated with ribonuclease lose protein synthetic activity as anticipated but that the treatment does not alter the enzymic hydroxylation activity.

The principal oxidative reactions that occur at the microsomes are presented in Fig. 3.

Demonstrations that remarkably high absolute levels of P-450 exist raise very interesting questions. Cytochrome–P-450 is present in much greater amount than one would predict from the apparent stoichiometry of the reactions sequences. Estabrook *et al.* (1969) have postulated, therefore, that this represents a tie-in point with other electron flow systems.

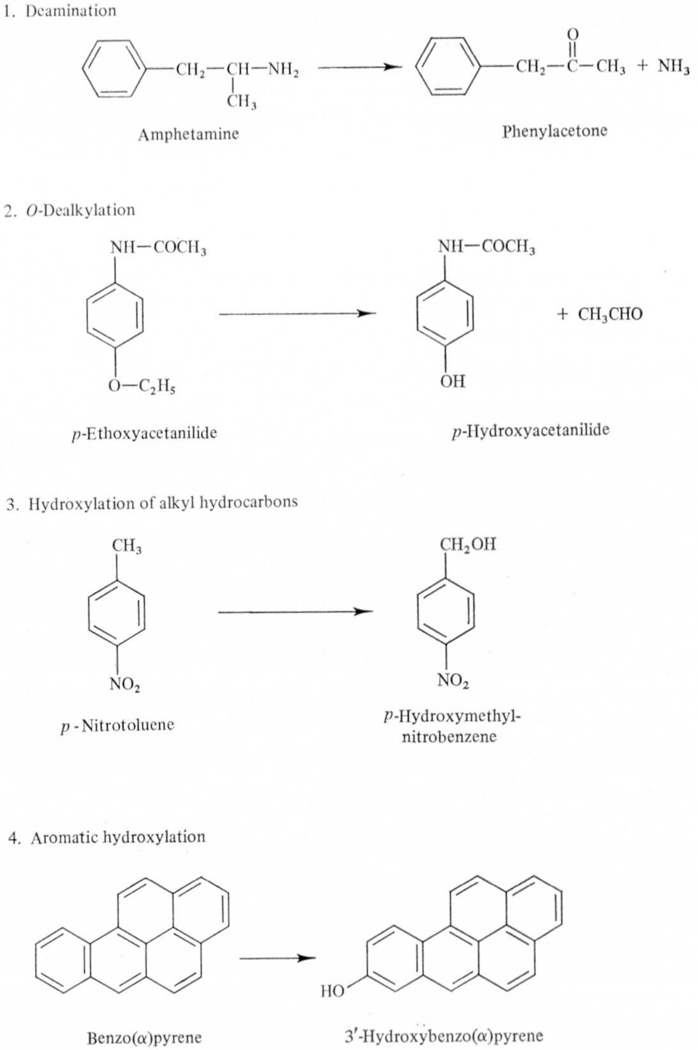
1. Deamination
$CH_2-CH-NH_2$
CH_3
$CH_2-C-CH_3 + NH_3$
O
Amphetamine
Phenylacetone
2. O-Dealkylation
$NH-COCH_3$
$O-C_2H_5$
$NH-COCH_3$
OH
$+ CH_3CHO$
p-Ethoxyacetanilide
p-Hydroxyacetanilide
3. Hydroxylation of alkyl hydrocarbons
CH_3
NO_2
CH_2OH
NO_2
p-Nitrotoluene
p-Hydroxymethyl-nitrobenzene
4. Aromatic hydroxylation
HO
Benzo(α)pyrene
3′-Hydroxybenzo(α)pyrene

FIG. 3. Principal oxidative reactions that occur at the endoplasmic reticulum.

It is presumed that NADH provides reducing equivalents. Indeed, although NADPH must be present, the extent of hydroxylation reactions exceeds that which can be computed on the basis of total NADPH available.

The quantity of P-450 is so much greater than that required for the re-action sequence that functionality within the system must be considered. It may be that the large reservoir of the cytochrome may allow it to function not only as a tie-in point with other electron flow systems but as a

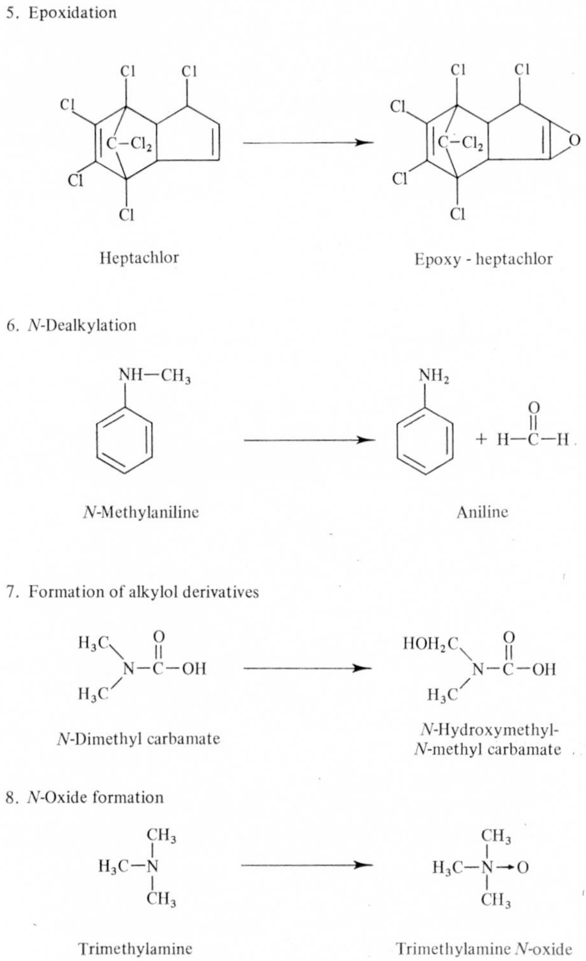

5. Epoxidation

Heptachlor   Epoxy - heptachlor

6. N-Dealkylation

N-Methylaniline   Aniline

7. Formation of alkylol derivatives

N-Dimethyl carbamate   N-Hydroxymethyl-
                               N-methyl carbamate

8. N-Oxide formation

Trimethylamine   Trimethylamine N-oxide

FIG. 3 (Continued).

holding position for electrons during formation or change of other components. Experimental studies of MFO induction may indicate the need for an energy storage form as essential to the oxidation system. Changes that may be genetically or cytoplasmically controlled as induction-like phenomena may require some finite period of time during which the buildup of reducing equivalents at the "level" of P-450 serves as an economy of value to the cell's metabolism.

These speculations about the role of P-450 are, in a way, a vignette of

9. *N*-Oxidation of primary and secondary amine and their derivatives

Primary

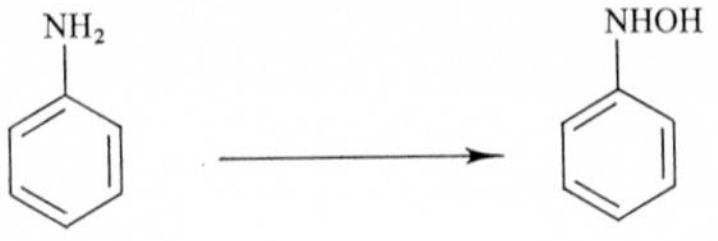

Aniline *N*-Hydroxyaniline

Secondary

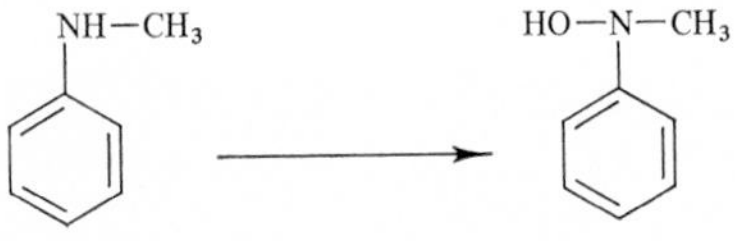

N-Methylaniline *N*-Hydroxy-*N*-methylaniline

10. S-Demethylation

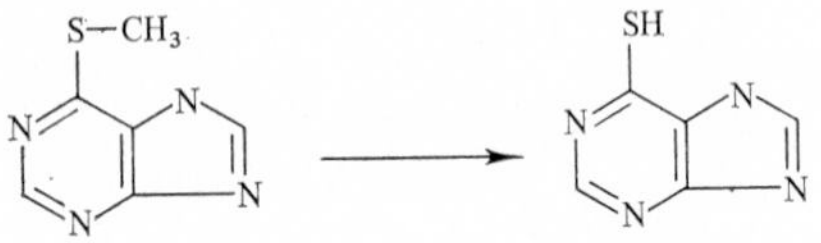

6-Methylmercaptopurine 6-Mercaptopurine

11. S-Oxidation

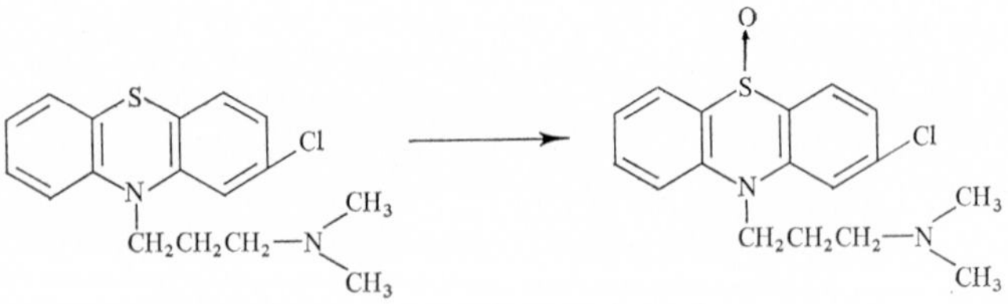

Chlorpromazine Chlorpromazine sulfoxide

FIG. 3 (*Continued*).

thought about the entire problem of the electron flow components. Because P-450 of mammals is tightly bound to a supporting membrane, pure preparations have not been obtained. Where purification has been reported, the characteristic Soret band of the purified material (presumably lipid free) has been found at about 420 mμ. Although the assumption is widely made that the P-420 represents denatured P-450 and that both forms are hemoproteins, it is by no means clear that this transformation is stoichiometric in the expected manner. The problem here as it is throughout the

12. Phosphothionate oxidation

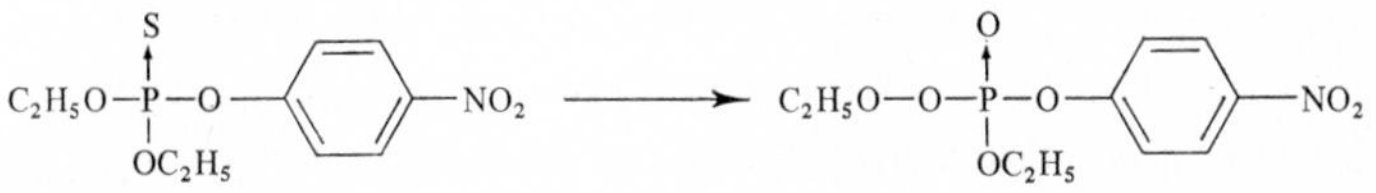

$$S \uparrow \\ C_2H_5O-P-O-\!\!\bigcirc\!\!-NO_2 \longrightarrow C_2H_5O-O-\overset{O\uparrow}{P}-O-\!\!\bigcirc\!\!-NO_2 \\ \underset{OC_2H_5}{\,} \qquad\qquad \underset{OC_2H_5}{\,}$$

Parathion   p-Nitrophenol-diethylphosphate

13. Conversion of thiobarbiturates

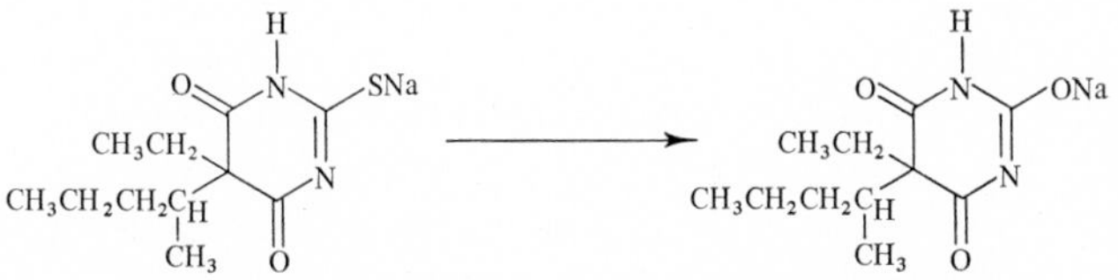

Sodium thiopental   Sodium pentobarbital

14. Dehalogenation

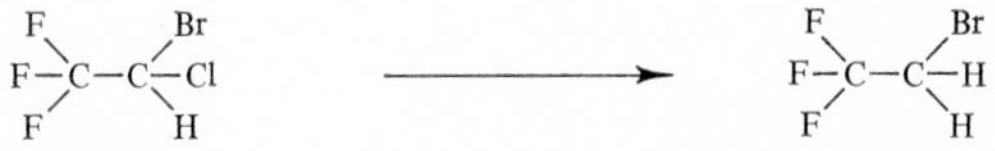

Halothane   1-Bromo-2-trifluoroethane

15. Dealkylation of metalloalkanes

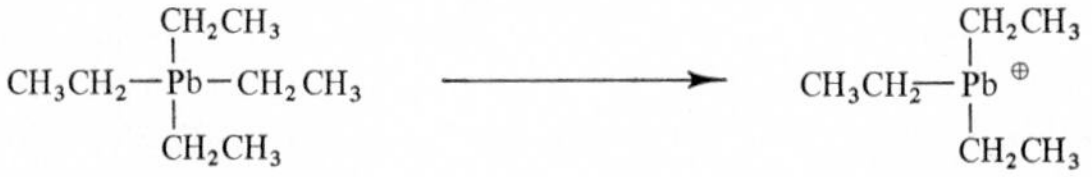

Tetraethyl lead   Triethyl lead

FIG. 3 (*Continued*).

entire sequence is that characterization depends upon the association of components with a lipid membrane. So complex are the interactions at the membrane, that conflicting and often apparently unrelated data have developed and frustrated attempts at interpretation. Hence an attempt has been made to develop models which by virtue of their spectral similarity could be used as a reference base to indicate the nature of the P-450–CO complex. This complex was originally described (Omura and Sato, 1962) in terms of a Soret peak at 450 mμ. Imai and Sato (1966a) showed that reduced P-450–ethyl isocyanite complex has dual peaks at 428 and 455 mμ. For the formation of the protoheme–ethyl isocyanite complex, two or more of the protoheme molecules are required. Thus, since the protoheme–ethyl isocyanite complex and the P-450–ethyl isocyanite complex in common show Soret absorption maxima at 455 mμ, they suggest that the spectral properties of the P-450 complexes are due to the interaction of two hemes.

In another model system which may have reference to P-450, Holden and Lemberg (1939) reported that reduced protoheme–pyridine–CO has an absorption maximum at 440 mμ. Jefcoate and Gaylor (1969) have extended this work and shown that complexes of these components (protoheme–pyridine–CO) are formed which have a variety of characteristics depending upon formation conditions. Very interestingly, those agents that affect the stability of these model complexes are identical with those that affect the conversion of P-450 into P-420 (e.g., Na deoxycholate, and alcohols). Conversely, increased ionic strength favors association (i.e., a higher-wavelength Soret band) suggesting that the association energy in these complexes derives from a hydrophobic interaction of a pair of hemes. An ultimate interpretation of the models must await the availability of pure P-450 for comparative study.

Nishibayashi and Sato (1968) have undertaken a spectral study of a purified material from biological membranes and have published absolute spectra of this material. They showed Soret maxima at 360, 416, 535, 570, and 650 mμ, similar to the spectra of (other) protoheme-containing proteins; but the reduced spectrum they have published with absorption maxima at 412 and 555 mμ is anomalous for hemoproteins. Elucidation of the heme interactions with the protein or the heme interactions with a multiple complex, if indeed this is involved, may provide us with a useful reference point for the further description of the NADPH–electron flow–hydroxylase systems.

Definitive enzymological characterization has not been possible for the component enzymes of the mixed-function oxidation systems. Though some of the classic parameters have been obtained for NADPH oxidase (Gillette et al., 1957; Gillette, 1963; Kato and Takanaka, 1968; H. D. Brown et al., 1969e), ferricyanide reductase (C. H. Williams and Kamin, 1962; H. D. Brown et al., 1969e), and the hydroxylases (Gillette, 1966; Imai and Sato, 1968a,b; Levin et al., 1969; Gurtoo et al., 1968), systems in which these entities have been observed are extremely complex. In some studies these systems have involved the entire cycle of events including regeneration (reduction) systems for the oxidized nucleotide. It would certainly appear that definitive elucidation of the enzymology of the components of this system will require some degree of separation of the individual activities or identification of multiple sites or multiple activities associated with single proteins active in the pathway.

3. Adenyl Cyclase and Cyclic Nucleotide Synthesis

Cellular processes, to a very large extent, are influenced by cyclic nucleotides. Cyclic adenosine monophosphate (AMP) has been shown to affect an incredibly large number of phenomena presumably by a catalytic level effect upon enzymes, though this has not been directly demonstrated in most systems. These actions of cyclic AMP have been summarized in Table III. Figure 4 is a visualization of hypothesized mechanisms for the interaction between catecholamines, polypeptide hormones, and cellular metabolic events. Although the mechanisms of these interrelationships remain speculative, extensive experimental data have been brought to bear upon the problems. It has been suggested, for example, that the catecholamines facilitate the synthesis of cyclic AMP through a steric interaction of the hormone amine with the enzyme itself (Bloom and Goldman, 1966). Reactions involving cyclic AMP are, in the chemical sense, little known.

The adenyl cyclase activity is found to be associated with the internal membrane reticulum widely throughout the animal kingdom and in many microorganisms as well. Somewhat surprisingly, nuclear membranes are often highly active, and, indeed, much of the existing literature describes activity derived from the nuclear envelope. To as great an extent as possible, the discussion which follows will draw upon examples of studies which have used endoplasmic reticulum-derived activity. However so much of the work has been done upon particulate homogenates or nuclear materials that we need, for the moment, to make the questionable assumption that generalizations about the activity in various particulate sources can be extrapolated to endoplasmic reticulum specifically. Fortunately it does appear that adenyl cyclase (Pulsinelli and Eik-Nes, 1970) is distributed throughout the particulate matter and that the enzymic character of the materials from the several particulate sources are not distinguishable by conventional kinetic parameters or by their relationships to inhibitors and stimulatory agents.

Elucidation of the subcellular distribution of adenyl cyclase is complicated by the existence of what are apparently exceptional tissues. The rat brain cortex, for example, is at variance with many other source tissues studied. In this material (de Robertis et al., 1967) adenyl cyclase is largely in the mitochondrial fraction. These authors further report that

TABLE III

PHENOMENA INFLUENCED BY CYCLIC ADENOSINE $3',5'$-MONOPHOSPHATE

Enzyme or process	Reference	Effect
Amino acids → liver protein	Pryor and Berthet (1960)	−
Uterine protein synthesis	Szego (1965)	+
	Creange and Roberts (1965)	
	Hechter et al. (1967)	
Lac messenger ribonucleic acid (mRNA) synthesis	de Crombrugghe et al. (1970)	+
Uterine RNA synthesis	Hechter et al. (1967)	+
Parotid gland deoxyribonucleic acid synthesis	Malamud (1969)	+
Serine dehydratase synthesis	Jost et al. (1969)	+
	Jost et al. (1970)	
Biosynthesis of progesterone—bovine granulosa cells	Cirillo et al. (1969)	+
Synthesis of progesterone—immature rat ovary	Sulimovici and Boyd (1968)	−
Morphogenesis aggregation—acrasin activity, slime mold	Chassy et al. (1969)	+
	Barkley (1969)	
Cell growth—HeLa and Strain L cells	Ryan and Heidrick (1968)	−
Cerebellar Purkinje cells (discharge frequency)	Siggins et al. (1969)	−
Platelet adhesiveness	Zieve and Greenough (1969)	−
HCl secretion—gastric mucosa	J. B. Harris and Alonso (1965)	+
Release of protein from polysomes	Khairallah and Pitot (1967)	+
Pigmentation—frog skin	Bitensky and Burstein (1965)	+
	Noveles and Davis (1967)	(Darkening)
	Abe et al. (1969)	
	Hadley and Goldman (1969)	
Pigmentation—lizard skin	Hadley and Goldman (1969)	−
		(Lightening)
Permeability—toad bladder	Orloff and Handler (1967)	+
Permeability—rat uterus	Hechter et al. (1967)	+
Reduced nicotinamide adenine dinucleotide oscillations—yeast extract	Chance and Schoener (1964)	+
Secretion of amylase—rat parotid	Babad et al. (1967)	+
	Schramm (1967)	
Secretion of insulin	Sussman and Vaughan (1967)	+
	Turtle et al. (1967)	
Glycogen synthetase	Larner (1966)	−
Protein kinase (phosphorylation of histone)	Walsh et al. (1968)	+
	Kuo and Greengard (1969)	
	Miyamoto et al. (1969)	
	Langan (1969)	
Kinase	J. S. Bishop and Larner (1969)	+
Tyrosine aminotransferase	Tryfiates and Litwack (1964)	+
	Wicks (1969)	
Phosphofructokinase	Mansour (1966)	+
	Stone and Mansour (1967)	
Fructose-1,6-diphosphatase	Mendicino et al. (1966)	−
Tryptophan oxygenase	G. D. Gray (1966)	+
	Chytil (1968)	
Phosphorylase	Haugaard and Hess (1965)	+
	E. G. Krebs et al. (1966)	
Lipolysis	Chytil and Skrivanova (1963)	+
	Rizack (1964)	
	Butcher (1966)	
	Butcher and Sutherland (1967)	
Steroidogenesis	Haynes et al. (1960)	+
	Karaboyas and Koritz (1965)	
	P. F. Hall and Koritz (1965)	
	N. M. Kaplan (1965)	

TABLE III (*Continued*)

Enzyme or process	Reference	Effect
Steroidogenesis—*Continued*		
	Marsh *et al.* (1966)	
	Graham-Smith *et al.* (1967)	
Gluconeogenesis	Exton and Park (1966)	+
	Exton *et al.* (1966)	
Ketogenesis	Bewsher and Ashmore (1966)	−
Glucose oxidation	Pastan (1966)	+
	Blecher (1967)	
Acetate → liver fatty acids	Berthet (1960)	−
Cholesterol → pregnenotoney	Sulimovici and Boyd (1968)	+
Urea formation	Exton and Park (1968)	+
Contractile responses (by oxytocin and Ca²⁺ in isolated rat uterus)	Mitzmegg *et al.* (1970)	+

the mitochondrial fraction, when subfractionated, had an adenyl cyclase activity distribution indicating it is compartmentalized within the organelle. Hence, if this work can be generalized to include other organisms, it would appear that the localization of adenyl cyclase in the central nervous system, although particulate, is somewhat different than in other tissues (Sutherland and Rall, 1960; Sutherland *et al.*, 1962, 1965; Klainer *et al.*, 1962; Davoren and Sutherland, 1963).

Little can be said with certainty about the chemistry of the adenyl cyclase reaction. The level of cyclic AMP in any system so far described depends upon the rates of activity of at least two enzymes and quite possibly of others as well [most systems contain ATPases as well as phosphodiesterase (cyclic AMP → AMP)]. The literature in general has been built upon methods in which inhibitors were used to allow the adenyl cyclase activity a competitive advantage. Kinetics in consequence require complex interpretation, and final description must await a more nearly satisfactory *in vitro* reaction system. The assays themselves are difficult because of the conditions under which these reactions are run; adenyl cyclase activity is feeble when compared with some of the enzymic activities which are competitive for substrate (e.g., the ATPase ubiquitously present with it). Though the enzyme is relatively stable and purification procedures which include a lipid extraction have been successful, the reports of characteristics of the activity vary greatly from source to source. It is certainly to be suspected that characterization will require greater success at isolation.

It remains a probability that adenyl cyclase activity, while described by the type reaction, ATP → cyclic AMP + 2 P_i, is a multiple phenomenon that may differ from organism to organism or even cell site to cell site. Since the chemistry of the reaction per se is not available to us, we can only speculate upon the interrelationship of catecholamines with the adenyl cyclase. A number of pharmacologists have offered interesting suggestions along this line. Typical is that of Belleau (1960) ; see Fig. 5.

Cyclic AMP has loomed large in endocrinological studies because more than any other single entity yet described in the biochemical literature, its activity is under hormonal control. Epinephrine, at least, appears very

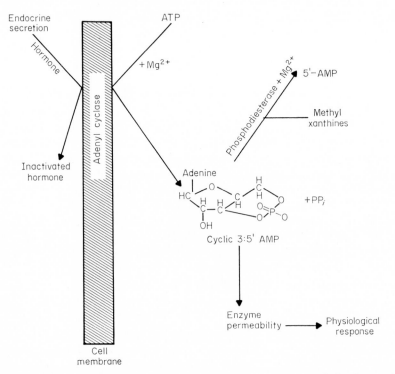

Fig. 4. Hypothesized mechanisms for the interaction among catecholamines, polypeptide hormones, and cellular metabolic events.

likely to exhibit its effect by target action directly on adenyl cyclase which may be a single or multiple enzyme activity. Sutherland and associates (Sutherland et al., 1965; Sutherland and Robinson, 1966; Butcher et al., 1968) have called cyclic AMP a possible "second messenger" system in hormonal regulation. In their scheme the first messenger is the hormone that travels to the cell from its site of production where it excites the formation of the second messenger, cyclic AMP. This messenger then accomplishes the final effect by altering in some way membrane permeability and/or the activities of enzymes.

Distribution of the enzyme throughout the plant and animal kingdom appears to be ubiquitous, even including the slime mold *Dictyostelium discoideum* where 3′,5′-AMP has been reported to be the long-sought aggregating chemotaxic medium, acrasin (Chang, 1968).

4. Transport Proteins

Proteins associated with transport phenomena have been described. In general, based upon the concept of Osterhout (1933), binding to or reaction

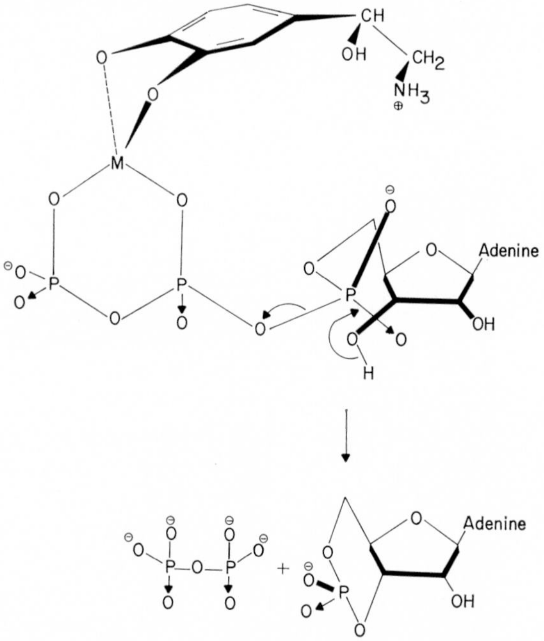

Fig. 5. Catecholamine catalysis of adenyl cyclase. [Redrawn from B. Belleau (1960). *In* "Ciba Foundation Symposium on Adrenergic Mechanisms," p. 223, by permission.]

with transported substrates (Fig. 6) has been the most direct criterion used in the identification of these components. A host of indirect indications have also been brought to play. Though problematic, such components may conceivably constitute a separate category of membrane proteins. It has been possible to isolate protein components of membranes that specifically bind transported moieties.

Both column adsorbants and equilibrium dialysis have been used to demonstrate substrate binding by proteins. In the column method the protein, either purified or as a complex mixture, is held in the column medium either by adsorption or by lattice entrapment. Substrate mixtures are then passed through the column, and the time of elution of substrates is measured by direct assay or by radiolabel detection. In equilibrium dialysis, proteins are isolated within a cellulose dialysis bag; this causes an equilibrium solution of substrate to increase its concentration inside the bag.

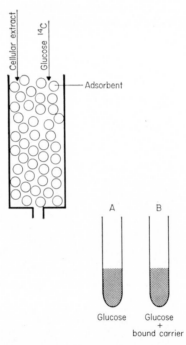

Fig. 6. Researchers have attempted to demonstrate specific transport agents in cellular extracts by isolating bound "complexes." Adsorbent columns have been used to separate specific binding agents in this way.

This technique has been used by Oxender (1968) to isolate a leucine-binding protein from *Escherichia coli*. The protein has been crystallized (Fig. 7), has a molecular weight of 36,000, and has a low content of sulfur-containing amino acids; its heat stability and other physical characteristics have also been described (Oxender, 1968; Nichoalds and Oxender, 1970). The protein is specific for L-isomers of branch chain amino acids (presumably binding occurs in nature at the surface of the membrane prior to transport across the bacterial wall).

Physiological experiments employing osmotic shock of the bacterium from which the protein has been isolated are also interpreted to favor the concept of a transport protein (J. H. Weiner *et al.*, 1970). In osmotic shock, *E. coli* is rapidly carried from a sucrose solution of high osmotic strength to a dilute salt solution that results in the release of about 5% of total cellular protein. The transport protein was isolated from the shock

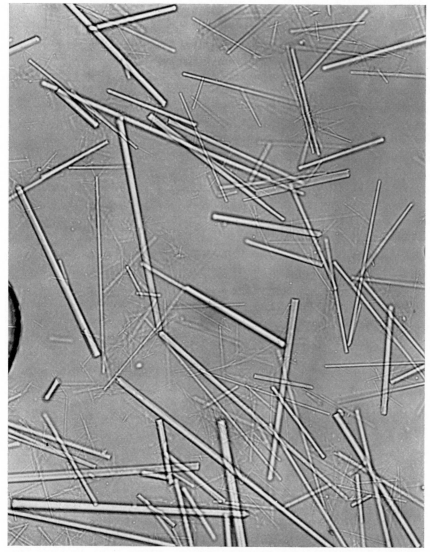

Fig. 7. A crystalline membrane protein identified by specific binding to amino acids is thought to participate in a transport phenomenon. (Photograph courtesy of Dale L. Oxender.)

fluid, and it shows similar kinetic constants for binding as does the organism for the transport of leucine itself. Further synthesis of the protein is repressed when the cells are grown in culture medium containing leucine. The so-called transport protein has been shown to be localized in the cell membrane and does not exist elsewhere in the *E. coli* cell. Other proteins related by similar inferential evidence have been described from other sources, e.g., transport galactosidase (Cohen and Monod, 1957; Fox and Kennedy, 1965; Kepes, 1969); the concept was early given the name permease (Cohen and Monod, 1957). Indeed, as S. G. Schultz (1969) points out, "it has become acceptible jargon to describe carrier mediated transport processes as 'enzyme-like' [when they] function ... to transfer a substance from one place to another." The conventional characterization of an enzyme based upon its catalysis of a transformation has been applied to only a few of the transport proteins described in the literature.

Whether all of these protein entities are solely involved in transport, whether they are carriers in the traditional sense of mobile components, or whether or not they have a catalytic role essential to an energy-requiring phenomenon remains to be clarified (Figs. 8 and 9). By far the greatest number of transport-associated proteins have been identified in microbial systems though there have been interesting experiments performed with animal materials (A. M. Stein *et al.*, 1967).

In higher animals (Skou, 1957, 1965, 1967; Post *et al.*, 1965, 1969; H. D. Brown, 1966c) and higher plants (H. D. Brown *et al.*, 1965, 1967b; Epstein and Rains, 1965; J. L. Hall, 1969), the most elaborately studied protein presumed to be associated with the transport phenomenon is so-called "transport" ATPase, an enzyme which was originally described by Skou (1957, 1965, 1967). By analogy to a Na^+ transport phenomenon in isolated crab nerves, an enzyme catalyzing the hydrolysis of ATP was hypothesized, and when nerve tissue was studied, an activity which met the predicted criteria was found.

The reasoning by which the existence of transport ATPase was predicted was straightforward and admirably simple. Skou knew that Na^+ transport per se would occur in isolated crab nerve only if both Na^+ and K^+ were present together. Transport also required Mg^{2+} and energy (conventionally associated with adenosine triphosphate hydrolysis) and was inhibited by the heart drug ouabain. This description of the physiological event was reminiscent of metal-requiring enzymes known in many systems. Hypothetically, a transport enzyme could be described in these same terms: an ATPase activated by Mg^{2+}, requiring $Na^+ + K^+$, and inhibited by ouabain. Fortunately for the development of this important area of knowledge, the entire system was found to be contained within isolated cell particulates. Hence, membrane ATPase literature presumably related to transport phenomena has developed and now numbers, literally thousands of papers, covering virtually every family of higher animals (Skou, 1957, 1965, 1967; Post *et al.*, 1965, 1969; H. D. Brown, 1966a,b,c; Albers, 1967; Glynn, 1968; Sen *et al.*, 1969; Jorgensen and Skou, 1969; Tobin and Sen, 1970), many microorganisms (Bonting and Caravaggio, 1966; Hafkenscheid and Bonting, 1968; Munoz *et al.*, 1968a,b; Gainor

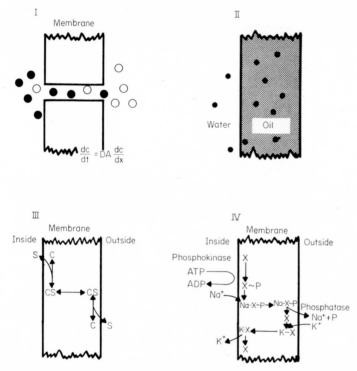

FIG. 8. These diagrams illustrate aspects of the thinking which has contributed to the present understanding of membrane phenomena. I—diffusion, II—lipid solubility, III—passive carrier facilitated diffusion, and IV—active cation transport.

and Phillips, 1969), and some higher plants (H. D. Brown *et al.*, 1965, 1967b; Epstein and Rains, 1965; J. L. Hall, 1969).

Thus, transport ATPase has been defined as a Mg^{2+}-activated $Na^{+}+K^{+}$-requiring ATPase which is specifically inhibited by the cardiac glycosides, and particularly by ouabain. In addition, the enzyme preparations which are still quite impure (despite a great and widespread labor which has been expended upon purification) have other properties which by inference relate it to the transport phenomenon. The precise role of such enzyme systems as this in the transduction of chemical energy to do osmotic work is not known. Useful hypotheses range from conformational changes which literally propel the substrate (D. E. Green, 1954; D. E. Green *et al.*, 1968a) to more conventional chemical interactions (E. C. Slater, 1967a) with mobile carriers to a vectorial distribution of substrate and product of a coupled reaction (Mitchell, 1963, 1968).

A number of papers have provided information about endoplasmic reticulum $Na^+ + K^+$ ATPase. Hokin and Reasa (1964) have interpreted the inhibition of ATPase by diisopropyl fluorophosphate (DFP) to indicate that a serine residue is part of the protein and is involved in its cata-

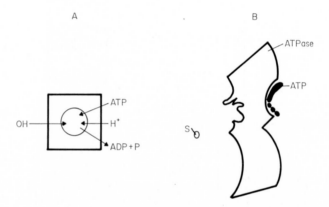

Fig. 9. A. Anisotropic distribution of ions as a function of membrane geography. [Redrawn from Peter Mitchell (1963) *in* "Cell Interface Reactions" (H. D. Brown, ed.), p. 33, Scholar's Library, New York, by permission.] B. Anisotropic distribution resulting from hypothesized relationship among ATPase, ATP, and a transported ion (S).

lytic center. Mercury inhibitors affect the activity of the enzyme, indicating that a sulfhydryl group is at or near the active site (Skou, 1963; Taylor, 1963). The enzyme also appears to require association with a phospholipid (Schatzmann, 1962; Hokin and Yoda, 1965; Hegyvary and Post, 1969; Israel, 1969). Hokin and his associates have proceeded along a route toward chemical characterization of the protein using the DFP- and halogen-substituted strophanthan as tagged reagents. In this way a tagged though denatured (and hence catalytically inactive) protein can be followed through purification. These workers reason that the difficulties of purification reported by others have come about in consequence of attempts to maintain the activity of the protein. Thus, if the ATPase is tagged in a specific manner, it may be identified during a purification process.

Another approach to studies of the metal-activated ATPase employs pulse labeling of the enzyme with AT^{32}-P (Albers *et al.*, 1963, 1968; Post *et al.*, 1965; Post and Sen, 1967a,b). This physiological orientation may ultimately serve to describe the transport phenomenon itself rather than provide chemical characterization of the protein participants. Results indicate that a labeled phosphorylated intermediate is functionally an acyl phosphate and that one step in the reaction is an acyl phosphate hydrolysis which requires potassium and is inhibited by ouabain (Skou, 1969).

Peptic digestion of these same relatively impure membrane preparations has provided some physical information about fragments. From electrophoretic mobility, Jean and Bader (1967) calculated pK values which are taken to be an indication of possible identity of ionized groups in the fragment.

Purification methods for $Na^+ + K^+ ATPase$ as an active entity have been conventional—centrifugation methods, salt fractionations, or chromatographic procedures applied to protein (Nakao *et al.*, 1963, 1965; H. D. Brown *et al.*, 1965, 1967d; Matsui and Schwartz, 1966). To date the purified materials have not made it possible to determine whether the acyl phosphatase and kinase activities are functions of more than one or of a single protein. In any case, the functional distinctness of the phosphatase–kinase activities makes it possible to "construct" a reaction scheme upon a rationale using ionic radii. Such a mechanism has been outlined in general terms by Schoner *et al.* (1967). Similar schemes for other enzymes are well established. For example, peptidase hydrolysis has been described by Vallee and Coleman (1965). Ouabain is almost always much less than 100% effective as an inhibitor; and though rigorous demonstration is lacking, it is thought to be active at an allosteric site. We have suggested that the variability in the enzyme's response to inhibitors may relate to conformation and local environment changes which alter the interaction of allosteric and catalytic sites (H. D. Brown, 1966c; H. D. Brown *et al.*, 1966, 1967d, 1968c,d). Schrier *et al.* (1969) have, from similar data, concluded that the ouabain effect is not directly upon the ATPase. Barnett (1970), based upon a study of the kinetics of ouabain binding, concludes that more than one mode of ouabain binding occurs. This fact may also contribute to the variation in reported observations.

An additional physical approach to description has been radiation inactivation. This technique, based upon statistical treatment of X-ray "hits," has made the functional ATPase appear to be a much larger unit (mol wt 1,000,000) than are the purified enzymes.

Definitive description of the membrane ATPase or the individual enzymes which compose the complex has been frustrated by the failures to obtain pure components. The effort to accomplish purification has been unsuccessful in the sense that preparations which have been obtained, although not lacking ATPase activity, did lack one or more of the characteristics by which the transport enzyme is recognized. Metal activation, or cardiac glycoside sensitivity, changes importantly as a function of protein purification procedures; some authors have assumed, therefore, that a large family of related enzymes exists. Heinz (1967) has stated another interpretation of some of these data, "Purification of the ATPase as a rule does not mean a clear-cut separation but rather a selective depression, inhibition or deterioration of the apyrase component." (Apyrase is usually defined as an ATPase–ADPase activity associated with membrane preparations. It is generally assumed that these activities are quite separate from the $Na^+ + K^+ATPase$.) We have interpreted our data and some of that in the literature in terms of an hypothesis (H. D. Brown et al., 1965, 1966, 1967a,b, 1968a,b; H. D. Brown, 1966c) that a single protein may be altered by these rather harsh procedures to show the various properties.

Interpretations aside, it is a fact that "purifications" of ATPases have resulted in compounds with higher percentages of the ion-sensitive activity than in the original crude preparations. This has been accomplished by prolonged storage (Hokin and Reasa, 1964), heating (Somogyi, 1964), detergents (Jarnefelt, 1964a; Uesugi et al., 1969), deoxycholate (Nakao et al., 1965), NaI (Nakao et al., 1963; Schwartz, 1965), or inhibitors such as histone-like substances.

Table IV illustrates the variety of characteristics associated with ATP-hydrolyzing preparation obtained from membranes.

In microorganisms, similar lines of study have led to the description of an energy-supplying protein, HPr, and to the existence of a transport-related P-enolpyruvate P-transferase (Kundig et al., 1964; Kaback, 1969).

5. Carbohydrate Transformations

A number of the reactions of carbohydrate intermediary metabolism are catalyzed by endoplasmic reticulum enzymes (Table V). Evaluation of the specific contribution of these membrane site events to metabolism cannot be made upon present knowledge, though it appears probable that control mechanisms are associated with the membrane activities. Many such pathways exist, particularly in plant materials, but have seen little intensive or extensive study. Many of the reactions occur only in part at the endoplasmic reticulum, and the possibility exists that reactions involve "soluble" systems adsorbed to the membranes. (It is great fun to conjecture about the difference in function between a membrane enzyme tightly bound and one adsorbed, perhaps in equilibrium with the solution form.

However, in the absence of information of solidity we are not able to offer satisfactory generalities.)

Alternative routes to a final product, in this instance ascorbic acid and xylulose, by particulate systems were considered by Kanfer et al. (1959). Details of synthesis of L-ascorbate, particularly as the synthetic events may be contrasted in various organisms, have been the subject of numerous studies (Chatterjee et al., 1960). The interrelationship of cytoplasmic and microsomal systems is illustrated by the results of Strominger et al. (1957). These workers showed that a cytosol enzyme catalyzed transformation of uridine diphosphoglucose to uridine diphosphoglucuronic acid with NAD as cofactor.

The literature dealing with cellular compartmentalization of LDH isozymes is another area of chemistry of carbohydrate transformation which points to a broadly applicable concept of specialization of classes of catalysis with respect to cellular geography (Güttler, 1967; Güttler and Clausen, 1967; H. D. Brown et al., 1969a,b).

In recent literature the attempt—much needed—has been made not only to localize enzymes associated with carbohydrate transformation but to characterize them. Examples are studies of hexosphosphate dehydrogenase (Beutler and Morrison, 1967; Srivastava and Beutler, 1969), lactose synthetase (Barra et al., 1969), and glycogen synthetase (Parodi et al., 1969).

6. Lipid Synthesis

a. Sterols. In no subject area of biochemistry has the relationship of cell substructure to biosynthetic mechanism been better illustrated than in the literature of the steroid. This class of molecule appears not to be present in bacteria or in cyanophytic (blue-green) algae. In these forms, in which cellular organization is much less complex than in higher forms, the sterol is not synthesized. Hence, based upon this lesson of phylogeny, we may suppose that the elaboration of membrane-delineated compartments which provide physical discreteness to biosynthetic reactions is essentially related to sterol biosynthesis. The evolution of the subcell morphology can be traced in outline by reference to sterol evolution.

Not only the synthetic machinery but the sterol molecule itself is not randomly distributed within the cell; rather, it is ubiquitously associated with membrane systems. It is easy to rationalize this observation when one recalls that the sterol molecule, by virtue of its chemical properties, provides a suitable and quite probably essential building block for the membrane.

The literature of sterol biosynthesis is voluminous and presents a diversity as great as the number of organisms that have been studied. Research

TABLE IV

Characteristics of Membrane Adenosinetriphosphatases (ATPases)

Source	Treatment	ATPase(s)	Comment	Reference
Pig brain	NaI	Four ouabain-sensitive activities: (a) Mg^{2+} stimulated ATPase, (b) K^+ stimulated ATPase, (c) $(Na^+ + K^+)$-stimulated ATPase, (d) Na^+ stimulated ATPase	All have identical substrate specificity. Activities are inhibited by low concentration of ouabain. Enzymes inactivated by N-ethylmaleimide. Mg^{2+}-sensitive enzyme activated by Ca^{2+}; others inhibited by Ca^{2+}. Temperature coefficient, 6.6 for K^+-stimulated ATPase activity, 3.7 for $(Na^+ + K^+)$-stimulated ATPase, and 2.6 for Na^+-stimulated ATPase	Fujita et al. (1968)
Rabbit brain, kidney intestine, stomach, and heart	NaI	(a) Mg^{2+} ATPase, (b) $Mg^{2+} + Na^+ + K^+$ ATPase	Activity inhibited by fairly low concentration of ouabain. pH plotted vs. activity showed a peak at pH 7.7. Adenosine triphosphate and cytidine triphosphate more readily hydrolyzed than other triphosphates	Nakao et al. (1963, 1965)
Guinea pig brain	Lubrol WX	(a) Ouabain-sensitive $Na^+ + K^+$ ATPase (b) Ouabain-insensitive ATPase	$Na^+ + K^+$ ATPase (ouabain-sensitive), mol wt 670,000; ouabain-insensitive ATPase, mol wt 775,000. Solubilized enzyme remains stable at 0° to 4°C if either K^+ or Na^+ is present in the preparation	Medzihradsky et al. (1967)
Beef brain	NaI-Lubrol	(a) Ouabain-sensitive $Na^+ + K^+$ ATPase (b) Ouabain-insensitive ATPase	$Na^+ + K^+$ ATPase has a molecular weight of 670,000. It is a lipoprotein which can be solubilized and separated from other protein in the membrane and from unbound solubilizing detergent	Uesigo et al. (1969)
Rabbit and rat skeletal muscle	Deoxycholate	(a) ATPase	The Mg^{2+}-moderated soluble microsomal ATPase is activated 10–20-fold by 10^{-5} M Ca^{2+} and inhibited by ethylene glycol tetraacetate	Martonosi (1968)
Bacillus metaterium	Dialysis and treatment with alkali	(a) Soluble Ca^{2+}-activated ATPase	The membrane ATPase is activated by Mg^{2+} or Ca^{2+} and stable in cold, but the solubilized activity is greatly activated by Ca^{2+}, not by Mg^{2+}, and inactivated by cold. The solubilized ATPase was stimulated by SO_2^{2-}. This enzyme does not resemble $Na^+ + K^+$ ATPase	Ishida and Mizushima (1969a)
Beef brain	NaI–salt and isoelectric precipitation	(a) $Na^+ + K^+$ ATPase	Partially purified enzyme (10 times higher specific activity); inhibited by ouabain	Kahlenberg et al. (1969)
Guinea pig heart	NaI	(a) $Na^+ + K^+$ ATPase, ouabain-sensitive	Specific activity is higher than membrane. The enzyme is more stable than the membrane preparation	Schwartz (1965)
	Digitonin	(a) $Na^+ + K^+$ ATPase (b) Ouabain-stimulated	Purified after digitonin solubilization and by gel chromatography in active fractions of varying molecular weight. When soluble enzyme was bound to synthetic matrix certain characteristics associated with membrane preparations were obtained.	H. D. Brown et al. (1966)

	Butanol	(a) Ouabain-sensitive Na$^+$ + K$^+$ ATPase (b) Ouabain-insensitive ATPase		H. D. Brown et al. (1967d)
Pig brain	Digitonin	(a) Mg^{2+} + Ca^{2+} ATPase (b) Mg^{2+} ATPase	g-Strophanthin-stimulated, Mg^{2+} + Ca^{2+} ATPase activity. The Mg^{2+} + Ca^{2+} ATPase inhibited by ethyleneglycol-bis(β-aminoethylether) N,N'-tetraacetate	Nakamaru and Konishi (1968)
Barley root	Digitonin	(a) Ouabain-sensitive ATPase (b) Ouabain-insensitive ATPase		H. D. Brown et al. (1967b)
Bacillus metagoerium	Ammonium sulfate fractionation, protamine sulfate treatment and diethylaminoethyl-cellulose column chromatography	(a) ATPase I (b) ATPase II	Both I and II ATPase require the presence of Ca^{2+} or Mg^{2+}	Ishida and Mizushima (1969b)
Calf kidney	Salting-out and deoxycholate	(a) Na$^+$ + K$^+$ ATPase		Rendi and Uhr (1964)
Thiobacillus thioxidans	Osmotic salt solution washing (without Mg^{2+})	(a) Mg^{2+} ATPase (b) Mg^{2+} + SO$_3^{2-}$ ATPase		Marunouchi and Mori (1967) Marunouchi (1969)

113

TABLE V
CARBOHYDRATE TRANSFORMATIONS CATALYZED BY MICROSOMAL ENZYMES[a]

E. C. number	Enzyme (trivial name)	Reaction	Comment
1.1.1.19	Glucuronate reductase	L-Gulonate + NADP → D-glucuronate + reduced NADP	Also reduces D-galacturonate
1.1.1.20	Glucuronolactone reductase	L-Gulono-γ-lactone + NADP → D-glucurono-γ-lactone + reduced NADP	
1.1.1.22	UDPG dehydrogenase	UDP glucose + 2 NAD + H_2O → UDP glucuronate + 2 reduced NAD	
1.1.1.27	Lactate dehydrogenase	L-Lactate + NAD = pyruvate + reduced NAD	Also oxidizes other L-2-hydroxymonocarboxylic acids. NADP also acts more slowly
1.1.1.47	Glucose dehydrogenase	β-D-Glucose + NAD(P) → D-glucono-δ-lactone + reduced NAD(P)	Also oxidizes D-xylose
1.3.2.-	Gulonolactone dehydrogenase	L-Gulonolactone + NAD(P) = L-ascorbic acid + reduced NAD(P).	
2.4.1.1	α-Glucan phosphorylase, glycogen phosphohydrolase	$(α-1,4-Glucosyl)_n$ + orthophosphate → $(α-1,4-glucosyl)_{n-1}$ + α-D-glucose 1-phosphate	The mammalian enzyme contains pyridoxal phosphate
2.4.1.11	UDP glucose–glycogen glucosyltransferase, glycogen–UDP glucosyltransferase	UDP glucose + $(glycogen)_n$ → UDP + $(glycogen)_{n+1}$	Activated by D-glucose 6-phosphate and other hexose phosphates
2.4.1.17	UDP glucuronyltransferase	UDP glucuronate + acceptor → UDP + acceptor glucuronide	A wide range of phenols, alcohols, amines, and fatty acids can act as acceptors
2.4.1.23	UDP galactose–sphingosine galactosyltransferase, psychosine–UDP galactosyltransferase	UDP galactose + sphingosine → UDP + psychosine	
2.7.1.1	Hexokinase	ATP + D-hexose → ADP + D-hexose 6-phosphate	D-Glucose, D-mannose, D-fructose, and D-glucosamine can act as acceptor; ITP and deoxy-ATP can act as donors
3.1.1.19	Uronolactonase	D-Glucurono-δ-lactone + H_2O → D-glucuronate	
3.1.3.9	Glucose-6-phosphatase	D-Glucose 6-phosphate + H_2O → D-glucose + orthophosphate	Also acts on D-glucosamine 6-phosphate
3.2.1.1	α-Amylase	Hydrolyzes α-1,4-glucan links in polysaccharides containing three or more α-1,4-linked D-glucose units	Acts on starch, glycogen, and related polysaccharides and oligosaccharides
3.2.1.2	β-Amylase	Hydrolyzes α-1,4-glucan links in polysaccharides so as to remove successive maltose units from the nonreducing ends	Acts on starch, glycogen, and related polysaccharides and oligosaccharides, producing β-maltose by an inversion
3.2.1.31	β-Glucuronidase	β-D-Glucuronide + H_2O → alcohol + D-glucuronate	Also catalyzes glucuronotransferase reactions
4.1.2.7	Ketose 1-phosphate aldolase (aldolase)	Ketose 1-phosphate → dihydroxyacetone phosphate + aldehyde	Wide specificity

[a] Abbreviations: NADP, nicotinamide adenine dinucleotide phosphate; UDPG, uridine diphosphate glucose; ADP, adenosine diphosphate; ATP, adenosine triphosphate; ITP, inosine triphosphate.

impetus to a significant degree has been directed toward those organisms and reactions that promised utility to the drug industry. Thus many of the enzyme systems studied are derived from microbial participants in commercial fermentations. Where sufficient work has been done to describe these biosyntheses, they have proven to be associated with microsomes, and for the most part NADPH has been required as a source of reducing equivalents. Literature on the enzymology of cholesterol biosynthesis presents the most fully developed illustration of this disciplinary area (Bucher and McGarrahan, 1956; Tchen and Bloch, 1957; Olson et al., 1957; Bloch, 1965; Dean et al., 1967).

Similar to other sterol-biosynthesizing organisms, the tissues known to synthesize cholesterol in man (liver, adrenal cortex, arterial wall, etc.) are capable of using the two-carbon acetate residue of acetyl-CoA to generate cholesterol. Conventional techniques have been used to isolate from membranes (microsomes) many of these complex enzyme systems. They have proven active when dispersed in water in *in vitro* studies though NADPH and metal ions have been required. A complication in interpretation has resulted from the fact that regenerating systems have frequently been used to replenish the reducing agent.

Details of the transformations have been described in recent reviews: Bloch (1965), Frantz and Schroepfer (1967), and in the primary literature, e.g., Scallen et al. (1968), Etemadi et al. (1969), and Akhtar et al. (1969a,b).

b. *Triglycerides and Phospholipids.* Triglyceride synthesis has been studied intensively in the intestinal epithelium of higher animals (Johnston, 1959; J. L. Brown and Johnston, 1964; Brindley and Hübscher, 1965, 1966; Bickerstaffe and Annison, 1969). Two major pathways of triglyceride synthesis have been associated in part or wholly with the endoplasmic reticulum. In these, glycerol 3-phosphate or monoglyceride acts as acceptor of activated fatty acids. At least in some tissues, all of the enzymes necessary for the two pathways are located in the microsomes. However, phosphatidate phosphohydrolase (M. E. Smith et al., 1967; Johnston et al., 1967) has been identified with a particle-free supernatant of cat and hamster intestinal epithelium, although the enzyme is microsomal in pig and sheep (M. E. Smith et al., 1967). Bickerstaffe and Annison (1969) found in the sheep intestinal epithelium that only glycerol 3-phosphate pathway (Fig. 10) could be demonstrated unequivocally; the 1-monoglycerides showed only a very slight capacity to accept activated fatty acids.

Subcellular localization of the enzymes involved in the glycerol 3-phosphate pathway, though varying not only as a function of tissue but of species, is generally found in the mitochondrial and microsomal fractions

(Y. Stein and Shapiro, 1957; Y. Stein *et al.*, 1957; Senior and Isselbacher, 1962; J. L. Brown and Johnston, 1964; Rao and Johnston, 1966; M. E. Smith and Hübscher, 1966; M. E. Smith *et al.*, 1967). Although for the majority of systems studied the pathway is a microsomal function (Hübscher *et al.*, 1963, 1964; Brindley and Hübscher, 1965), some participation of cytosol is involved since the supernatant fraction must be added back

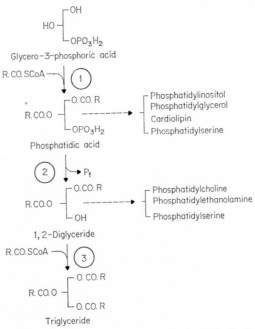

Fig. 10. Glycerate pathway to triglyceride. ① Acyl-CoA: L-glycerol-3-phosphate *O*-acyltransferase (glycerophosphate acyltransferase) (2.3.1.15) ② L-phosphatidate phosphohydrolase (phosphatidate phosphatase) (3.1.3.4) ③ Acyl-CoA 1,2-diglyceride *O*-acyltransferase (diglyceride acyltransferase) (2.3.1.20).

to the reaction systems (M. E. Smith *et al.*, 1967; Johnston *et al.*, 1967). Contribution of cytosol has been assigned to nonenzymic factors—probably a protein and lipoprotein (Tzur and Shapiro, 1964) and long-chain fatty acids—although a phosphatidate phosphohydrolase (M. E. Smith *et al.*, 1967; Johnston *et al.*, 1967) is also involved. Enzymes that have been demonstrated in microsomal fractions to participate in this pathway are lipase, phospholipase A, diglyceride acyl transferase, and cholinephosphotransferase.

In the first position of the alternative pathway, glycerol 3-phosphate is acylated by CoA esters of fatty acids to give the corresponding phosphatidic acid. This is dephosphorylated to yield a 1,2-diglyceride which is, in turn, acylated to yield the triglyceride. Throughout, the L-configuration is maintained. Phosphatidic acid and 1,2-diglyceride are intermediates in the biosynthesis of phosphoglycerides.

When labeled fatty acids are fed to experimental animals, they are rapidly incorporated into liver phospholipids (Possmayer *et al.*, 1969). Incorporation is in the 2-position of glycerol in phosphatidylcholine and phosphatidylethanolamine; saturated fatty acids are incorporated in the 1-position. It is not known if the incorporation of fatty acids represents a *de novo* synthesis of the entire phospholipid molecule for an exchange of components of the fatty acid within the phospholipid.

The presence of a saturated fatty acid in the 1-position and an unsaturated fatty acid in the 2-position would be explicable if the enzymes catalyzing the acylations preferentially localized the saturated fatty acids in the 1-position and the unsaturated fatty acids in the 2-position. In the phosphatidic acid pathway, glycerol 3-phosphate is acylated at the 1- and 2-positions. The phosphatidic acid thus formed is dephosphorylated to yield 1,2-diglyceride which reacts with cytidine diphosphocholine to form phosphatidylcholine. The cytidine diphosphocholine was formed from cytidine triphosphate and phosphocholine (E. P. Kennedy, 1961). Phosphatidylethanolamine is formed similarly in a reaction involving cytidine diphosphocholine-ethanolamine. Palmitate and oleate appear more rapidly in phosphatidic acid moiety than in phosphatidylcholine or phosphatidylethanolamine. Linoleate is more rapidly incorporated into phosphatidylcholine and phosphatidic acid but only slowly into phosphatidylethanolamine. Stearate, in the experiments of Possmayer *et al.* (1969), was incorporated only very slightly into phosphatidic acid and appeared mostly in phosphatidylethanolamine.

There is evidence that an exchange of acyl groups in phospholipids is the mechanism for the introduction of more highly unsaturated fatty acids.

To interpret these data about positional specificity of fatty acids, we must consider that there exists either an enzyme specificity in the acylation of glycerol phosphate or random incorporation of the fatty acids followed by hydrolysis and specific reacylation with another fatty acid. Mudd *et al.* (1969) undertook to test these alternatives in a study of the substrate specificity of cholinephosphotransferase. In the reaction yielding phosphatidylcholine from diglycerides and cytidine diphosphocholine, the catalyst was found to be relatively nonspecific, and, hence determination of a positional specificity, at least in this instance, did not appear to be controlled by sharply determined enzyme specificity. The somewhat more complex alternative of more-or-less random incorporation followed by hydrolysis and specific reacylation is, hence, favored by these and other studies (Van den Bosch et al., 1967, 1968, 1969).

Progressive acylation of monoglyceride to diglyceride and triglyceride also occurs, probably representing a nonstereospecific shunt which functions to produce large amounts of di- and triglycerides during absorption of the fatty meal.

7. Protein and Amino Acid Metabolism

a. *Protein Synthesis.* Drain upon the cell's economy imposed by the energy requirements of protein synthesis is substantial, especially in a heavily synthetic tissue. Although it is difficult to assign percentage values because of wide variation even within a particular tissue, the magnitude of the negative entropy term accompanying protein synthetic activities is certainly imposing.

In addition, the synthesis of protein is under an extremely elaborate set of control mechanisms which are not only interdependent and correlated with the cells' metabolic activity but are also influenced by the availability of the amino acids essential for the synthetic process (Christensen, 1964).

Protein must be supplied for renewal of the endoplasmic reticulum membranes themselves and for transfer to the membranous structure of other organelles. Soluble protein must be synthesized at the reticulum surface for its further transfer to mitochondria and to the so-called "soluble" or, at least, " less-structured" areas of the cytoplasm. Major synthetic tissues, such as liver, produce proteins which are then released to the blood.

Protein synthesis has been associated primarily with the microsomal fraction though Howell et al. (1964) have shown that the synthetic activity is not exclusively associated with this organelle. However, most of the ribosomes are bound to endoplasmic reticulum membranes, and the membrane ribosome combination appears the major site of synthetic activity. Although Wettstein et al. (1963) described synthesis in isolated polysomes, later authors (Lawford et al., 1966; Blobel and Potter, 1966) presented the "majority conclusion" that the structures are best described as "bound polysomes." Diversity of view is not made less by recent recognition that other cell areas are also capable of some protein synthesis. The literature remains much involved with its technology, however, and some of the unsettled questions must await more effective ways of handling these complex systems. Figure 11 illustrates a representative method employed for separation of ribosomes from other particles (Campbell and Lawford, 1968). Much of the literature, however, has been based upon *in vitro* systems (Nirenberg, 1964) composed of complex particulate fractions.

It may be that the specific site of synthesis will prove to be associated with the ultimate geographic destination of the product although some tests of this thesis have given negative results (Peters, 1962; Sargeant and Campbell, 1965; Campbell and Lawford, 1968). The large number of components in the several biologically unrelated systems that have been studied appear to vary greatly in their character. There is an enzyme-catalyzed cycle of peptide bond formation. [Reaction mechanisms have been discussed in the extensive review of Lengyel and Söll (1969).] Hence, a series of enzymes essential to the chain of events leading to specific pro-

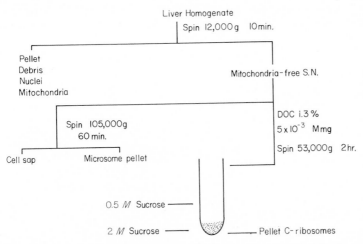

Fig. 11. Method of separation of ribosomes from other particles. [Redrawn from P. N. Campbell and G. R. Lawford (1967) *in* "Structure and Function of the Endoplasmic Reticulum in Animal Cells" (F. C. Gran, ed.), p. 61, Academic Press, New York.]

tein synthesis are intimately related to and controlled by the genetic mechanism *per se*. The basic transfer reaction involves aminoacyl (AA)–ribonucleic acid (RNA), ribosomes, and messenger RNA (mRNA) which can result in peptide bond formation. The transfer reaction in higher animal systems requires two enzyme fractions, a source of bond energy, and various salts functioning presumably as enzyme cofactors.

Prior to synthesis of the protein chains, AA–transfer RNA (tRNA) synthetases are involved in the binding of the correct amino acid to the 3′-hydroxyl of the terminal adenosine of a specific tRNA molecule. A variety of such synthetases have been described and their physical and enzymological properties have been studied.

It is thought that the first step in protein chain synthesis is the binding of aminoacyl–RNA to the mRNA–ribosome complex. Moldave (1965) has concluded that it is a nonenzymic binding. The binding appears to be non-covalent. When the enzyme components (two enzyme fractions) and guanosine triphosphate (GTP) are added, the bound amino acid is incorporated into protein. The two enzyme fractions required for polypeptide synthesis were described by J. Bishop and Schweet (1961) and by Fessenden and Moldave (1961). The most extensively studied system has been that purified from *Escherichia coli*. Specific assignment of function to the fractions has not been accomplished though it appears that one of the fractions is active for GTP hydrolysis in the presence of ribosomes (Allende *et al.*, 1964). Conway and Lipmann (1964) and Arlinghouse *et al.* (1964) have described this reaction as a hydrolysis yielding guanosine diphosphate (GDP) and inorganic phosphate. Peptide bond synthesis has been shown to require a second enzymic activity. The second fraction is more heat labile than is the GTP hydrolase itself. The exact nature of the interrelationship between the hydrolase activity and the transferase activity *per se* is not well established. Whether these activities could be associated with separate sites on a single protein, of course, is speculative. Lipmann (quoted by Schweet and Heintz, 1966) has indicated a one-to-one stoichiometry between GTP hydrolysis and peptide bond formation. Hence, if indeed, the transferase is a separate activity, then the coincidence of a similar rate function exists. A number of attempts to purify the fractions have been reported. Rat liver was used as a source of the two components, and separations were made on Sephadex G-200 (Gasior and Moldave, 1965a,b). One of the fractions which Gasior and Moldave (1965a,b) call transferase I had an apparent molecular weight above 300,000. This material appears to be equivalent to the *E. coli* heat-labile binding fraction. It now appears that, in addition to the nonenzymic binding, fraction B of *E. coli* or transferase I of liver is essential for an enzymic binding. Too, an enzyme-requiring GTP binding has been reported in rat liver ribosomes by Bont *et al.* (1965). The functions of the two enzyme complexes and their chemical nature as well as the number of actual steps in the reaction have not yet been made clear, though a number of laboratories are using specific inhibitors to illuminate further the mechanisms involved. It presently appears that the *E. coli* system and that of higher organisms have essential differences. Studies in the area have provided substantial lists of enzyme properties associated with the various tRNA's (Holley *et al.*, 1965; Zachau *et al.*, 1966; Henley *et al.*, 1966; Lindahl *et al.*, 1966).

It is interesting to note that, in the context of the thesis that membrane structures interrelate in a functional way with enzymic activities, several authors have emphasized the importance of membrane constituents in the enzymic synthesis of proteins (Howell *et al.*, 1964; Cammarano *et al.*,

1965; Tsukada and Lieberman, 1965). The role of membranes in protein synthesis has scarcely been approached and will undoubtedly prove to be very important. Association of isolated bacterial ribosomes with various enzymic activities has been described by Raacke and Fiala (1964), and Dresden and Hoagland (1965) have considered the stabilization of ribosomes by membranes. Although the subject area of the relationship of protein synthesis components to membrane has been considered by a number of researchers, it undoubtedly remains a major problem area.

b. Enzymes with Amino Acids as Substrates. Because of possible atypical distribution of amino acid metabolizing enzymes in neoplastic cells, subcellular distribution of several has been studied. Carruthers *et al.* (1959) have reported that arginase and glutamate dehydrogenase activity were bound to microsomes in Ehrlich ascites cells. L. T. Myers and Worthen (1961) have described cysteine reductase in animal tissues (rat, rabbit, and man) as being present exclusively in the microsomal fraction. Binkley (1961) has associated a glutathionase, hydrolyzing the release of cysteinyl glycine from glutathione, with the lipoprotein particles of microsomes.

C. Plasma Membrane

The plasma membrane, though shown in many electron micrographs to be interconnected with the endoplasmic reticulum proper, is functionally a distinct structure, and hence *prima facie* one would expect that its enzymic complement should be demonstrably different from that of other cellular reticulum areas. This has been studied by Benedetti and Emmelot (1968), Wattiaux-De Coninck and Wattiaux (1969), and Perdue and Sneider (1970). These authors list activities associated with rat liver plasma membrane preparations (Table VI). Of these the following activities are intrinsic to the plasma membrane: Mg^{2+} (or Ca^{2+})-dependent ATPase, $Na^++K^++Mg^{2+}$ATPase, K^+-p-nitrophenylphosphatase (NPPase), K^+-acetylphosphatase (AcPase), 5'-mononucleotidase, NAD-pyrophosphatase, alkaline glycerolphosphatase and phosphodiesterase, cytidinetriphosphatase (CTPase), inosinediphosphatase (IDPase), ADPase, and L-leucylnaphthylamidase (leucine aminopeptidase). The other enzymes of Table VI are found also in other subcellular fractions and, hence, are not categorically distinct to the plasma membrane. It may be that some are in dynamic equilibrium with other subcell populations. The converse hypothesis that such activities, though having a common catalytic character, may nonetheless be chemically and physically different has been suggested.

TABLE VI

Specific Enzyme Activities of Liver Plasma Membranes[a]

Enzymes	Rat[b]	Product	Mouse[b]
Adenosinetriphosphatase	46.4 ± 7.8	P_i	32.4 ± 3.7
Na-K-adenosinetriphosphatase	11.7 ± 2.3	P_i	17.4 ± 1.7
5'-Nucleotidase	51.0 ± 6.7	P_i	13.2 ± 3.0
Glycerolphosphatase		P_i	
Alkaline	0.0 (1.0)[c]		2.2 ± 0.8
Acid	0.4 ± 0.1		—
p-Nitrophenylphosphatase		p-Nitrophenol	
Alkaline[d]	1.5 ± 0.5		3.0 ± 1.0
Alkaline K+-activated	1.5 ± 0.5		1.5 ± 0.3
Acid	5.8 ± 1.0		5.3 ± 1.0
Acetylphosphatase	11.4 ± 2.2	Acetylphosphate	13.6 ± 2.5
K+-activated AcPase	10.8 ± 1.9		8.2 ± 0.7
Phosphodiesterase		p-Nitrophenol	
Alkaline	3.6 ± 0.4		2.0 ± 0.3
Acid	0.7 ± 0.1		1.3 ± 0.2
Ribonuclease		A_{444} (mµ)	
Alkaline	1.1 ± 0.25		2.4 ± 0.5
Acid	0.27 ± 0.12		0.29 ± 0.07
Glucose-6-phosphatase	1.4 ± 0.2	P_i	2.0 ± 0.1
Esterase		α-Naphthol	
α-Naphthyl laurate	0.75 ± 0.02		1.03 ± 0.08
α-Naphthyl caprylate	34.0 ± 1.6		38.1 ± 1.1
Triose-3-phosphate dehydrogenase	2.04	Reduced nicotinamide adenine dinucleotide	—
Reduced nicotinamide adenine dinucleotide–cytochrome c reductase	7.68 ± 0.5	Reduced cytochrome c	—
Adenosinediphosphatase	0–04	P_i	—
Inosinediphosphatase	30 ± 3	P_i	—
Nicotinamide adenine dinucleotide pyrophosphatase	5.72 ± 0.2	Nicotinamide adenine dinucleotide disappeared	—
Nicotinamide adenine dinucleotide nucleosidase	0.5 ± 0.1	Nicotinamide adenine dinucleotide disappeared	—
Leucyl-β-naphthylamidase	3.7 ± 0.1	β-Naphthylamine	3.5 ± 0.2

[a] From Emmelot and Bos (1966) by permission.

[b] Data given in micromoles product per milligram protein per hour.

[c] See text.

[d] About half of this activity is due to nitrophenylphosphatase activity which is not dependent on Mg^{2+} ions.

Success in assigning active proteins to specific subcell organelles has led to still greater resolution in the distinction of membrane activities with identifiable morphological areas. E. L. Smith and Hill (1960) demonstrated a leucyl-β-naphthylamidase activity associated with membrane knobs which were released from membranes by papain treatment. The knobs themselves are specialized areas (perhaps microvilli) of the plasma membrane (Benedetti and Emmelot, 1965).

As this example illustrates, fine resolution of enzyme activities is being made with quite minute subcellular entities, and no doubt chemical and structural specificities are resolvable into relationships that may portend only a relatively few molecules. Whether simplifying interpretations (e.g., a particular morphological association imposes special catalytic activity) may be made useful or whether we must presume that each distinct catalytic pattern represents a genetically determined entity remains for elucidation by methodology which is yet to be developed.

D. Golgi Apparatus (Dictyosomes)

The Golgi apparatus has a history of controversy in the scientific literature which may surpass the fiction writer's best story. Beginning with the description of a silver-staining organelle by Camillo Golgi in 1898, its existence and function have been constantly debated. Electron microscopic studies have now brought conclusive support to the indirect evidence which had earlier been amassed to establish the existence of Golgi's subcellular apparatus. Function—though presumably secretory—is not known with completeness.

The specialized membrane area called by Sjöstrand (1962) "Golgi membranes" or "α-cytomembranes" is universal in distribution, and lipid-soluble membranes of the apparatus appear to be associated with vitamin C localization in some cells. The characteristic metal affinity (silver and osmium) is not well understood.

Assignment of enzyme activity to the Golgi apparatus has depended greatly upon microchemical techniques. In consequence, it has to this date proven difficult or impossible to know whether enzyme activities of the Golgi apparatus are membrane-bound or present within the cysternal cavities of the apparatus. [This point is further developed in the review of Beams and Kessel (1968).] Novikoff and Goldfischer (1961) and Goldfischer (1964) have associated a nucleoside diphosphatase activity with the Golgi apparatus.

Thiamine pyrophosphatase appears to be the most consistent marker available for the apparatus (Fig. 12) in both animal and plant cells

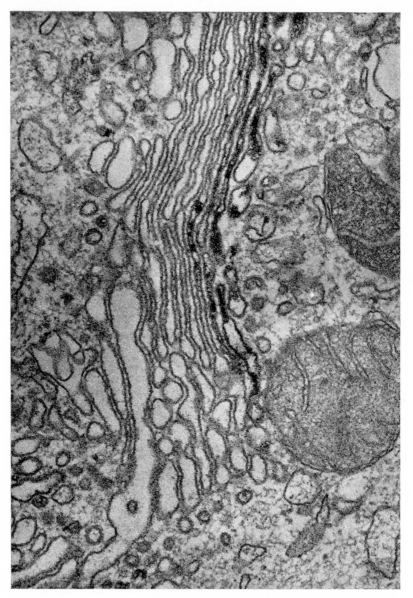

Fɪɢ. 12. Thiamine pyrophosphatase marking Golgi apparatus in rat epididymis. (Photograph courtesy of Daniel S. Friend.)

(Shanthaveerappa and Bourne, 1965; Dauwalder *et al.*, 1966) though it is not universal in all organisms which have been studied. Both acid and alkaline phosphatases have been associated with some Golgi complexes (Kitada, 1959; Dominas *et al.*, 1963; Novikoff *et al.*, 1965). Succinic dehydrogenase from the rat prostate (Likar and Rosenkrantz, 1964) and esterase and β-glucuronidase activity isolated from rat pituitary and thyroid (Sobel, 1962) have been associated with Golgi structures.

Definition of enzyme localization with reference to Golgi membranes must await strict separation of endoplasmic reticulum membranes from Golgi saccules. Endoplasmic reticulum and Golgi membranes frequently are sufficiently interwoven so that this may be accomplished with difficulty, and for the moment our categories may be set up only by the exercise of considerable arbitrariness. Hence, functionality of the Golgi enzymes is for the most part still to be worked out.

III. Mitochondrial Enzymes

A. Structure and Function of Mitochondria

The mitochondria have been the subject of elaborate morphological studies. Comparatively regular particles, varying in length from 0.5 to 7.0μ, they are present in great numbers in some cells (e.g., as many as 2500 in liver cells).

The overall shape and size of the mitochondria and the numbers and internal organization of their cristae vary with respect to the tissue in which they occur (Palade, 1952). Aerobic oxidation and the coupling of oxidation with the principal energy storage compounds occurs largely within the mitochondria. Hence, because their function is paramount in the cell's economy and the technology for their isolation is available, they have been intensively and extensively studied. Thus structures, so recently visualizable only as Janus green-stained dots, now are recognized by their double-membrane highly compartmented form, and much of the intermediary metabolism associated with their complex substructure has been elucidated (Figs. 13–15).

Light scattering and extinction changes using light and electron microscopy have been employed to study the relationship of structural changes to metabolism.

Mitochondria may show changes in their structure and in their enzymology in response to changed physiological conditions. The descriptive literature is closely tied to the technology of materials preparation for electron microscopy—isolated and fixed organelles differ characteristically in appearance from those that are fixed *in situ*. The peripheral and intercristal spaces are enlarged and the matrix is decreased in isolated structures. Uncoupling agents can be used, however, to render mitochondria *in situ* in appearance much like those isolated before fixation (Weinbach

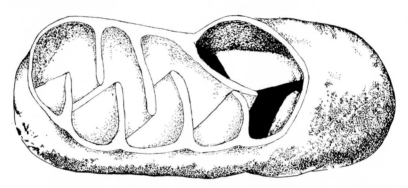

FIG. 13. Schematic representation of mitochondrion showing outer and inner walls and infolded cristae.

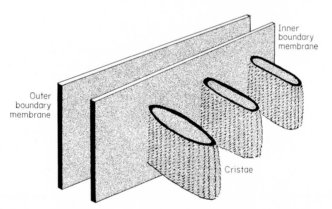

FIG. 14. Interrelationships of the mitochondrial membranes. [Redrawn from J. T. Penniston *et al. Proc. Nat. Acad. Sci. U.S.* **59**, 624–631 (1968), by permission.]

and Garbus, 1966; Weinbach *et al.*, 1967; Hackenbrock, 1966, 1968; Deamer *et al.*, 1967).

Green has related similar appearance changes to a series of configurational states theoretically related to cell energetics (D. E. Green and Baum, 1970). Their subunit theory of membrane composition relates to the reported particulate components of the inner membrane (Fernández-Morán, 1962, 1963). These appear to be stalked (Fernández-Morán,

1963; Parsons, 1963; Stoeckenius, 1963; D. S. Smith, 1963) and to have a base piece (D. S. Smith, 1963) which is continuous with the membrane proper. Head pieces of the particles have ATPase activity (coupling factor I of Racker and Horstman, 1967); the stalks have been associated with an oligomycin-sensitivity-conferring protein (MacLennan and Asai, 1968), and some of the base pieces appear to be involved in the electron transfer chain (D. E. Green and Tzagoloff, 1966). Most authors associate succinic dehydrogenase and ferrochelatase with the inner membrane (M.

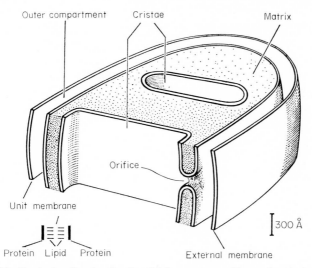

Fig. 15. Idealized detail of mitochondrial structure. [Reproduced from V. P. Whittaker (1966) *in* "Regulation of Metabolic Processes in Mitochondria" (J. M. Tager, S. Papa, E. Quagliariello, and E. C. Slater, eds.), Vol. 7. p. 1, Elsevier (BBA Library), Amsterdam, by permission.]

S. Jones and Jones, 1968), monoamine oxidase (Sottocasa *et al.*, 1967a), and palmitate–CoA ligase (Norum *et al.*, 1966b) with the outer membrane solely. Many of the citric acid cycle enzymes (reviewed by Greville, 1969) are in the intercompartment solution within the inner membrane or between the inner and outer membranes. Much of this literature, however, has been invested with ingenious, stimulating, and heated disagreement.

Mitochondria have been dissected by a variety of mechanical and digestive techniques. These methods, briefly discussed below, have served as biochemical tools brought to bear upon questions concerning the association of series of chemical reactions with the several major structural components. Microchemical procedures have been added recently to this approach. It is no longer adequate to associate enzymic phenomena with a mitochondrion; rather we would hope that the suborganelle site might be specified. When this has been accomplished, then the qualifications which apply to the description of enzymic activities in terms of site-specific populations may be referred—at an even higher resolution—to the discrete populations of catalysts within the single mitochondrion. Although the task has hardly been begun, certainly structure–activity relations and the interaction of active proteins in specific suborganelle compartments have been better described in this organelle than in any other.

As with questions of structure and function of the mature organelle, the nature of the development of the mitochondrion remains only incompletely elucidated. Most probably the mitochondria are derived from preexisting incomplete "promitochondria" and grow—at least in part—by the activity of a protein-synthesizing system intrinsic to the structure itself (Pullman and Schatz, 1967). Mitochondrial deoxyribonucleic acid (DNA) appears to represent a functional entity apart from the control of nucleus-contained genetic material.

However, it appears that most mitochondrial proteins are coded for by nuclear genetic material and assembled on cytoplasmic ribosomes. Hence the formation of the mitochondria involves two separate genetic systems and the interaction of a formidable array of cellular components.

B. The Enzyme Systems

1. Introduction

Mitochondria isolated by differential and gradient centrifugation have been the object of organelle biochemistry to a far greater extent than any other cell particle (Table VII). The various aspects of the biochemistry of mitochondria have been extensively reviewed (Lehninger, 1965; E. C. Slater, 1967b; Pullman and Schatz, 1967; Lardy and Ferguson, 1969; Lenaz, 1969).

Localization of enzyme activities within substructures has been accomplished at least with reference to several of the reactive sequences. In order to separate the discrete areas of the mitochondrion, various degenerative procedures have been used. Phospholipase (Bachman et al., 1966; Allmann and Bachmann, 1967) from snake venom has been used to hydrolyze lipid elements of the outer membrane, and, thus, under carefully controlled conditions, gradually to peel it away. Similar results have been reported using deoxycholate. Detergents have been employed in a similar way (C. Schnaitman et al., 1967; Levy et al., 1967; Hoppel and Cooper, 1968; C. Schnaitman and Greenawalt, 1968), and mechanical methods of various design have also served to allow a more-or-less controlled detachment of the outer from the inner membrane (Parsons, 1965; Parsons *et al.*, 1966, 1967; Parsons and Williams, 1967; Sottocasa *et al.*, 1967a,b).

Separations, whether disruption has been accomplished by mechanical or detergent–bile salt or phospholipase dispersion of the membrane, ultimately have been obtained by differential or density gradient centrifugation (though presumably other methods may also prove to be appropriate). As the nature of the separation technology would lead one to expect, the identification of enzymic activities with the several suborganelle fractions is not a matter of absolute agreement. Table VIII (A and B) represents a synthesis of these studies.

The technical literature remains at this writing controversial. Assignment of functional systems to the outer and inner membranes has been discussed by Pullman *et al.* (1960), Parsons *et al.* (1966), D. E. Green *et al.* (1966), Kagawa and Racker (1966a,b), Sottocasa *et al.* (1967a,b), C. Schnaitman *et al.* (1967), Sottocasa and Sandri (1969), Lenaz (1969), Lowenstein (1969), and D. E. Green and Baum (1970). Though laudable first attempts at the development of a methodology which will allow suborganelle differentiation have been made, results are only in part satisfactory because criteria for identification of the fractions after separation are inadequate and by all existing techniques damage to some of the particles is the rule. Electron microscopy with its own inherent limitations is further restricted by the fact that the separation methods themselves tend to break down the structure of the fractions and their visual identification becomes more than ordinarily difficult.

2. A Terminal Respiratory Chain and Oxidative Phosphorylation

Transfer of energy via a respiratory chain assembly in the mitochondrion and the bioenergetic phenomena associated with it have been studied by many scientists. With the exception of the genetic mechanisms, the power of "major programs" has been brought to bear more upon this subject than upon any other area of biochemistry. Much of the accomplishment has been discussed in frequent and intensive reviews (e.g., C. Ernster and Lee, 1964; Lehninger, 1965; T. E. King *et al.*, 1965; Racker, 1965; Sanadi, 1965; Quagliariello *et al.*, 1967; E. C. Slater, 1966; Tager *et al.*, 1966; Boyer, 1967; Pullman and Schatz, 1967; Lardy and Ferguson, 1969; D. E. Green and MacLennan, 1969; D. E. Green and Baum, 1970). Individual enzyme activities have been elucidated; for the most part, however, these have been studied in relatively impure systems. In fact, the overall mechanism of the energy flow and its transduction are a matter of intellectual and, unfortunately, frequently emotional, dispute. The volumes of conjecture are ingenious and stimulating; at this moment they are

TABLE VII

Enzymes Associated with Mitochondria

E. C. number	Enzyme (trivial name)	Comments and references
1.1.1.27	Lactate dehydrogenase (LDH)	Although found essentially in cytoplasmic fraction, it is also found in subcellular particles. Electrophoretic studies on cellular compartmentalization of LDH isozymes indicate a cathodic band which could be identified as LDHs in rat heart and liver mitochondria. Güttler and Clausen have reported an anodically migrated isozyme, LDHt in purified mitochondrial fraction. Faihimi and Karnowsky (1966); Güttler (1967); calves heart, liver, kidney, and skeletal muscle; Güttler and Clausen (1967); Alessio et al. (1967), rat heart and liver; H. D. Brown et al. (1969a), Morris hepatoma; H. D. Brown et al. (1969b), rat breast tissue; Pokrovskii and Karovnikore (1969), liver
1.1.1.30	β-Hydroxybutyrate dehydrogenase	Reported to be bound to the outer membrane of mitochondria. Beaufay et al. (1959a), rat liver; Lehninger et al. (1960), rat liver; Norum et al. (1966a)
1.1.1.37	Malate dehydrogenase, nicotinamide adenine dinucleotide (NAD)-dependent	This enzyme has both cytoplasmic and mitochondrial distribution. Nicotinamide adenine dinucleotide phosphate (NADP)-dependent malate dehydrogenase (malic enzyme) is mostly cytoplasmic in distribution but in pigeon brain, nerve, and heart, part of the activity is particulate. Found in multimolecular forms; probably attached to the inner membrane of the mitochondrion. Abood et al. (1952), rat brain; Christie and Judah (1953), rat liver; Sacktor (1953), housefly flight muscle; Abood and Gerard (1954), nerve; Davies and Kun (1957), ox heart; Mahler et al. (1958), chick embryonic liver and chick embryo; Rutter and Lardy (1958), pigeon liver; Beaufay et al. (1959a, b), rat liver; Wieland et al. (1959–60), ox heart; Solomon (1959), chick embryonic liver; Delbrück et al. (1959a), rat liver; Delbrück et al. (1959b), Locusta migratoria muscles; W. Vogell et al. (1959–1960), Locusta migratoria muscles; L. Siegel and Englard (1960), ox heart; Thorne (1960), rat liver; L. Siegel and Englard (1961), beef heart; Grimm and Doherty (1961), beef heart; L. Siegel and Englard (1962), beef heart; Kitto et al. (1966), chicken and pig heart; Kitto and Kaplan (1966), chicken heart; Henderson (1966), mouse tissues; Moorjani and Lemonde (1967), Tribolium confusum; Sottocasa (1967), rat liver; Schechter and Epstein (1968); Dupourque and Kun (1968), ox kidney; Salganicoff and Koeppe (1968), rat brain; Guha et al. (1968), beef heart; Henderson (1968), mouse liver, heart, and other tissues; Mann and Vestling (1969), rat liver; Fahien and Strmecki (1969a), bovine liver; Vary et al. (1969); Silverstein and Sulebele (1969a,b), pig and bovine heart; Silverstein and Sulebele (1970), pig heart; Sulebele and Silverstein (1970), pig heart; Kitto et al. (1970), chicken heart
1.1.1.41–42	Isocitrate dehydrogenase	Subcellular distribution of NAD- and NADP-dependent isocitrate dehydrogenase differ from each other. The NAD-dependent isocitrate dehydrogenase is localized in mitochondria. The NADP-linked enzyme is present both in mitochondria and in soluble fractions. Most pyridine nucleotide-linked dehydrogenases are attached to inner membrane of mitochondrion. L. Ernster and Navazio (1956), rat liver; Lowenstein and Smith (1962), rat and chicken liver; Bell and Baron (1964), rat liver and heart; Highashi et al. (1965), beef heart; A. M. Stein et al. (1967), Ehrlich ascites carcinoma; Cox and Davies (1967), pea; Sottocasa (1967), rat liver; Salganicoff and Koeppe (1968), rat brain; Henderson (1968), mouse liver, heart, and other tissues; Yamamota (1969), bean; LeJohn et al. (1969), Bastocladiella emersonii; Cox and Davies (1969), pea; Nicholls and Garland (1969), rat liver
1.1.1.43	Phosphogluconate dehydrogenase (6-phosphogluconic acid dehydrogenase)	Phosphogluconate dehydrogenase along with glucose-6-phosphate dehydrogenase were thought to be exclusively cytosol enzymes. However, recent work indicates that they are associated with mitochondria in a latent form. Zaheer et al. (1967), rat liver

130

1.1.1.49	Glucose-6-phosphate dehydrogenase	See comment for 1.1.1.43. Zaheer et al. (1967), rat liver
1.1.1.51	Δ⁵-3-β-Hydroxysteroid dehydrogenase	This enzyme and Δ^5-3-ketosteroid isomerase (5.3.3.1) catalyze the enzymic conversion of pregnenolone to progesterone in various animal tissues. The pathway is associated with microsomal and mitochondrial fractions. The mitochondrial dehydrogenase is mainly NAD-dependent whereas the microsomal enzyme can utilize NADP as well as NAD. Koide and Torres (1965), human placenta; Sulimovici and Boyd (1969), rat ovary
1.1.99.1	Choline dehydrogenase	Essentially mitochondrial. Kensler and Langemann (1951), rat liver; Christie and Judah (1953), rat liver; Rendina and Singer (1959), rat liver; Zebe (1959–1960), rat liver; J. N. Williams (1952a), rat liver
1.1.99.5	L-α-Glyceorophosphate dehydrogenase	Two forms of glycerophosphate dehydrogenase are found in mammalian cells. One type is localized in the cytoplasm fraction and requires NAD as cofactor, whereas the other form is found to be bound tightly to mitochondria and does not require NAD. This enzyme (cytochrome-linked) occurs in insect muscle mitochondria. Zebe et al. (1959) and Estabrook and Sacktor (1958) believe that these insects have very low level of LDH activity, and the enzymes in two different locations play a role in mito-chondrial reoxidation of glycolytic reduced NAD (NADH) with glycerophosphate as electron carrier between two cell phases. Sacktor (1955), housefly muscle; Zebe et al. (1959), fat body of Locusta migratoria; Fiala (1967), rat brain and liver; Tipton and Dawson (1968), pig brain; Y.-P. Lee and Hsu (1969), mouse liver, kidney, heart, brain, lung, and spleen; Dawson and Thorne (1969), pig brain
1.2.1.8	Betaine aldehyde dehydrogenase	Isolated from mitochondrial fraction. J. N. Williams (1952b), rat liver
1.2.2.2	Pyruvate dehydrogenase	Found in mitochondria. Silbert and Martin (1968), rat liver; Reinauer and Junger (1969), swine heart
1.3.2.–	Gulonolactone dehydrogenase	This enzyme has been mainly recovered with microsomal fraction in rat liver (Isherwood et al. 1960) and goat liver (Chatterjee et al., 1959). Associated with mitochondria in chicken kidney. Chatterjee et al. (1960), chicken kidney
1.3.3.–	Coproporphyrinogen oxidase (decarboxylating)	Predominantly mitochondrial. Sano and Granick (1961), guinea pig liver
1.3.99.1	Succinate dehydrogenase	Found in mitochondrial fraction along with cytochromic oxidase and used as an indicator for mitochondrial contamination in other subcellular fractions. A considerable activity has been reported in supernatant from rat mammary gland and chick embryo homogenates. Schneider (1946), rat azodye hepatoma and rat liver; Schneider et al. (1948), rat liver; Hogeboom et al. (1948), rat liver; Schneider and Hogeboom (1950a), rat liver; Schneider and Hogeboom (1950b), mouse liver; Schein et al. (1951), rat liver; J. A. Shepherd and Kalnitsky (1951), rat liver; Schneider et al. (1952), rat brain; Lipner and Barker (1953), rat liver; Novikoff et al. (1953b), rat liver; Maxwell and Ashwell (1953), mouse spleen; Chappel and Perry (1953), pigeon muscle; Sacktor (1953), housefly muscle; Cleland and Slater (1953), rat heart; McSham et al. (1953), rat anterior pituitary; Dianzani (1954b) rat liver; Paigen (1954), rat liver; C. F. Strittmatter and Ball (1954), rat liver; Kuff and Schneider (1954), mouse liver; Kretch-mer and Dickerman (1954), rat kidney; Straus (1954), rat kidney; Holter (1954), ameba Chaos chaos; de Duve et al. (1955), rat liver; Lowe and Lehninger (1955), rat liver; Blaschko et al. (1955), ox adrenal medulla; Nelson (1955), bull spermatozoa; Holter (1955) ameba Chaos chaos; Birbeck and Reid (1956), rat liver; Kuff et al. (1956), rat liver; Stevens and Reid (1956), rat liver; J. M. Foster (1956), lobster nerve and squid nerve; Ziegler and Melchior (1956), rat anterior pituitary; Masko et al. (1956), pancreatic islets from Pleuronectus; Thomson and Klipfel (1957), rat and mouse liver; Eichel (1957b), rat spleen; Blaschko et al. (1957), ox adrenal medulla; Scevola and de Barbieri (1957), chick whole embryos; Novikoff (1957), Novikoff hepatoma; Thomson and Klipfel (1958), rat liver; Krishnamurthy et al. (1958), chicken liver; La Bella and Brown (1958), pig anterior pituitary; R. L. Schultz and Meyer (1958), ox adrenal cortex medulla; Baillie and Morton (1958), cow mammary gland; Mahler

TABLE VII (Continued)

E. C. number	Enzyme (trivial name)	Comments and references
		et al. (1958), chick whole embryos and embryonic liver; Caravaglios and Franzini (1959), rat liver; Datta and Shepard (1959), rat liver; Kikuchi *et al.* (1959), *Ascaris lumbricoides* muscle; Aldridge and Johnson (1959), rat brain; Giuditta and Strecker (1959), rat brain; Whittaker (1959), guinea pig and rabbit brain; Carey and Greville (1959), chick whole embryos; Kawai (1960), snail hepatopancreas; N. Weiner (1960), human, cow, and dog brain; T. F. Slater and Planterose (1960), rat mammary gland and rat liver; Wakid and Needham (1960), rat myometrium; Samson *et al.* (1961), rat brain; Spiro and Ball (1961a), ox adrenal medulla; Spiro and Ball (1961b), ox cortex; Greenfield and Boell (1968), chick liver, heart, and skeletal muscle during embryonic development; Baginsky and Hatefi (1969); Aithal and Ramsarma (1969), rat liver
1.3.99.3	Acyl-CoA dehydrogenase	Principally in mitochondrial fraction. Waldschmidt (1967), rat liver, spleen, and thymus
1.4.1.2–4	Glutamate dehydrogenase	This enzyme is considered to be bound to inner membrane of mitochondria though its presence in other particulate and soluble fraction is not uncommon. Christie and Judah (1953), rat liver; Hogeboom and Schneider (1953), mouse liver; L. Ernster and Navazio (1956), rat liver; Allard *et al.* (1957), rat liver; de Lamirande and Allard (1957), rat liver; Beaufay *et al.* (1959a), rat liver; Beaufay *et al.* (1959b), rat liver; Carruthers *et al.* (1959), Ehrlich ascites cells; Delbrück *et al.* (1959b), muscles and fat body from *Locusta migratoria*; Delbrück *et al.* (1959a), rat uterine epithelioma and rat liver; Solomon (1959), chick embryonic liver; Bendall and de Duve (1960), rat liver; Novikoff (1960), Novikoff hepatoma; Sottocasa (1967), rat liver; LeJohn (1968), *Blastocladiella*; di Prisco *et al.* (1968), rat liver; Fahien and Strmecki (1969b), bovine liver; Fahien *et al.* (1969), bovine liver; Baudhuin *et al.* (1969), rat liver; LeJohn *et al.* (1969); T. L. Mason and Hooper (1969), chick embryo liver
1.4.3.4	Monoamine oxidase	This enzyme has been reported to be entirely mitochondrial in many animal tissues. However, microsomal distribution also is not uncommon. A significant part of the total activity is located in microsomal fraction in rat liver and vas deferens. Hawkins (1952), rat liver; Blaschko *et al.* (1955), ox adrenal medulla; Bogdanski *et al.* (1957), dog and cat brain; Blaschko *et al.* (1957), ox adrenal medulla; Zile and Lardy (1959), rat liver; N. Weiner (1960), human, dog and cow brain; de Duve *et al.* (1960), rat liver; C. Schnaitman *et al.* (1967), rat liver; Hellerman and Erwin (1968), bovine kidney; McEwen *et al.* (1968), human liver; Fuller (1968), rat liver and brain; Tipton and Dawson (1968), pig brain; Gorkin and Akopyan (1968), beef and rat liver; Jarrott and Iversen (1968), rat liver and vas deferens; Erwin and Hellerman (1968), bovine kidney; Yasunobu *et al.* (1968), beef liver; Sourkes (1968), rat liver; Allman *et al.* (1968); Brunner and Bygrave (1969), rat liver; McEwen *et al.* (1969a,b), human liver; C. G. S. Collins and Youdim (1969), human and rat liver; Hartman *et al.* (1969), ox liver
1.4.3.6	Diamine oxidase, histaminase	In pig kidney, approximately one-fourth of total enzyme activity is mitochondrial, the rest in the supernatant fraction. In rabbit liver the enzyme is largely associated with nuclear and mitochondrial fractions. Cotzias and Dole (1952) rabbit liver; Valette *et al.* (1954), pig kidney
1.5.1.2	Pyrroline-5-carboxylate reductase	Essentially associated with mitochondria in the tissues studied. Roche and Ricaud (1955), rat liver; Strecker and Mela (1955), rat liver; Sacktor (1955), flight muscle of housefly; Lang and Lang (1958), rat liver
1.5.1.3	Tetrahydrofolate dehydrogenase	Essentially in mitochondrial fractions when homogenized in sucrose medium; if saline medium is used, however, this enzyme is found in supernatant also. Noronha and Sreenivasan (1960), rat liver
1.5.1.4	Dihydrofolate dehydrogenase	Comment as for 1.5.1.3. Noronha and Sreenivasan (1960), rat liver

132

1.5.3.1	Sarcosine dehydrogenase	Reported to be present only in the soluble protein fraction of liver mitochondria. Frisell *et al.* (1965), rat liver; Honova *et al.* (1967), infant and adult rat liver
1.6.1.1	NAD(P) transhydrogenase	Associated with particles (N. O. Kaplan *et al.* 1953) mainly but not exclusively mitochondrial. N. O. Kaplan *et al.* (1953), animal tissues; Spiro and Ball (1961a), ox adrenal medulla; Spiro and Ball (1961b), cortex; Kramer *et al.* (1968), rat liver; Oldham *et al.* (1968), bovine adrenal cortex; Papa *et al.* (1968), rat liver
1.6.99.1	Reduced NADP cytochrome c reductase	This enzyme and NADH cytochrome c reductase have mitochondrial as well as microsomal distribution. Hogeboom and Schneider (1950), mouse liver; de Duve *et al.* (1955), rat liver; Beattie (1968a,b), rat liver; Brunner and Bygrave (1969), rat liver
1.6.99.2	Reduced NAD(P) dehydrogenase (menadione reductase)	T. E. King *et al.* (1969), heart; Bois and Estabrook (1969), beef heart
1.6.99.3	NADH cytochrome c reductase	Rotenone-insensitive NADH cytochrome c reductase activity has been described in outer mitochondrial membrane and it is considered by many investigators as a specific identifying characteristic of the outer membrane. Hogeboom (1949), rat liver; Brody *et al.* (1952), rat liver and brain; V. R. Potter and Reif (1952), rat liver; Abood *et al.* (1952), rat brain; Sacktor (1953), housefly flight muscle; Abood and Gerard (1954), rat nerve; de Duve *et al.* (1955), rat liver; Kuff *et al.* (1956), rat liver; Eichel (1957b), rat liver; Thomson and Klipfel (1957), rat liver; Baillie and Morton (1958), cow mammary gland; Giuditta and Strecker (1959), rat brain; Brand and Mahler (1959), chick embryo and chick embryonic heart; Spiro and Ball (1961a), ox adrenal medulla; Spiro and Ball (1961b), cortex; Parsons *et al.* (1966); Sottocasa *et al.* (1967a); Levy *et al.* (1967); Okamoto *et al.* (1967); D. E. Green *et al.* (1968b), beef and rat liver; Beattie (1968a,b), rat liver; Horgan and Singer (1968); Hatefi and Stempel (1969); Hatefi *et al.* (1969); Spiegel and Wainio (1969), beef heart; Abraham *et al.* (1969), rat liver
1.6.99.–	Reduced NAD diaphorase	Activity of this enzyme varies as a function of the acceptor used. When neotetrazolium is the acceptor, distribution is mitochondrial as well as microsomal. Baillie and Morton (1958), cow mammary gland; T. F. Slater (1959), rat liver; C. H. Williams *et al.* (1959), rat liver; Novikoff (1960), Novikoff hepatoma ascites
1.7.99.1	Hydroxylamine reductase	Bernheim (1969), liver
1.8.3.1	Sulfite oxidase	This enzyme is in all fractions in rat liver homogenate. Baxter *et al.* (1958), rat liver
1.9.3.1	Cytochrome oxidase	This enzyme, like succinate dehydrogenase, is thought to be entirely mitochondrial in a large number of tissues, although microsomal distribution in mouse liver prepared in a medium containing a high proportion of glycerol (Carruthers *et al.*, 1959) has been reported. Schneider (1946), rat liver and rat azodye hepatoma; Schneider and Hogeboom (1950b), mouse liver; Recknagel (1950), amphibian eggs; Brody *et al.* (1952), rat liver and brain; Hogeboom *et al.* (1952), rat liver; Abood *et al.* (1952), rat brain; Dounce *et al.* (1953), rat liver; Cleland and Slater (1953), rat and cat heart; Paigen (1954), rat liver; C. F. Strittmatter and Ball (1954), rat liver; Kretchmer and Dickerman (1954), rat kidney; Abood and Gerard (1954), rat nerve; Straus (1954), rat kidney; M. W. Woods (1954), mouse brain; Thomson and Mikuta (1954), rat liver; Appelmans *et al.* (1955), rat liver; de Duve *et al.* (1955), rat liver; Boell and Weber (1955), tadpole liver and amphibian eggs; Weber (1955), amphibian eggs; Nelson (1955), bull spermatozoa; Straus (1956), rat kidney; Wattiaux *et al.* (1956), rat liver; Eichel (1957b), rat liver; Thomson and Klipfel (1957), mouse liver; Straus (1957), rat kidney; Krishnamurthy *et al.* (1958), chicken liver; Fenwick (1958), fat body of the locust; Beaufay *et al.* (1959a,b), rat liver; Giuditta and Strecker (1959), rat brain; Carruthers *et al.* (1959), Ehrlich ascites cells and mouse liver; Woernley *et al.* (1959), Ehrlich ascites cells; Sellinger *et al.* (1960), rat liver; T. F. Slater and Planterose (1960), rat liver; Van Lancker (1960), regenerating rat liver; Sauer *et al.* (1960), Lettre-Ehrlich ascites cells; Sun and Jacobs (1967); Greenfield and Boell (1968), chick liver, heart, and skeletal muscle during embryonic development; Person *et al.* (1969a,b), beef heart; Birkmayer *et al.* (1969), *Neurospora crassa*

TABLE VII (Continued)

E. C. number	Enzyme (trivial name)	Comments and references
1.10.3	Ubiquinone oxidase	In rat liver ubiquinone oxidase has a mitochondrial distribution. Seshadri Sastri et al. (1961), rat liver
1.11.1.8	Iodinase	Mainly mitochondrial. De Groot and Carvalho (1960), sheep thyroid
1.14.1.2	Kynurenine hydroxidase	Cammer and Moore (1969), rat liver
1.14.1.6	Steroid 11-β-hydroxylase	This is one of the relatively few hydroxylases of this type associated with mitochondria rather than endoplasmic reticulum. Sweat (1951), ox adrenal; Yago and Ichii (1969), hog adrenal cortex
2.1.2.1	Serine hydroxymethyltransferase	Associated with soluble and mitochondrial fractions in rat liver. Noronha and Sreenivasan (1960), rat liver; Nakano et al. (1968), rat liver; Fujioka (1969), rabbit liver
2.1.3.2	Aspartate carbamoyltransferase	Reichard (1954). rat liver
2.1.3.3	Ornithine carbamoyltransferase	Cytochemically identified with mitochondrion. Leuthardt and Müller (1948), rat liver; Leuthardt et al. (1949), rat liver; Müller and Leuthardt (1950a), rat liver; Grisolia and Cohen (1953), rat liver; Siekevitz and Potter (1953), rat liver; Caravaca and Grisolia (1960), rat liver; Grillo and Bedino (1968), bovine liver; Mizutani (1968), rat and mouse liver; Mizutani and Fujita (1969), pig liver
2.3.1.6	Choline acetyltransferase	Associated with mitochondrial membrane in brain tissue. Hebb and Whittaker (1958), rabbit, sheep, and guinea pig brain tissue; Bellamy (1959), rat brain and pigeon brain tissue; Whittaker (1959), guinea pig brain; E. G. Gray and Whittaker (1960), guinea pig brain; de Robertis et al. (1963), rat brain; L. T. Potter and Glover (1968), rat brain
2.3.1.7	Carnitine acetyltransferase	Activity solely in the isolated mitochondrion. Barker et al. (1968), guinea pig liver
2.3.1.9	Acetyl-CoA acetyltransferase	This enzyme is located essentially in cytoplasmic particles, presumably mitochondria, in muscle from Locusta migratoria. Zebe (1959-1960), muscle from Locusta migratoria
2.3.1.13	Glycine acyltransferase synthesis of hippurate and p-aminohippurate	Catalyzes the final step of the detoxication pathway for benzoic acid; essentially mitochondrial in liver and kidney. Nielsen and Leuthardt (1949), rat liver; R. K. Kielley and Schneider (1950), mouse liver; Leuthardt and Nielsen (1951), rat liver; Bremer (1955), rat liver; I. K. Brandt et al. (1968), rat liver
2.6.1.1	Aspartate aminotransferase (glutamic oxalacetic transaminase)	Present in mitochondrial and soluble fractions. Two forms of mitochondrial aspartate aminotransferase can be distinguished by electrophoretic mobility and solubility characteristics. Hird and Rowsell (1950), rat liver; May et al. (1959), rat liver and brain; de Duve (1960b), rat liver; Gaull and Villee (1960), rat liver; Rosenthal et al. (1960), rat liver; McArdle et al. (1960), rat liver and brain; Eichel and Bukovsky (1961), rat liver; Boyd (1961), rat liver; Bengmark et al. (1967), rat liver; Bhargava and Sreenivasan (1968), rat liver and kidney; S. H. Lee and Torack (1968), rat liver; Waksman and Rendon (1968), rat heart, kidney, and brain; Rendon and Waksman (1969), rat liver; Michuda and Martinez-Carrion (1969), pig heart; Morino and Watanabe (1969), pig heart; S. H. Lee (1969), rat heart
2.6.1.2	Alanine aminotransferase (glutamic pyruvic transaminase)	Alanine aminotransferase activity is represented by two enzymes located in soluble and mitochondrial fractions. Enzyme associated with mitochondria represents only 5-20% of the total activity. Delbrück et al. (1959a), rat liver and rat uterine epithelioma; Kafer and Pollack (1961), rat liver and embryonic rat liver; Swick et al. (1965a, b), rat liver; Ziegenbein (1966), rat heart; Bengmark et al. (1967), rat liver
2.6.1.5	Tyrosine aminotransferase	65-70% of the total activity is tightly bound to mitochondrial fraction. Fellman et al. (1969), human and rat liver, heart, muscle, and kidney; J. E. Miller and Litwack (1969), rat brain

134

2.6.1.19	Aminobutyrate aminotransferase	Aminobutyrate transaminase exists in several molecular forms attached to mitochondrial and other cell fractions. Waksman and Bloch (1968), mouse and rat brain
2.7.1.1	Hexokinase	Hexokinase is found in both soluble and particulate fractions of animal and plant tissues. In animal tissues the greater part of the total enzyme activity is associated with mitochondria (in beef brain, 75–90%). Saltman (1953), plant tissue; Crane and Sols (1953), rat tissues; M. K. Johnson (1960), brain; A. Hernandez and Crane (1966), rat heart; J. E. Wilson (1967), rat brain; Bachelard (1967), rat and guinea pig brain; Newsholme et al. (1968), guinea pig cerebral cortex; C. L. Moore (1968) beef brain; H. D. Brown et al. (1968a) dystrophic duck muscle; Teichgräber and Biesold (1968), rat and guinea pig brain; Koskow and Rose (1968), ascites tumor; Craven and Basford (1969), beef liver; Craven et al. (1969), beef and rat liver; Thompson and Bachelard (1969), ox cerebral cortex
2.7.2.2	Carbamate kinase	This enzyme, together with aspartate carbamyltransferase is partly associated with mitochondrial fraction. Leuthardt and Müller (1948), rat liver; Leuthardt et al. (1949), rat liver; Grisolia and Cohen (1953), rat liver; Siekevitz and Potter (1953), rat liver; Reichard (1954), rat liver
2.7.3.2	Creatine kinase	75–80% of the total activity is found in supernatant; the mitochondria contain the largest percentage of total particulate population of the enzyme. Wood (1963) bovine brain; Wood and Swanson (1964), guinea pig cerebral cortex; P. D. Swanson (1967), guinea pig cerebral cortex; Sullivan et al. (1968), rat cerebral cortex; H. D. Brown et al. (1968a), dystrophic duck muscle.
2.7.4.3	Adenylate kinase	Variable in distribution frome one tissue to another. It is found predominantly in supernatant in heart tissue (Cleland and Slater, 1953); essentially particulate in rat nerve (Abood and Gerard, 1954) and mouse mammary tumor (L. A. Miller and Goldfeder, 1961). The total enzyme activity is equally distributed between mitochondria and supernatant in rat liver and human placenta. In rat liver most of the mitochondrial activity is associated with the outer membrane fraction. Novikoff et al. (1952), rat liver; Cerletti and Zichella (1960), human placenta; Sottocasa et al. (1967a), rat liver
2.7.4.6	Nucleoside diphosphate kinase	Studied extensively because of involvement in oxidative phosphorylation, has been considered exclusively mitochondrial but its presence in cytosol of mammalian cells has been reported. Goffeau et al. (1967), beef liver; Goffeau et al. (1968), bovine liver; Pedersen (1968), bovine liver; Colomb et al. (1969), beef heart
2.7.7.7	Deoxyribonucleic acid nucleotidyl transferase	E. Wintersberger (1968), yeast; R. R. Meyer and Simpson (1969), rat liver; V. Wintersberger and Wintersberger (1970)
2.7.7.16–17	Ribonuclease	Distribution remains controversial. Morais et al. (1967) and de Lamirande et al. (1967) have reported it to be essentially mitochondrial. Morais et al. (1967), rat liver; de Lamirande et al. (1967), rat liver
2.7.–.–	Adenyl cyclase	Adenyl cyclase has been shown to be localized in particulate fractions of liver. Sutherland et al. (1962). Most activity is in nuclear membrane and plasma membrane of liver cell though significant activity is characteristically in endoplasmic reticulum. Rodbell (1967) reported that adenyl cyclase occurs in fat cell ghosts. The enzyme also is associated with mitochondria in rat liver and rabbit psoas muscle homogenates. Rabinowitz et al. (1965) rabbit psoas muscle; de Robertis et al. (1967), rat liver
2.8.1.1	Thiosulfate sulfurtransferase	Nearly all of the activity is found in the mitochondrial fraction in rat liver. Ludewig and Chanutin (1950), rat liver; Rosenthal and Vars (1954), rat liver; de Duve et al. (1955), rat liver
2.8.1.2	β-Mercaptopyruvate sulfurtransferase	55% of the activity is found associated with nuclear fraction and the rest in mitochondria, supernatant, and possibly microsomes to a smaller extent. Kun and Fanshier (1959), rat liver
3.1.1.4	Phospholipase A	Found in mitochondria, microsomes, and lysosomes. In submitochondrial fractions, phospholipase A is largely in the outer membrane although the inner membrane also has some activity. Nachbaur and Vignais (1968), rat liver; Waite et al. (1969), rat liver
3.1.3.2	Acid phosphatase	Although in rat liver acid phosphatase is localized in the lysosome, it is not uncommon to find a nonspecific acid phosphatase in hepatic tissue associated with mitochondria. The mitochondrial enzyme has little or no activity toward added glycerophosphate at pH 5. Berthet and de Duve (1951), rat liver; Palade (1951), rat liver

135

TABLE VII (*Continued*)

E. C. number	Enzyme (trivial name)	Comments and references
3.1.3.4	Phosphatidate phosphatase	Sedgwick and Hübscher (1965)
3.1.4.2	Glyceerylphosphorylcholine diesterase	Highest activity is associated with microsomal fraction though lower levels of activities have been found in mitochondria and nuclear fractions. Baldwin and Cornatzer (1968), rat kidney
3.1.4.5	Deoxyribonuclease	Alkaline or neutral deoxyribonuclease has been found associated with mitochondria. It differs from acid deoxyribonuclease which is associated with lysosomes. Beaufay et al. (1959a,b), rat liver
3.2.2.5	NAD nucleosidase	Essentially a microsomal enzyme in Ehrlich ascites cells, but is found with mitochondrial fraction. In some cases the activity is found in the nuclear fraction. Kun et al. (1951), Ehrlich ascites cells
3.4.2.1-2	Carboxypeptidase A and B	The distribution of this enzyme is not clear. In bovine pancreas procarboxypeptidases A and B and in rat liver, a carboxypeptidase-like enzyme is thought to be mitochondrial. Rademaker (1959), rat liver; Keller and Cohen (1961), ox pancreas
3.4.4.9	Cathepsin C	Cathepsin C is a lysosomal enzyme; separates also with light mitochondria. Rademaker (1959), rat liver
3.4.4.–	Cathepsin A	Cathepsin A has a distribution pattern similar to that of carboxypeptidase A and B (3.4.2.1-2) and is essentially associated with mitochondria. Rademaker (1959), rat liver
3.4.4.–	Cathepsin B	The distribution of cathepsin B has been studied in various tissues of rat and found mainly in mitochondria. Maver and Greco (1951), rat liver, kidney, spleen, and azodye hepatoma; Finkenstaedt (1957), rat liver; Rademaker (1959), rat liver
3.4.4.–	"Cathepsin," cathepsin D, proteinase I	The enzyme has been found in lysosomes in some liver tissues and in kidney "droplets." In other tissues it is found in mitochondria. Maver and Greco (1951), rat liver, kidney, spleen, and rat azodye hepatoma; B. Weiss (1953), ox thyroid; Holter (1954), ameba Chaos chaos (substrate casein); Holter (1955), ameba Chaos chaos (substrate casein); Wattiaux et al. (1956), rat liver; R. K. Meyer and Clifton (1956), rat anterior pituitary tumors induced by stilbestrol; Allard et al. (1957), rat liver and Novikoff hepatoma; Beaufay et al. (1957), rat brain; La Bella and Brown (1958), pig anterior pituitary; Brachet et al. (1958), Müllerian ducts of chick embryo; Korner and Tarver (1958), rat liver; Greenbaum et al. (1960), rat mammary gland; Zalkin et al. (1961), mouse, rabbit, rat and chicken skeletal muscle; de Bernard et al. (1963), ox heart
3.5.1.2	Glutaminase	It has been known that rat liver homogenates contain two enzyme systems capable of deamidating glutamine; one accelerated by phosphate (glutaminase I) and the other by pyruvate (glutaminase II). The phosphate-activated glutaminase is associated entirely with mitochondria, whereas pyruvate-activated glutaminase is found mostly in supernatant. Errera (1949), rat liver; J. A. Shepherd and Kalnitsky (1951), rat liver; O'Donovan and Lotspeich (1966), rat kidney
3.6.1.1	Inorganic pyrophosphatase	Localized on the inside of the inner mitochondrial membrane. Schick and Butler (1969), rat liver
3.6.1.3-8	Adenosinetriphosphatase (ATPase)	Mitochondrial forms of ATPase activity have been described by Chantrenne (1947), mouse liver; Schneider et al. (1950), mouse liver; Allard and Cantero (1952), rat liver; Novikoff et al. (1952), rat liver; Perry (1952), rabbit skeletal muscle; Hecht and Novikoff (1952), rat azodye hepatoma and Walker 256 hepatoma; Christie and Judah (1953), rat liver; Novikoff et al. (1953a,b), rat liver; Maxwell and Ashwell (1953), mouse spleen; Sacktor (1953) housefly flight muscle; Cleland and Slater (1953), rat heart; Dianzani (1954b), rat liver; Thomson and Mikuta (1954), rat liver; Straus (1954), rat kidney; Abood and Gerrard (1954), rat nerve; Nelson (1954), bull spermatozoa; Birbeek and Reid (1956), rat liver; Reid et al. (1956), rat liver; Jordan and March (1956), rat brain; de Lamirande and Allard (1957), rat liver; Thomson and Klipfel (1957), rat liver; Scevola and de Barbieri (1957), chick embryos; Allard et al. (1957), Novikoff hepatoma and rat liver; Novikoff (1957), Novikoff hepatoma; Novikoff et al. (1958), rat liver; Kasbekar et al. (1959), rat liver; Thorell and Yamada (1959), bone marrow from both normal and leukemic hens; Wakid and Needham (1960), rat myometrium; Cerletti et al. (1960), human placenta; Blecher and White

136

EC number	Enzyme	Description
4.1.1.9	Malonyl-CoA decarboxylase	(1960), lymphosarcoma; Mills and Cochran (1967), American cockroach, thoracic muscle; Rechardt and Kokko (1967), rat spinal cord; MacLennan and Tzagoloff (1968), bovine heart; Amons et al. (1968), rat liver; Beattie and Basford (1968), rat brain; Drabikowski and Rafalowska (1968), calf liver; Veldsema-Currie and Slater (1968), rat liver; Vigers and Ziegler (1968), beef heart; Kagawa (1969); Sone et al. (1969), yeast; Takeuchi et al. (1969), castor bean
4.1.3.4	Hydroxymethylglutaryl-CoA lyase	This enzyme is found in mitochondrial fraction along with propionyl-CoA carboxylase (6.4.1.3) in rat liver. Scholte (1969), rat liver
4.1.3.5	Hydroxymethylglutaryl-CoA synthase	Essentially a mitochondrial enzyme. Bucher et al. (1960), rat liver
4.1.3.7	Citrate synthase	It is localized in mitochondria, although association with microsomes has been reported as well. Bucher et al. (1960), rat liver
4.1.3.8	ATP-citrate lyase	Predominantly mitochondrial. Zebe (1959-1960), Locusta migratoria muscle; Shepherd and Garland (1969b), rat liver Daikuhara et al. (1968), rat liver
4.2.1.2	Fumarate hydratase (fumarase)	Associated partly with mitochondria, although present in other fractions as well. Kuff (1954), mouse liver; J. A. Shepherd and Kalnitsky (1954), rabbit brain; J. A. Shepherd et al. (1955), human liver; de Duve et al. (1955), rat liver; Blaschko et al. (1957), ox adrenal medulla; Mahler et al. (1958), chicken liver; Fenwick (1958), fat body of the desert locust; Fahien and Strmecki (1969b), bovine liver; Vary et al. (1969), Saccharomyces cerevisiae
4.2.1.3	Aconitate hydratase	Activity located in mitochondria and supernatant. The mitochondrial enzyme has optimum activity at pH 5.8, cytosol form, at pH 7.3. Dickman and Speyer (1954), rat liver; J. A. Shepherd and Kalnitsky (1954), rabbit brain; J. A. Shepherd et al. (1955), human liver; Mahler et al. (1958), chick embryo
4.2.1.17	Enoyl-CoA hydratase	Has been obtained essentially from mitochondria, but its participation both in fatty acid oxidation and fatty acid synthesis indicates its presence in cytosol as well (Langdon, 1957). E. P. Kennedy and Lehninger (1948), rat liver; Schneider (1948), rat liver
4.99.1.1	Ferrochelatase	Exclusively a mitochondrial enzyme bound to the inner membrane (M. S. Jones and Jones, 1968). Nishida and Labbe (1959), liver; M. S. Jones and Jones (1968), rat liver; McKay et al. (1969), rat liver
5.3.3.1	Δ⁵ ⁴-Ketosteroid isomerase	This enzyme is present in mitochondria as well as in microsomes. Ewald et al. (1964); Kruskemper et al. (1964); Koide and Torres (1965), placenta
6.2.1.1	Acetyl-CoA synthetase	Aas and Bremer (1968), rat liver
6.2.1.2-3	Acyl-CoA synthetase	This enzyme catalyzes two distinct reactions possibly indicating a dual localization in rat liver tissue. Fatty acid oxidation is a mitochondrial phenomenon associated with the outer membrane. E. P. Kennedy and Lehninger (1948), rat liver; Schneider (1948), rat liver; Norum et al. (1966a); Farstad et al. (1967), rat liver; Galzigna et al. (1967), rat liver; Aas and Bremer (1968), rat liver
6.2.1.—	Benzoyl-CoA synthetase	Mitochondrial in mouse liver and probably in rat liver. Nielsen and Leuthardt (1949), rat liver; R. K. Kielley and Schneider (1950) mouse liver; Leuthardt and Nielsen (1951), rat liver
6.2.1.—	Octanolyl-CoA synthetase	Graham and Park (1969), ox liver
6.2.1.—	Palmityl-CoA palmityl-transferase	Norum (1966), human tissues
6.4.1.—	δ-Aminolaevulinate synthetase	Thought to be a mitochondrial enzyme. Sano and Granick (1961), guinea pig liver; McKay et al. (1969), rat liver
6.4.1.1	Pyruvate carboxylase	Located exclusively in mitochondria. Keech and Utter (1962), rat liver; H. A. Krebs et al. (1963), rat liver; Struck et al. (1966), rat liver; Böttger et al. (1969), rat liver; Kimmich and Rasmussen (1969), rat liver; Gul and Dils (1969), rat and rabbit mammary gland; Irias et al. (1969), chicken liver
6.4.1.3	Propionyl-CoA carboxylase	This enzyme, along with malonyl CoA decarboxylase, is associated with mitochondria. Scholte (1969), rat liver

137

TABLE VIII A
Submitochondrial Localization

E. C. number	Enzyme (trivial name)	Location and reference
1.1.1.37–39	Malate dehydrogenase	Outer membrane (S form). Bachmann et al. (1966); D. E. Green and Perdue (1966); Allmann et al. (1966a,b, 1967, 1968); D. E. Green (1967); D. E. Green et al. (1968b) Inner membrane and matrix. Parsons (1965); Parsons et al. (1966); Parsons and Williams (1967); Brdiczka et al. (1968); Hoppel and Cooper (1968)
1.1.1.41–42	Isocitrate dehydrogenase	Outer membrane (S form). Bachmann et al. (1966); D. E. Green and Perdue (1966); Allmann et al. (1966a,b, 1967, 1968); D.E. Green (1967); D. E. Green et al. (1968b) Inner membrane and matrix. Parsons (1965); Parsons et al. (1966); Parsons and Williams (1967); C. Schnaitman et al. (1967; Brdiczka et al., 1968); Hoppel and Cooper (1968); C. A. Schnaitman and Pedersen (1968)
1.2.2.2	Pyruvate dehydrogenase	Outer membrane. Bachmann et al. (1966); D. E. Green and Perdue (1966); Allmann et al. (1966a,b, 1967, 1968; D. E. Green (1967); D. E. Green et al. (1968b) Inner membrane and matrix. Sottocasa et al. (1967a,b); Sottocasa (1967); Tubbs and Garland (1968)
1.2.4.2	α-Oxoglutarate dehydrogenase	Outer membrane. Bachmann et al. (1966); D. E. Green and Perdue (1966); Allmann et al. (1966a,b,1967,1968; D. E. Green (1967); D. E. Green et al. (1968b) Inner membrane and matrix. Tubbs and Garland (1968)
1.3.99.1	Succinate dehydrogenase	Inner membrane and matrix. Parsons (1965); Bachmann et al. (1966); D. E. Green and Perdue (1966); Allmann et al. (1966a,b, 1967, 1968; Parsons et al. (1966); D. E. Green (1967); Parsons and Williams (1967); Sottocasa et al. (1967a,b); Sottocasa (1967); D. E. Green et al. (1968b), Brdiczka et al. (1968); Hoppel and Cooper (1968)
1.4.1.2	Glutamate dehydrogenase	Inner membrane and matrix. C. Schnaitman et al. (1967); C. Schnaitman and Greenawalt (1968); C. A. Schnaitman and Pedersen (1968)
1.4.3.4	Monoamine oxidase	Inner membrane and matrix. Bachmann et al. (1966); D. E. Green and Perdue (1966); Allmann et al. (1966a,b, 1967, 1968; D. E. Green (1967); D. E. Green et al. (1968b) Outer membrane. Parsons (1965); Parsons et al. (1966); Parsons and Williams (1967); C. Schnaitman et al. (1967); Beattie (1968a,b); Hoppel and Cooper (1968); C. A. Schnaitman and Greenawalt (1968); C. A. Schnaitman and Pedersen (1968)
1.6.99.3	Rotonone-insensitive NADH-cytochrome c reductase Rotonone-sensitive NADH-cytochrome c reductase	Outer membrane. Parsons (1965); Parsons et al. (1966); Parsons and Williams (1967); Sottocasa et al. (1967a,b); Sottocasa (1967); Beattie (1968a,b); Hoppel and Cooper (1968) Inner membrane. Sottocasa et al. (1967a); Whereat et al. (1969)
1.9.3.1	Cytochrome oxidase	Inner membrane and matrix. C. Schnaitman et al. (1967); C. Schnaitman and Greenawalt (1968); C. A. Schnaitman and Pedersen (1968); Whereat et al. (1969)
1.14.1.2	Kynurenine hydroxylase	Outer membrane (S form). C. Schnaitman et al. (1967); C. Schnaitman and Greenawalt (1968); C. A. Schnaitman and Pedersen (1968); Beattie (1968a,b); Cammer and Moore (1969)
2.1.3.3	Ornithine transcarbamylase	Inner membrane and matrix. Sottocasa et al. (1967a,b); Sottocasa (1967)

138

EC number	Enzyme	Localization and references
2.3.1.7	Palmitoyl-CoA: carnitine palmityltransferase	Outer membrane. Bachmann *et al.* (1966); D. E. Green and Perdue (1966); Allmann *et al.* (1966a,b, 1967, 1968); D. E. Green (1967); D. E. Green *et al.* (1968b)
2.6.1.1	Aspartate-amino transferase	Inner membrane and matrix. Norum *et al.* (1966b); Beattie (1968a,b); Garland *et al.* (1969)
2.7.–.–	Oligomycin-sensitive adenosine diphosphate-adenosine triphosphate exchange activity	Inner membrane and matrix. C. Schnaitman *et al.* (1967); C. Schnaitman and Greenawalt (1968); C. A. Schnaitman and Pedersen (1968)
2.7.1.1	Hexokinase	Inner membrane and matrix. Sottocasa *et al.* (1967a,b); Sottocasa (1967)
2.7.4.3	Adenylate kinase	Outer membrane. Craven *et al.* (1969)
		Intracristal space and intermembrane space. Parsons (1965); Bachmann *et al.* (1966); D. E. Green and Perdue (1966); Allmann *et al.* (1966a,b, 1967, 1968); Parsons *et al.* (1966); D. E. Green (1967); Parsons and Williams (1967); C. Schnaitman *et al.* (1967); Sottocasa *et al.* (1967a,b); Sottocasa (1967); D. E. Green *et al.* (1968b); Brdiczka *et al.* (1968); Hoppel and Cooper (1968); C. Schnaitman and Greenawalt (1968); C. A. Schnaitman and Pedersen (1968)
2.7.4.6	Nucleoside diphosphokinase	Intracristal space and intermembrane space or outer membrane. C. Schnaitman *et al.* (1967); C. Schnaitman and Greenawalt (1968); C. A. Schnaitman and Pedersen (1968)
3.6.1.3–5	Mg²⁺ Adenosinetriphosphatase	Inner membrane and matrix. Bachmann *et al.* (1966); D. E. Green and Perdue (1966); Allmann *et al.* (1966a,b, 1967, 1968); D. E. Green (1967); Sottocasa *et al.* (1967a,b); Sottocasa (1967); Racker and Horstman (1967); D. E. Green *et al.* (1968b)
	Ca²⁺ Adenosinetriphosphatase	Inner membrane and matrix. Bachmann *et al.* (1966); D. E. Green and Perdue (1966); Allmann *et al.* (1966a,b, 1967, 1968); D. E. Green *et al.* (1968b)
4.1.1.9	Malonyl-CoA decarboxylase	Matrix. Scholte (1969)
4.1.3.7	Citrate synthetase	Inner membrane and matrix. Tubbs and Garland (1968)
4.2.1.1–2	Fumerase	Outer membrane (S form). Parsons (1965); Bachmann *et al.* (1966); D. E. Green and Perdue (1966); Allmann *et al.* (1966a,b, 1967, 1968); Parsons *et al.* (1966); D. E. Green (1967); Parsons and Williams (1967); C. Schnaitman *et al.* (1967); D. E. Green *et al.* (1968b); Brdiczka *et al.* (1968); Hoppel and Cooper (1968); C. A. Schnaitman and Pedersen (1968)
4.2.1.3	Aconitase	Outer membrane (S form). Bachmann *et al.* (1966); D. E. Green and Perdue (1966); Allmann *et al.* (1966a,b, 1967, 1968; D. E. Green (1967); D. E. Green *et al.* (1968b)
		Inner membrane and matrix. Parsons (1965); Parsons *et al.* (1966); Parson and Williams (1967); C. Schnaitman *et al.* (1967); C. Schnaitman and Greenawalt (1968); C. A. Schnaitman and Pedersen (1968)
6.2.1.2–3	AcylCoA synthetase	Outer membrane. Bachmann *et al.* (1966); D. E. Green and Perdue (1966); Allmann *et al.* (1966a,b, 1967, 1968); Norum *et al.* (1966b); D. E. Green (1967); D. E. Green *et al.* (1968b)
6.4.1.3	Propionyl-CoA carboxylase	Inner membrane and matrix. Beattie (1968a,b); Tubbs and Garland (1968); Aas and Bremer (1968)
	β-Oxidizing enzymes	Matrix and loosely bound on the inner side of the inner membrane. Scholte (1969)
		Outer membrane. Bachmann *et al.* (1966); D. E. Green and Perdue (1966); Allmann *et al.* (1966a,b, 1967, 1968); D. E. Green (1967); D. E. Green *et al.* (1968b)
		Inner membrane and matrix. Norum *et al.* (1966b); Beattie (1968a,b); Brdiczka *et al.* (1968); Tubbs and Garland (1968)

discussed in terms of several general concepts which describe the event of energy conservation. The first possibility is a now classic visualization of an electrical flow through a series of intermediates at stepwise diminishing reduction poten-

TABLE VIIIB

ENZYMES OF OXIDATIVE METABOLISM ON MITOCHONDRIAL MEMBRANE[a]

E. C. number	Enzyme (trivial name)	Reaction
		Oxidoreductases acting on CH—OH group of donors with NAD or NADP as acceptor
1.1.1.37	Malate dehydrogenase (L-Malate–NAD oxidoreductase)	L-Malate + NAD = Oxalacetate + $NADH_2$
1.1.1.41	Isocitrate dehydrogenase (*threo*-D$_s$-isocitrate–NAD oxidoreductase)	*threo*-D$_s$-Isocitrate + NAD = 2-oxoglutarate + $NADH_2$ + CO_2
		Acting on the aldehyde or keto group of donors with a cytochrome as an acceptor
1.2.2.2	Pyruvate dehydrogenase (pyruvate–ferricytochrome b$_1$ oxidoreductase)	Pyruvate + ferricytochrome b$_1$ = acetate + ferrocytochrome b$_1$ + CO_2
		With lipoate as acceptor
1.2.4.2	Oxoglutarate dehydrogenase [2-oxoglutarate–lipoate oxidoreductase (acceptor-acylating)]	2-Oxoglutarate + oxidized lipoate = 6-S-succinylhydrolipoate + CO_2
		Acting on the CH—NH$_2$ group of donors with oxygen as acceptor
1.4.3.4	Monoamine oxidase [monoamine–oxygen oxidoreductase (deaminating)]	Monoamine + H_2O + O_2 = aldehyde + NH_3 + H_2O_2
		Acting on reduced NAD or NADP with other acceptor (= flavoprotein)
1.6.99.3	Reduced NAD dehydrogenase; cytochrome c reductase [reduced NAD–(acceptor) oxidoreductase]	Reduced NAD + acceptor = NAD + reduced acceptor
		Acting on paired donors with incorporation of oxygen into one donor (hydroxylases), with reduced NAD, or NADP as one donor
1.14.1.2	Kynurenine hydroxylase [L-kynurenine, reduced NADP–oxygen oxidoreductase (3-hydroxylating)]	L-Kynurenine + $NADPH_2$ + O_2 = 3 hydroxy-L-kynurenine + NADP + H_2O
		Lyases, carbon–oxygen lyases, hydrolyases
4.2.1.2	Fumerase; fumarate hydratase (L-malate hydrolyase)	Fumarate + H_2O = L-Malate
4.2.1.3	Aconitase; aconitate hydratase [citrate (isocitrate) hydrolyase]	Citrate = *cis*-aconitate + H_2O Isocitrate + H_2O
		Ligases forming C—S bonds acid–thiol ligases
6.2.1.2	Acyl-CoA synthetase [acid–CoA ligase (AMP)]	ATP + acid + CoA = AMP + pyrophosphate + acyl − CoA

[a] Abbreviations: NAD, nicotinamide adenine dinucleotide; NADP, nicotinamide adenine dinucleotide phosphate; $NADH_2$, reduced NAD; AMP, adenosine monophosphate; ATP, adenosine triphosphate.

tials from $NADH^+$ or $NADPH^+$ with the formation of a number of high-energy intermediates (Chance and Spencer, 1959). Figure 16 relates the "carriers" to some of the enzyme-catalyzed reactions. A physical change, presumably a conformational change in a protein, has also been suggested (Boyer, 1965; Baum *et al.*, 1967a,b).

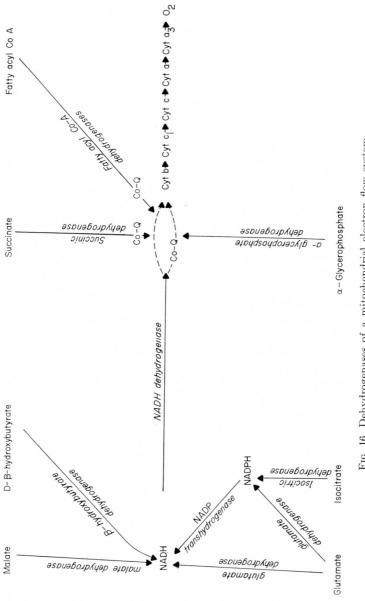

Fig. 16. Dehydrogenases of a mitochondrial electron flow system

An alternative view is the concept of chemiosmotic coupling proposed by Mitchell (1968; Mitchell and Moyle, 1969). This has been described as a membrane phenomenon in which the disparity of distribution of charged particles builds a transmembrane potential that is responsible for the electron flow. This concept implies that the vectorial enzyme be oriented in the biochemical complex so that the channel of entry of the group donor to the active center is spatially separate from the channel of exit of a group acceptor (Mitchell, 1961, 1963, 1968, 1969). An extensive literature has developed in which these concepts have been discussed (E. C. Slater, 1967a). To these major alternatives hybrid versions have been added and, indeed, the several concepts are not mutually exclusive. Much of this literature which has, conveniently, been very well reviewed is outside the scope of our present discussion and we shall consider only those topics that are most immediately pertinent to the theme of this chapter. We shall consider certain essential aspects of the enzymology of the mitochondrial electron transport system by reference to the functional involvement of mitochondrial dehydrogenases, using the following taxonomy:

A. Enzymes serving to transfer reducing equivalents from substrates to NAD
   1. Glutamate dehydrogenase
   2. Malate dehydrogenase
   3. β-Hydroxybutyrate dehydrogenase
B. Transferring reducing equivalents to NADPH
   1. Glutamate dehydrogenase
   2. Isocitric dehydrogenase
C. Transferring reducing equivalents from NADPH to NAD
   1. Reduced NADP transhydrogenase
D. Oxidation of NADH
   1. Reduced NAD dehydrogenase
E. Transferring reducing equivalents to an electron flow intermediate
   1. Succinic dehydrogenase
   2. Fatty acyl-CoA dehydrogenase
   3. α-Glycerophosphate dehydrogenase

Glutamate dehydrogenase is ubiquitous. Interrelationships of the dehydrogenase to other mitochondrial enzymes and their kinetic properties in an isolated complex system have been described by Fahien and Strmecki (1969a,b). The enzyme activity has been studied in intact mitochondria by Papa et al. (1967). Several authors have reported that glutamate dehydrogenase has a higher affinity for NADP than for NAD (Klingenberg and Slenczka, 1959; Klingenberg and Pette, 1962; Papa et al., 1967); though this property appears to be a function of the tissue source (see, for example, Corman et al., 1967).

Relative contribution of the glutamate–dehydrogenase reaction to electron flow will require evaluation after resolution of the functioning system of the organelle is accomplished. However, enzyme profile studies serve to provide some concept of level in the organelle (e.g., Pette *et al.*, 1962).

Malate dehydrogenase is distributed in both cytosol and mitochondria (Salganicoff and Koeppe, 1968), but catalytic properties of the enzyme isolated from mitochondria differ from the cytosol enzyme (Davies and Kun, 1957; Englard and Brieger, 1962; Berkes-Tomasevic and Holzer, 1967; Dupourque and Kun, 1968) e.g., the mitochondrial enzyme is sus-

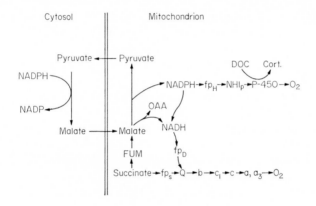

$$Malate + NADP^+ \rightleftharpoons$$
$$pyruvate + CO_2 + NADPH + H^+$$

Fig. 17. Malate shuttle. [Redrawn from E. R. Simpson and R. W. Estabrook, *Arch. Biochem.* **126**, 977–978 (1968), by permission.]

ceptible to substrate inhibition by oxalacetate and activation by L-malate and to variation in ionic strength whereas the cytoplasmic form is not. Kinetic properties of the two enzyme forms differ markedly in the presence of the synthetic substrate fluorooxalacetate. Malate dehydrogenase has been postulated to perform as a shuttle (E. R. Simpson and Estabrook, 1968; E. R. Simpson *et al.*, 1969) in a mechanism which serves to explain mitochondrial utilization of the reducing equivalents of NADPH. This scheme is illustrated in Fig. 17.

Physical properties of cytosol versus analogous mitochondrial enzymes have been described by Grimm and Doherty (1961), by Mann and Vestling (1969), and by Guha *et al.* (1968).

Lehninger *et al.* (1960) reported that D-β-hydroxybutyrate dehydrogenase is present in a large number of mammalian tissues. The enzyme is tightly bound and presumably capable of oxidation of a number of aromatic substrates (McCann, 1957). Its lipid requirement probably explains the tight binding of the enzyme to organelle structure (Sekuzu et al., 1961, 1963; Jurtshuk et al., 1963). The enzyme has been detergent-solubilized (Gotterer, 1964) and "freed" by digestion with phospholipase A (Fleischer et al., 1966). Its properties and a discussion of methods for purification of this enzyme have been given by Gotterer (1967).

Isocitric dehydrogenase also exists in multiple forms. Bell and Baron (1964) have shown that both isozymes were present in mitochondria of rat liver but only one (the more rapidly migrating isozyme) was present in the soluble fraction. A description of the intra- and extramitochondrial isocitric dehydrogenase activities has also been published by Lowenstein and Smith (1962). Kinetic properties of a soluble preparation of isocitric dehydrogenase of Ehrlich ascites cells mitochondria have been published by A. M. Stein et al. (1967).

N. O. Kaplan et al., in 1953, described an NAD(P) transhydrogenase. Its role in nature has been a matter of interesting conjecture, and it has consequently been extensively investigated by a series of workers. Klingenberg and Schollmeyer (1963), Estabrook and Nissley (1963), and Danielson and Ernster (1963) have shown that the reduction of NADP is regulated by the concentration of ATP in the mitochondria. Oldham et al. (1968) have also shown that steroid hydroxylation is affected by the addition of NADP to an in vitro system, and the existence of transhydrogenase was, therefore, postulated (Sweat and Lipscomb, 1955).

Purified NADH dehydrogenase from cardiac mitochondria has been studied by T. E. King et al. (1969) using optical rotatory dispersion (ORD). Their study was a test of the concept (T. E. King et al., 1965) that the electron transport phenomenon can be explained in terms of respiratory enzyme conformation and particularly in the geometric relationships between the prosthetic group and apoprotein, changes which are brought about by a shift of the oxidation state. The ORD spectrum obtained supported this concept. Chance and co-workers (1967) have described the NADH dehydrogenase as being composed of two flavoprotein components ($F_{p_{D1}}$ and $F_{p_{D2}}$). Bois and Estabrook (1969) have evaluated the effect of rotenone upon the nonheme iron protein associated with the dehydrogenase by electron paramagnetic resonance (EPR) spectroscopy and have found a positive correlation.

Succinic dehydrogenase, which is firmly bonded to the organelle structure, in detergent-solubilized preparations is associated with flavin in a 1:1 stoichiometry. The activity is within the intercristal space (Tsou et al., 1968; Kalina et al., 1969). Racker and associates have used purified preparations in reconstitution of elements of the electron transport chain (Bruni and Racker, 1968; Yamashita and Racker, 1968, 1969).

Four fatty acyl-CoA dehydrogenases have been purified from liver mitochondria. The several enzymes are quite distinct in their physical properties, in their chain lengths, and in their response to radiation (Waldschmidt, 1967).

L-α-Glycerophosphate dehydrogenase activities are found in the cytosol as well as tightly bound to mitochondria. The two forms are electrophoretically distinct (Fiala, 1967); as for many other mitochondrial flavoproteins, the mechanism of the interaction of glycerol phosphate with the electron transport chain remains problematic.

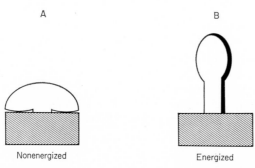

FIG. 18. Nonenergized (A) and energized (B) forms of head-piece-stalk sector of membranes. These changes are thought to occur as part of a conformation change energy transduction. [Redrawn from D. E. Green and H. Baum (1970) in "Energy and the Mitochondrion," Chapter IV, p. 45–65, Academic Press, New York, by permission.]

The coupled synthesis of ATP (oxidative phosphorylation) has been extensively investigated (Racker, 1965, 1967). In relatively early studies, a number of authors described the ATPase activity of mitochondria (M. A. Swanson, 1950a; Novikoff et al., 1952; W. W. Kielley and Kielley, 1953; V. R. Potter et al., 1953). D. E. Green and MacLennan (1969) identify the mitochondrial ATPase system with a "head-piece-stalk-sector" of the membrane and theorize that ATP synthesis is "powered by" conformation changes (Fig. 18) which result from electron transfers in the membrane (R. A. Harris et al., 1968; Penniston et al., 1968; D. E. Green et al., 1968a; D. E. Green and Baum, 1970). Other authors have suggested that the phosphorylation can be described in purely chemical terms (E. C. Slater, 1967b) or in terms of an osmotic differential (Mitchell and Moyle, 1967). As we stated earlier, the mechanisms of the final phosphorylation and of the electron flow phenomena associated with it have not been given a definitive interpretation. The literature is extremely interesting; certainly many of the contributors have oriented themselves toward the association of the catalytic activity with a specific membrane site so that the studies are, in fact, essential to our consideration of organelle biochemistry. We shall hence comment upon certain of the aspects of this literature though this present discussion will in no way substitute for the many excellent reviews, some of which have been summarized in Table VIIIA.

Much of the attempt to describe mitochondrial energetics has related to so-called coupling factors, compounds derived from mitochondria which when added to a depleted mitochondrial membrane preparation cause a simulation of one or more of the reactions associated with energy conservation. Such factors can, in thesis, participate in chemical reactions which are directly or indirectly necessary for ATPase synthesis; they can bring about conformational changes or in some manner inactivate inhibitors of energy conservation enzymes. A number of such preparations have been described. These may be categorized as having ATPase activity or lacking it: F_1 and A described, respectively, by Pullman *et al.* (1958) and by D. R. Sanadi *et al.* (1962) are the best known of those having ATPase activity; F_1X (Vallejos *et al.*, 1968) has been interpreted to be a combination of factor F_1 and another coupling factor (Groot and Meyer, 1969). Lardy and Wellman (1953) and Selwyn (1967) have described other soluble mitochondrial ATPase preparations. Oligomycin sensitivity is known, and intact mitochondria and ATPase preparations inhibited by this antibiotic have been described by Kagawa and Racker (1966a,b).

Mitochondrial ATPase activity has been demonstrated in a considerable variety of tissues (Lazarus and Barden, 1962; Ashworth *et al.*, 1963; Essner *et al.*, 1965; Schulze and Woolenberger, 1965; Moses *et al.*, 1966). The enzyme has been characterized using inhibitors (Robertson and Boyer, 1955; Penefsky *et al.*, 1960; Bogucka and Wojtczak, 1966; Vigers and Ziegler, 1968; Veldsema-Currier and Slater, 1968; Reddi and Nath, 1969). It is sensitive to metals (Lardy and Wellman, 1953; Amons *et al.*, 1968; Beattie and Basford, 1968).

The enzymological literature describing the ATPase activity of mitochondria varies in much the way that the endoplasmic reticulum ATPase literature does (see Section II, B, 4). Dinitrophenol stimulation, metal activation, azide, and oligomycin inhibition, etc., are the type of characteristics variously assigned to a wide variety of preparations. Whether each represents a distinct protein or rather characteristics associable with the mode of preparation and/or the physiological or chemical state of the material at the time of study remains to be established. It is known that several kinds of mitochondrial fragments exhibit a variety of activities which differ in one or more regards from the character generally assigned to the ATPase of intact mitochondria. Sonication (Bronk and Kielley, 1957; Lardy *et al.*, 1958), mechanical blending (W. W. Kielley and Kielley, 1953), freezing and thawing (D. K. Myers and Slater, 1957), deoxycholate treatment (Watson and Siekevitz, 1956; Siekevitz *et al.*, 1958; Ulrich, 1963, 1964, 1965), and lipid peroxidation (McKnight and Hunter, 1966; Drabikowski and Rafalowska, 1968) have all been used to isolate the activity from larger membrane pieces. Hence, although widely promulgated theses involving mitochondrial ATPase make the assumption, implied or overt, that only a single activity exists, direct experimental evidence is not available to support this. Whether there is a single or multiple mitochondrial ATPase, the consensus is that the activity has a central role in oxidative phosphorylation, and a number of speculative mechanisms, which are consonant with the available experimental evidence (Selwyn, 1968), have been presented to describe the hydrolysis.

An ample literature describes coupling factors that do not themselves have enzymic activity but which enhance reactions that are catalyzed by particles fractionated from mitochondria. Such affected phenomena are reduction of NAD using either succinate or ascorbate and requiring ATP, the nicotinamide nucleotide transhydrogenase reaction, and ATP $\rightarrow$ P$_i$ exchange reaction. Some of the nonenzymic factors (F$_c$, OSCP, F$_5$, F$_3$, and F$_4$) appear to relate to the oligomycin sensitivity of the mitochondrial ATPase. Beechey (1970) suggests that comparison of the properties of factors A, B, F$_1$, F$_2$, F$_3$, F$_4$, F$_5$, F$_c$, F$_1$ − X, OSCP indicates that either "there are common components in these different preparations or that there are different proteins with similar functional properties."

3. Aerobic Oxidation

Elucidation of the reactions of the trichloracetic acid (TCA) cycle and their relationship to other events represents a major accomplishment of the discipline of biological chemistry. Reactions that comprise the cycle and its interplay with cytosol intermediary metabolism and mitochondrial phosphorylation have been studied in a variety of systems. Enzymes that catalyze these reactions are collectively as well known as any group of metabolic constituents. The extensive literature provides an excellent example of the success of the classic biochemical approach of removing physiologically active proteins from structural elements of cells and characterizing them physically and chemically in isolated reaction systems. In addition, however, the relationship of these mitochondrial enzymes to structure has gained the attention of many workers (reviewed by Greville, 1969), and these studies continue with considerable if abated force. It remains, however, a matter of conjecture whether the isolated enzymes are identical in their reaction characteristics with those of the crude preparations and, of course, more importantly with those in the living system. If, indeed, these solubilized enzymes are identical in their reaction characteristics with the original cellular form, then it is certainly of great importance that we analyze the difference between these proteins and the membrane-bound catalysts that are changed markedly when removed from the supporting structure.

The TCA cycle enzymes have been reported to be associated with the outer membrane of the mitochondrion (Allman et al., 1968), although a large number of studies report most of the characteristic activities present, to a degree, in the cytosol as well. If correct, the "economic" significance of this will require explanation. Green and associates (reviewed by D. E. Green and Baum, 1970) have striven to identify not only the particular boundary (Fig. 11) but specialized areas of the membrane. They have assigned the dehydrogenases acting upon pyruvate, α-ketoglutarate, and β-hydroxybutyrate to base pieces of the outer membrane. The other enzymes of the cycle have been localized in a sector of the membrane-repeating units (Fig. 15) described by these authors, which is outward of the base piece and which can be detached from the base pieces. Much of the methodology (and consequently the conclusions) employed in the localization of enzymes with suborganelle particles is disputed by proponents of various techniques.

The cycle is frequently described in terms of eight summary reactions. Some of these are multiple events and involve more than a single catalytic site. The citrate synthetase reaction forms a carbon–carbon bond between a dicarboxylic acid and a derivative of a monocarboxylic acid (oxalacetate and acetyl-CoA). Transformation of citrate to isocitrate is a dehydration catalyzed by aconitase (aconitate hydratase). Isocitrate dehydrogenase and α-ketoglutarate dehydrogenase reactions ultimately lead to the formation of a 4-carbon dicarboxylic acid, succinyl—S—S—CoA. This compound is then oxidized (succinate thiokinase) to yield succinate. Succinate dehydrogenase catalyzes the dehydration to form fumarate. An additional dehydration (fumarase reaction) to malate is the penultimate event of the cycle. The oxidation catalyzed by malate dehydrogenase transforms malate to oxalacetate and the cycle is complete.

Citrate synthetase, though fluoroacetyl-CoA (Brady, 1955) and fluorooxalacetate (Fanshier et al., 1962) are substrates, is relatively substrate specific. Reaction mechanisms have been suggested (Chen and Plaut, 1963) which require that citryl-CoA is an intermediate (Eggerer and Remberger, 1963). It is hydrolyzed, presumably, to citrate and CoA, then cleaved to acetyl-CoA and oxalacetate. The synthetase of rat liver has a high affinity for both acetyl-CoA and oxalacetate with K_m values of 14 and 4 mmoles, respectively. In some other species, however, much lower values have been calculated. Rat liver enzyme has an 87,000 mol wt. Physical properties have been studied and reviewed by Srere (1969), D. Shepherd and Garland (1969a), Parvin (1969), Bogin and Wallace (1969), Weitzman (1969), and Plaut (1969).

The aconitase hydrate dehydration is substrate specific. Labeling experiments have made it apparent that the chemistry is not the conventional hydration–dehydration, but rather involves formation of an enzyme-bound ion (Dickman, 1961).

Isocitric dehydrogenase, like other NAD-linked mitochondrial enzymes, is unstable and "peculiar" in molecular and kinetic properties. This fact may relate to the control function that it exerts in intermediary metabolism. The enzyme has been described by Plaut (1969). The isocitric dehydrogenase literature is complicated by the existence of an NADP- as well as an NAD-specific dehydrogenase. The two enzymes are considerably different in subcellular localization. This extensive literature has been reviewed by Greville (1969).

The α-ketoglutarate dehydrogenase reaction represents the second of two successive decarboxylation–dehydrations. These events catalyzed by an enzyme complex result in the oxidative decarboxylation of α-ketoglutarate to succinyl-CoA. In some preparations the complex has been purified with a resultant separation of dihydrolipoic dehydrogenase activity from a flavoprotein fraction (Ishikawa et al., 1966). Succinyl-CoA generated in the reaction preponderantly continues in the cycle.

Since the conversion of succinyl-CoA to succinate in the reaction catalyzed by succinic thiokinase is highly exothermic, the energy of the thiolester bond is available for the endothermic synthesis of a nucleotide triphosphate. Chemical and physical properties of this enzyme from a variety of animal sources have been described by Cha (1969). Phosphorylation of guanosine represents energy "storage" in a "potentially energetic" triphosphate.

Intimate association of succinic dehydrogenase with structural elements of the mitochondrion structure (presumably an adaptation which brings it in close relationship to the enzymes of the respiratory chain) is attested to by the frustrating difficulties which have been experienced in the attempt to solubilize an active material. Although this has been accomplished (Singer and Kearney, 1955) the similarity of the solubilized purified dehydrogenase to the enzyme in its native physiological state is still moot. Surprisingly, succinic dehydrogenase has also been reported as an activity of the cytosol (T. F. Slater and Planterose, 1960; Mahler, et al., 1958). Attempts to elucidate the relationships between the enzyme and its flavin cofactor [flavin adenine dinucleotide (FAD)] and its interaction with

nonheme iron have not provided a complete picture, though a large literature has developed.

The fumarase reaction is a hydration–dehydration between fumarate
and L-malate. No cofactors are required but basic and acidic residues have
been shown to be essential. Based upon pK values, it appears probable that
two histidine residues, imidazolium and imidazole, are essential to the re-
activity of the enzyme. Cardiac fumarase of the pig has a 194,000 mol wt
and exists as four identical polypeptide chains. There are twelve half-
cysteine residues indicating twelve thiol groups per molecule. Further study
with sulfhydryl reagents indicates that all of these are present as free
sulfhydryls and that the enzyme exists without disulfide bridges (Hill and
Bradshaw, 1969). A reaction mechanism has been postulated by Hansen
et al. (1969).

L-Malate is dehydrogenated in a reaction that requires NAD and yields
oxalacetate. In this way the TCA oxidative cycle series of events is com-
pleted. Beef heart malate dehydrogenase has a 62,000 mol wt, appears to
have twelve sulfhydryl groups too, and lacks disulfide bridges. The amino
acid composition has been determined; this and other of its physical
characteristics have been described by L. Siegel and Englard (1962).

There is evidence (Pette and Brdiczka, 1966), based upon differential
and density gradient centrifugation, that mitochondria are not homogeneous but rather that two types exist in which specific activities of the
TCA cycle enzymes differ from each other.

4. Oxidation of Fatty Acids

Isolated mitochondria synthesize fatty acids from acetate or from
acetyl-CoA (Harlan and Wakil, 1962, 1963; Baron, 1966; Whereat *et al.*,
1967, 1969). Lynen and co-workers (1967) have worked out the series of
events in yeast (Fig. 19). The reaction is apparently an elongation of
fatty acid primers, though in rat and beef liver mitochondria there may
be a genuine *de novo* synthesis. Elongation serves to add two carbon units
at a time to long- and intermediate-chain lengths fatty acids. In mammals
the most significant pathway to stearic acid probably uses as its precursor palmitic acid derived from synthetic events outside the mitochondria. Fatty acid syntheses at the least require NADP and an oxidizable
substrate derived from the TCA cycle (Hulsmann, 1960, 1962; Crist and
Hulsmann, 1962). It has also been reported in some systems that both
NADH and NADPH are required for chain formation using acetate or
acetyl-CoA. Adenosine triphosphate is conventionally supplied in test
systems as a "driving" element for *in vitro* synthesis if acyl-CoA primer
is present. These condensation reactions result in the formation of longer-

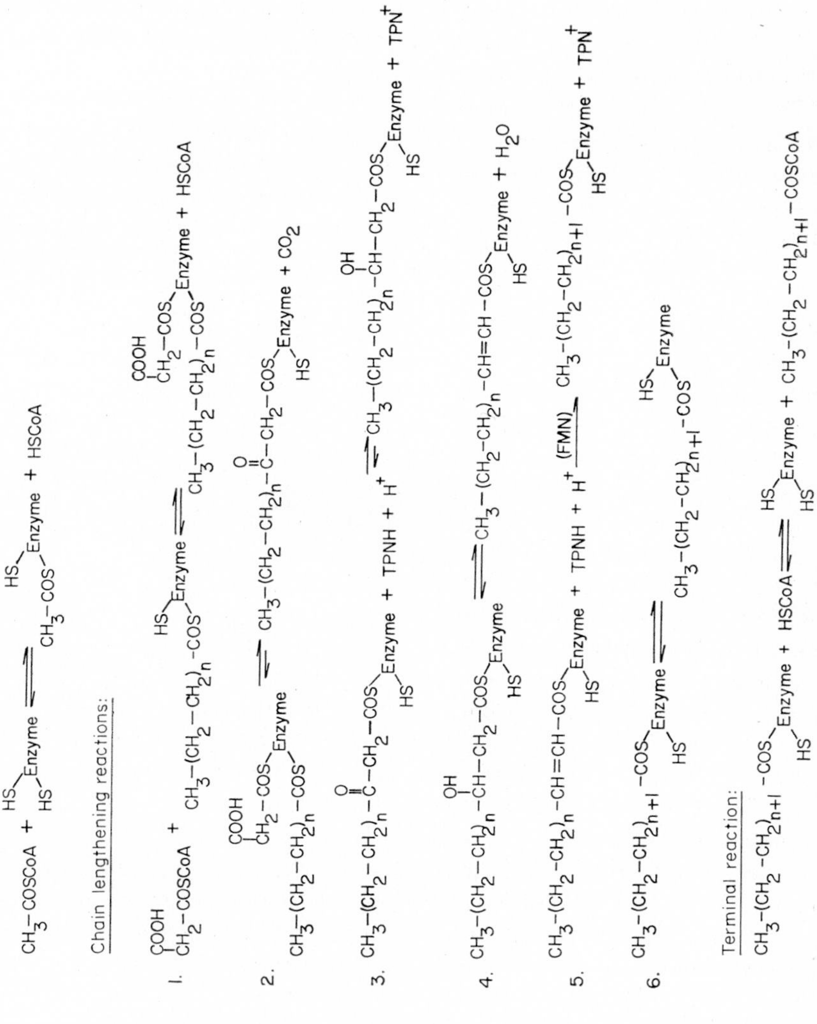

Priming reaction:

$$CH_3-COSCoA + \ ^{HS}_{HS}\!\!\diagdown Enzyme \rightleftharpoons \ ^{HS}_{CH_3-COS}\!\!\diagdown Enzyme + HSCoA$$

Chain lengthening reactions:

1. $$^{COOH}_{CH_2-COSCoA} + CH_3-(CH_2-CH_2)_n-COS \diagdown^{HS}\!\!Enzyme \rightleftharpoons \ _{CH_3-(CH_2-CH_2)_n-COS}^{COOH}\!\!\diagdown^{CH_2-COS}\!\!Enzyme + HSCoA$$

2. $$^{COOH}_{CH_2-COS}\!\!\diagdown Enzyme \rightleftharpoons \ ^{O}_{CH_3-(CH_2-CH_2)_n-C-CH_2-COS}\!\!\diagdown^{HS}\!\!Enzyme + CO_2$$

3. $$^{O}_{CH_3-(CH_2-CH_2)_n-C-CH_2-COS}\!\!\diagdown^{HS}\!\!Enzyme + TPNH + H^+ \rightleftharpoons \ _{OH}^{}CH_3-(CH_2-CH_2)_n-CH-CH_2-COS\!\!\diagdown^{HS}\!\!Enzyme + TPN^+$$

4. $$^{OH}CH_3-(CH_2-CH_2)_n-CH-CH_2-COS\!\!\diagdown^{HS}\!\!Enzyme \rightleftharpoons CH_3-(CH_2-CH_2)_n-CH=CH-COS\!\!\diagdown^{HS}\!\!Enzyme + H_2O$$

5. $$CH_3-(CH_2-CH_2)_n-CH=CH-COS\!\!\diagdown^{HS}\!\!Enzyme + TPNH + H^+ \xrightarrow{(FMN)} CH_3-(CH_2-CH_2)_{2n+1}-COS\!\!\diagdown^{HS}\!\!Enzyme + TPN^+$$

6. $$CH_3-(CH_2-CH_2)_{2n+1}-COS\!\!\diagdown^{HS}\!\!Enzyme \rightleftharpoons \ ^{HS}\!\!\diagdown Enzyme + CH_3-(CH_2-CH_2)_{2n+1}-COS$$

Terminal reaction:

$$CH_3-(CH_2-CH_2)_{2n+1}-COS\!\!\diagdown^{HS}\!\!Enzyme + HSCoA \rightleftharpoons \ ^{HS}_{HS}\!\!\diagdown Enzyme + CH_3-(CH_2-CH_2)_{2n+1}-COSCoA$$

FIG. 19. Mechanism of fatty acid synthesis. [Redrawn from F. Lynen (1967). *In* "Organizational Biosynthesis" (H. J. Vogel, J. O. Lampen, and V. Bryson, eds.), pp. 243–266. Academic Press, New York, by permission.]

chain fatty acids from intermediate-chain length compounds in the equivalent of the Claisen ester condensation. Pyridoxal phosphate is an essential cofactor in these reactions and is thought to be involved in the activation of acetyl-CoA through the formation of a Schiff-base-type intermediate. An impressive variety of long-chain fatty acids both saturated and unsaturated has been produced with these models.

The fatty acid synthesis enzyme complex has been isolated from the outer membrane of the mitochondrion (Bachmann et al., 1966). Attempts to purify the preparation have given only a tightly interacting group of enzymes that have not yielded to separation by chromatography or centrifugation. The synthesis complex in probability consists of a number of distinct activities. Failure to obtain isolated proteins that collectively possess the ability to synthesize fatty acids may indicate structural interaction in nature, as in other membrane systems, which is disrupted in the isolation.

Enzymological characterizations of the complex are, of course, tentative since we may suspect that reported properties are as much a reflection of substrate competitions and organizational hindrances as of intrinsic properties of the proteins themselves. Lynen and his co-workers have approached the problem of gaining knowledge of the molecular architecture of structured elements by studying the multienzyme complex of fatty acid synthetase in yeast. This particulate system is, relatively, simple and amenable to study. Hence, their results are illuminating, offering insight into the method of function of these multiple systems. However, even in this model which has been chosen for its "simplicity," the interrelationships are difficult to resolve. Components directly interact with each other and with substrates as well. Lynen has shown that the transformation of malonyl-CoA into fatty acids is essentially a continuous phenomenon in which the intermediates are covalently bound to sulfhydryl groups of the synthetase (Lynen, 1961).

Figure 20 illustrates the hypothetical structures of the multienzyme complex fatty acid synthetase. This has been developed as a plausible physical arrangement of enzymes serving to catalyze the seven reactions which are known to occur. In order to demonstrate the binding of substrates to sulfhydryl groups, these workers produced synthetic substrates in which the carboxylic acid intermediates of fatty acid synthesis were bound to pantetheine or N-acetylcysteamine (Lynen, 1961, 1962). Because the strong covalent bond to the central sulfhydryl group of the natural substrates is lacking, the affinity of these synthetics for the component enzymes is much smaller.

Dahlan and Porter (1968) have contrasted the beef heart, mitochondrial, fatty acid synthesizing enzyme complex with those described by other authors for rat and beef liver mitochondria. Beef heart enzyme, for example, is inhibited by NADPH which is, however, required by the liver enzyme systems. Substrate requirements and specificities are different. Hence, tentatively, on the basis of a very limited number of studies we have to expect that fatty acid synthesis systems may have characteristics that are common only in the general nature of substrates and products. The fascinating possibility that the "broad primer specificity and the low activity of the enzyme system suggest that its function is probably the remodeling of fatty acids synthesized by other systems" has been sug-

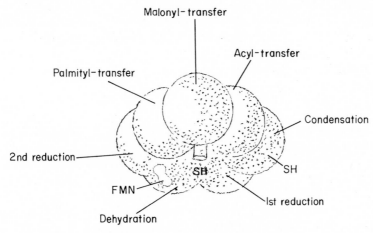

FIG. 20. Hypothetical structure of the multienzyme complex of fatty acid synthetase. [Redrawn from F. Lynen (1967). *In* "Organizational Biosynthesis," (H. J. Vogel, J. O. Lampen, and V. Bryson, eds.), pp. 243–266. Academic Press, New York, by permission.]

gested (Dahlan and Porter, 1968). This concept of reassembly points widely distributed and widely varying in chemical character does nothing to simplify our view of intermediary metabolism, but it certainly makes further study of the fat-synthesizing complex of broad interest. Whereat *et al.* (1969) have segmented the inner and outer mitochondrial membranes of rabbit and identified the fatty acid synthetic capacity in each of the subfractions. A *de novo* synthesis of C_{14} to C_{18} fatty acids was associated with the inner membrane. The outer mitochondrial membrane was capable of a chain elongation mechanism which added one or two acetates to previously formed acyl compounds to make fatty acids 14 to 24 carbon atoms in chain length. These authors reported that the two systems differed in their cofactor requirements and in responsiveness to added NADPH. Upon the basis of their study, they have offered an interesting scheme for possible functional roles of the two fatty acid synthetic systems.

C. F. Howard (1970) has further defined the fatty acids synthetic activities of the inner and outer mitochondrial membranes. He has also provided a useful review of the points of controversy in this literature particularly concerning the extent of de novo synthesis.

5. Other Enzymes

Studies of the subcellular localization of monoamine oxidase have been published which indicate that this enzyme is predominantly found in the mitochondria (McEwen et al., 1968; Yasunobu et al., 1968; Sourkes, 1968; Jarrott and Iversen, 1968; Hellerman and Erwin, 1968) though it is associated with microsomes as well (Bachmann et al., 1966; Jarrott and Iversen, 1968). There appears to be reasonable unanimity further that the enzyme is located in the outer mitochondrial membrane (Parsons and Williams, 1967; C. Schnaitman et al., 1967; Beattie, 1968a,b; Brdiczka et al., 1968), but it has been reported, in some systems, to be associated with the inner membrane (Lardy and Ferguson, 1969). The interesting possibility that the enzyme may move from inner to outer membrane has been raised by Waksman and Renden (1968). Monoamine oxidase has been purified and characterized in a number of systems (D. Fisher et al., 1967; Igaue et al., 1967; Sourkes, 1968; Yasunobu et al., 1968). Blaschko (1963) reviewed the literature of monoamine oxidase up to that time and Sourkes (1968) and Yasunobu et al. (1968) reviewed specialized aspects of the later literature. Lardy and Ferguson (1969) deal with additional aspects of this literature.

Okamato et al. (1967) described kynurenine hydroxylase activity—an outer mitochondrial membrane entity useful as a marker for the outer membrane fraction.

Several fatty acid-activating enzymes (e.g., acyl-CoA ligase) have been described in mitochondria (Webster et al., 1965; Norum et al., 1966a,b; Farstad et al., 1967; Galzigna et al., 1967; Aas and Bremer, 1968). Norum et al. (1966b) assigned the CoA ligase to the outer mitochondrial membrane.

Deoxyribonucleic acid and RNA synthetic and hydrolytic enzymes (Heppel et al., 1956; Curtis et al., 1966; Morais et al., 1967; R. R. Meyer and Simpson, 1968, 1969) and a number of other synthetic activities have been associated with mitochondria. Transaminase activity was described by Hird and Rowsell (1950). Müller and Leuthardt (1950a,b) reported glutamate aspartate transaminase in liver mitochondria. Two groups of aspartate amino transferase isozymes are present in different animal tissues (Borst and Peeters, 1961; Hook and Vestling, 1962; Morino *et al.*, 1963; Nisselbaum and Bodansky, 1963; Waksman and Rendon, 1968); one of them is mitochondrial, the other a cytosol form (Bhargava and Sreenivasan, 1968; Michuda and Martinez-Carrion, 1969). An amino butyrate transaminase has been described in brain mitochondria by Waksman and Bloch (1968). Swick *et al.* (1965a,b) have confirmed the localization of an alanine aminotransferase in rat liver; mitochondrial fractions of liver and kidney have glycine acyltransferase activity (an enzyme significant in the detoxication of benzoic acid and related compounds) (I. K. Brandt *et al.*, 1968). Higgins and Barnett (1970) assigned carnitine acetyltransferase to the inner surface of the outer mitochondrial membrane and/or the outer surface of the inner mitochondrial membrane. Rendon and Waksman (1969) in a localization study of aspartate aminotransferase have presented the interesting thesis that some of the controversy concerning localization of mitochondrial enzymes may result from a movement of the enzymes as a function of the energetic state of mitochondria. Enzymes involved in the synthesis of citrulene and related compounds have been discussed by Schneider (1959) in his review. Mizutani (1968) in a recent paper described the cytochemical localization of ornithine carbamyltransferase activity in liver mitochondria and has briefly outlined this area of the literature.

Other enzymes associated with the mitochondrion are listed in Table VII.

IV. Nuclear Enzymes

A. INTRODUCTION

By acclamation, the nucleus has been assigned as the principal genetic controller, though many details and mechanisms remain to be elucidated. Enzyme-catalyzed reactions related to the genetic machinery and perhaps to other functions of nucleoplasm have for good and obvious reason fascinated many investigators. The burgeoning literature has been reviewed by Hogeboom *et al.* (1953, 1957), Allfrey *et al.* (1955), Roodyn (1959), Mirsky and Osawa (1961), Siebert and Humphrey (1965), and Siebert (1968).

Here, as in the evaluation of the enzymological literature of other cell particulates, improvements in isolation techniques have made continuous reevaluation of the literature important. Methodology has been extensively discussed by Roodyn (1963), and aspects of the various methods have been evaluated by the several reviewers cited above. Establishment of an enzyme's inclusion within the nucleus is a problem which presents its own unique difficulties that are perhaps more imposing than establishing the fact of an association between an enzyme and other particles. This is especially true for soluble enzymes of the nucleus which frequently are identical with soluble proteins of the cytoplasm. Intact isolation is a practical necessity. Electron microscopy is widely used to observe the integrity of the fraction. The RNA/DNA ratios have also been useful criteria of preparation quality. For soluble enzymes, Siebert and Humphrey (1965) have suggested that an *identical specific activity* of an enzyme in nucleus and in cytoplasm can be taken as an additional point of evidence that the enzyme is a true nuclear component.

If soluble enzymes are freely exchanged between the nucleoplasm and the cytoplasm, then one would expect that the characteristics of both molecular pools would be identical. Siebert and Hannover (1963) have found lactate dehydrogenase isozyme distribution, heat stability, and other characteristics studied to be identical in nucleoplasm and in cytoplasm. However, other traditional enzymological criteria for identity, kinetic parameters (coenzyme influence upon reaction rate, substrate affinity, and rate as a function of temperature), were fundamentally different. These authors interpreted their results in terms of aggregation or complexing of the LDH in the nucleus which served to alter the catalytic properties of the enzyme. However, the possibility of inherent differences in the population averages, even if free diffusion into and out of the nucleus occurs, cannot be excluded, and it appears that, overall, soluble proteins of the cytoplasm and the nucleus are similar in character and percent distribution (Bakay and Sorof, 1964).

Enzymic constituency of the nucleus is as remarkable for what appears to be absent as for what is present. Although some of the individual enzymes are present, the phenomenon of oxidative metabolism is absent from the nucleus (Roodyn, 1959). The one validated exception is the nucleated erythrocyte. These nuclei contain the oxidative systems of these cells and are, hence, entirely exceptional. Mirsky et al. (1956) and Allfrey and Mirsky (1957), however, have reported that oxidative phosphorylation occurs in pituitary cell nuclei so that this too appears a noteworthy exception.

The nucleus is by no means an isolated organelle. Systems which are complete—glycolysis, nuclear biosynthesis, NAD synthesis—are significant as much for their interrelationships with cell structures outside the nucleus as for their presence within the nucleus. In glycolysis, glucose must enter the nucleus. This, in ascites cells, occurs within a span of a few seconds (Kohen et al., 1964). Similarly lactic acid or pyruvic acid must leave the nucleus for oxidation as part of a cytoplasmic respiratory sequence. The NADH, formed during nuclear glycolysis, must be reoxidized outside the nucleus and presumably even so large a molecule as mRNA must pass the nuclear membrane. Siebert and Humphrey (1965) state that nine-tenths of the NAD formed leaves the nucleus to play a cytoplasmic or mitochondrial role.

B. THE ENZYME SYSTEMS

1. Glycolytic Enzymes

With the exceptions discussed, oxidative phosphorylation appears to be lacking in the nucleus. Thus we make the assumption that energy-storage molecules are synthesized primarily via glycolysis. There is acceptable experimental support for this thesis; aldolase, enolase, triosphosphate dehydrogenase, and lactic dehydrogenase (Roodyn, 1959) have been measured in isolated nuclei. Manometric measurements of glycolysis in isolated plant nuclei have been made (Stern and Mirsky, 1953). Hexokinase, phosphofructokinase, phosphoglyceric kinase, and pyruvic kinase exist in nuclei, and their activity is sufficient for formation of enough ATP to meet the energy requirements of the nuclei. Table IX compares the activity of cytoplasmic vs. nucleoplasmic enzymes from liver, kidney, and brain cells.

Published data (Siebert, 1961; Siebert et al., 1961; Siebert and Humphrey, 1965) indicate that the specific activities of nuclear glycolytic enzymes (Table IX) are slightly higher than are comparable cytoplasmic activities. Despite the absence of oxidative metabolism in nuclei of most cells, soluble enzymes of the citric acid cycle have been described (malate dehydrogenase and isocitrate dehydrogenase) (Siebert et al., 1961; Mandel et al., 1962).

2. Chromatin Space Enzymes

Siebert and associates (reviewed by Siebert, 1968) have defined a nuclear region in which enzymes are tightly bound to nuclear proteins and by presumption to chromosomal materials (F. Fischer et al., 1959). It is thought that these are genuinely particulate enzymes and that their activities are spatially restricted. Nicotinamide adenine dinucleotide pyrophosphorylase and nucleoside triphosphatase D are examples of such enzymes. It has also been suggested by Leslie (1961) that the histones themselves may carry some enzymic activity; some supporting evidence has been developed (Martin et al., 1963). Interpretation of these present data does not preclude the possibility that other proteins bound to the histones are active although the enzymic activities are closely associated with the histones which may themselves be the catalysts. It is, of course, a matter of conjecture whether those enzymes that are closely associated with chromatin perform genetic tasks; although the nature of some of the enzymes would indicate this, the economy has not been established.

TABLE IX

Nuclear–Cytoplasmic Ratios of Activities of Soluble Enzymes in Different Tissues[a]

Activity ratio[b]	Rat liver[c]		Pig kidney[c]		Beef brain[c]	
	T	N	T	N	T	N
Lactate dehydrogenase/triose-phosphate dehydrogenase	7.7	6.7	2.1	1.5	1.2	1.1
Lactate dehydrogenase/glucose 6-phosphate dehydrogenase	—	—	47	57	—	—
Lactate dehydrogenase/phos-phofructokinase	1220	1140	460	310	—	—
Triosephosphate dehydrogenase/ 6-phosphogluconate dehydro-genase	13	11	19	24	143	106
Pyruvate kinase/aldolase	—	—	0.64	0.52	4.2	3.3
Lactate dehydrogenase/pyru-vate kinase	48	38	66	53	1.9	3.2
Lactate dehydrogenase/6-phos-phogluconate dehydrogenase	96	75	39	36	—	—
Lactate dehydrogenase/malate dehydrogenase	3.2	2.9	0.63	0.90	3.1	3.5

[a] From Siebert and Humphrey (1965) by permission.
[b] Activity expressed as μmole substrate per gram dry weight.
[c] T—nonfractionated tissue; N—nuclei.

Other enzymes that appear to be bound to the chromatin include ribonuclease (H. Busch, 1965; Reid et al., 1964), triphosphatase A, polyadenylate-synthesizing enzyme (Chambon et al., 1963), a terminal DNA polymerase (Krakow et al., 1962), a DNA-degrading enzyme (Bernardi et al., 1961), RNA methylase (Birnstiel et al., 1963), and amino acid-activating enzyme (E. E. Brandt and Finamore, 1963).

Adenosinetriphosphatases are strongly bound to chromatin; it would appear that nucleoside diphosphokinase and related enzymes are absent. Nuclear ATPases are apparently distinct; conventional inhibitors of mitochondrial ATPase or myosin are without effect, and the metal responses and kinetic characteristics are quite different from those of the endoplasmic reticulum plasma membrane. It is further likely that there are separate nucleoside triphosphatases with specificity, respectively, toward adenosine, guanosine, cytosine, and uridine. The nature of these enzymes remains, of course, to be elucidated, although there has been some speculation that they may be themselves the histones of the chromatin complex.

There is some evidence that chromatin-bound enzymes exhibit modified properties by virtue of this bound state. Irradiation does not seem to inactivate NAD pyrophosphorylase in nuclei from rat liver and thymus (Adelstein and Bigg, 1962) and tumors (Hirsch-Hoffman *et al.*, 1964), but the same enzymic activity in regenerating rat liver is radiation sensitive (Mandel *et al.*, 1962). During the regenerative process itself NAD pyrophosphorylase is decreased (Stirpe and Aldridge, 1961) ; under other conditions, D. K. Myers (1962) measured little or no change. The regenerative process enhances the activity of RNA polymerase (S. Busch *et al.*, 1962).

V. Enzymes Associated with Other Particles

A. Chloroplasts

A number of reviews dealing with the distribution of enzymes in organelles of plant cells have been published (Bonner, 1965a,b; Hallaway, 1965; Lieberman and Baker, 1965; Mollenhauer and Morré, 1966; D. O. Hall and Whatley, 1967; Pridham, 1968). Though there are, of course, characteristic elements which differ from their counterparts in animal cells, the problems are much the same. In general the conclusions of Hallaway (1968) that "precise intracellular distribution of many (plant) enzymes is unknown ... [and] requires the use of cytochemical techniques as well as the analysis of subcellular particles" for assignment of cellular geography are as true for animal material as for that from plants. The chloroplast subcellular structure is complex, and much has been accomplished in elucidation of the nature of the structure and its relationship to function. The carbon cycle events are located in the stroma, and the light reactions in the membranes proper (Leech, 1968). Weier *et al.* (1967) postulated that energy transfer from the membrane occurred by movement of ATP, ADP, NADP, and NADPH from the membrane to the site of the energy-consuming reaction.

In addition to the light and dark reactions of photosynthesis proper, which are catalyzed by the appropriate enzymes (Goodwin, 1966; Leech, 1968), the metabolism of lipids is a complex function of chloroplasts (Nichols and James, 1968).

B. Microbodies

Rhodin (1954) used the term *microbody* to describe inclusions of renal tubular cells of the mouse. Similar particles have since been described in many other organisms and in other tissues. That the microbodies are, indeed, distinct from lysosomes is supported by demonstration of the absence in these particles of acid phosphatase and by the fact that microbodies do not, as lysosomes do, accumulate material that enters the cells by pinocytosis and phagocytosis. Most importantly, the complement of enzymes shown to be associated with the microbody is quite distinct from that associated with lysosomes. Microbodies are defined, then, as cytoplasmic particles characterized by catalase activity and by that of one of the several oxidative enzymes: urate oxidase, d-amino acid oxidase, or L-α-hydroxy acid oxidase having slightly alkaline pH optima. The particles are surrounded by a single smooth membrane which contains a finely granular matrix and in some species a central core or marginal plates or both.

Catalase is probably the only enzyme present uniformly in all microbodies. D-Amino acid oxidase is absent in liver (Shack, 1943; B. R. Endahl and Kochakian, 1956; Meister *et al.*, 1960). Uricase is absent in kidney of some animals (Straus, 1956; Baudhuin *et al.*, 1965), and L-α-hydroxy acid oxidase too is limited in species distribution (Blanchard *et al.*, 1944).

Catalase is a hemoprotein and uricase a metaloprotein. Both amino acid oxidase and L-α-hydroxy acid oxidase are flavoproteins. It is not unlikely that other flavoprotein enzymes and enzymes that produce hydrogen peroxide are also present in microbodies though conclusive demonstration of this has not been made.

Catalase activity was, until recently, thought to be associated with mitochondria. Thomson and Klipfel (1957), however, observed that, in their gradient centrifugations, catalase and uricase sedimented together in a nonmitochondrial fraction. The catalase reactions:

$$H_2O_2 + H_2O_2 \rightarrow 2\ H_2O + O_2 \quad \text{and} \quad H_2O_2 + RH_2 \rightarrow 2\ H_2O + R$$

are ubiquitous in distribution and are associated (though not exclusively so) with microbodies throughout the plant and animal kingdoms. Rechcigl and Price (1968) have shown that catalase activity is embodied by at least two enzymes.

Catalase is found in particulate as well as in soluble forms, the relative proportion depending upon the species. Isozymes have been demonstrated not only by conventional electrophoretic separations but by immunological variations in inhibitor responses (King and Gutmann, 1964) as well. Isozymes of catalase show tissue distributions in the manner similar to that reported for other multiple-form enzymes. Paul and Fottrel (1961) related molecular variation relative to species. They have pointed to differences in heat stability, pH versus activity curves, and resistance to chemical denaturation (by urea, formamide, and other agents). The extensive literature of catalase heterogeneity has recently been reviewed (Higashi and Peters, 1963a,b; Patton and Nishimura, 1967; Rechcigl and Price, 1968; Rechcigl et al., 1969).

Enzymic oxidation of uric acid by extracts of liver, kidney, and muscle were reported as early as 1905 by Schittenhelm. Hogeboom et al. (1957) identified the activity with cytoplasmic particles and the enzyme was purified by Mahler et al. (1955).

Microhistochemical and biochemical techniques have been used to demonstrate the association of urate oxidase activity with the crystaloid of microbodies, whereas catalase and amino acid oxidase activities appear to be associated with the soluble matrix of the particle. This literature has been reviewed by Hruban and Rechcigl (1969) and by Mahler (1963).

The urate oxidase reaction which accomplishes the oxidation of uric acid to products, which have not been identified with certainty, has been found to occur in a large number of species. This together with the fact that uric acid is a principal end product of purine metabolism strongly suggests that urate oxidase is an important enzyme in the catabolism of nitrogenous compounds, particularly of purines.

Intracellular distribution of D-amino acid oxidase from rat liver was worked out by Dianzani (1954a) who established its distinctness from the mitochondria with which it has been, by convention, associated. De Duve (1960a) and de Duve et al. (1960) have further localized the activity to the microbody. The enzyme is a flavoprotein; the coenzyme is identified as FAD. Although microbody D-amino acid oxidases in all species and tissue studied show an absolute specificity for D-amino acid, the rates and other kinetic parameters varied widely and the possibility exists that more than a single protein entity may exhibit this activity.

Spectral properties of D-amino acid oxidase, as many other flavoprotein enzymes, have been extensively studied (Shiga et al., 1968). Absorption, ORD, circular dichroism, and nuclear magnetic resonance spectra have been obtained and compared with those of free flavins (Harbury and Foley, 1958; Shiga and Piette, 1964; Miyake et al., 1965; Tsuchiyama and Osaka, 1965; Kondo, 1966; Massey and Curti, 1966; Aki et al., 1966; Takagi et al., 1966). Some of these data establish that the enzyme-bound flavin differs spectroscopically from free flavins. Spectral studies, in addition, have provided information which allows us to induce information about the electronic structures of the molecule.

Baudhuin *et al.* (1965) and J. M. Allen and Beard (1965) demonstrated the presence of α-hydroxy acid oxidase in microbodies from rat kidney. It was suggested by de Duve and Baudhuin (1966) that this enzyme might catalyze not only the L-α-hydroxy acids but L-lactate and L-amino acids as well. Only indirect evidence has been brought to bear, however, and this point remains to be substantiated. Two different α-hydroxy acid oxidases have been identified by J. C. Robinson *et al.* (1962) in kidney cortex and heart. The FMN proteins apparently have protein moieties which may be specific for either short-chain or long-chain substrates, but in both the general reaction is

$$R—CHOH—COOH + O_2 \rightarrow R\text{-}CO\text{-}COOH + H_2O_2$$

Like others of the microbody enzymes, these oxidases appear to exist in more than a single molecular form (Domenech and Blanco, 1967).

C. Lysosomes and Other Particles

De Duve and Wattiaux (1966) have characterized lysosomes by their heterogeneousness, "lysosomes stand out in a unique fashion against all other cellular constituents by polymorphism and by variety of processes both physiological and pathological in which they are implicated." These organelles contained a large number of hydrolytic enzymes having acidic pH optima. They are relatively large, about 0.04 μ in diameter, though with considerable variation, and are enclosed by a unit membrane. Substrates enter the lysosomes by endocytosis and are hydrolyzed by a broad spectrum of activities capable of acting upon many proteins, polysaccharides, and nucleic acids (Dingle and Fell, 1969).

Lysosome-like particles are very nearly ubiquitous in distribution. They were found first in the liver and have since been shown to exist in high concentration in kidney, spleen, intestine, leukocytes, and in nervous tissue. Lysosomes have been associated with a variety of physiological events including metamorphic changes, tissue regression differentiation, catabolic disorders, and inflammatory processes (G. Weissman, 1965; de Duve and Wattiaux, 1966; Tappel, 1968). Enzymes reported to be associated with lysosomes are outlined in Table X. Their distribution is universal throughout the animal kingdom from Protozoa to vertebrates. Similar bodies have been described in higher plant cells, although fine points of definition may be invoked to classify these particles separately.

It would appear, thus, that lysosomal particles are quite variable in the enzymic composition though high concentrations of certain hydrolytic enzymes certainly represent a characteristic property of these bodies.

TABLE X

Enzymes Associated with Lysosomes

E. C. number	Enzyme (trivial name)	Comments and references
2.7.7.16–17	Acid ribonuclease	Acts on ribonucleic acid (RNA) at pH 6.0–6.5, specifically hydrolyzes phosphoester linkages of RNA. De Duve et al. (1955); Anfinsen and White (1961); Josefson and Lagerstedt (1962); Bernardi (1966a); Buchanan and Schwartz (1967); Jacques (1968); Stagni and de Bernard (1968); Shamberger (1969); Dingle and Dott (1969); Futai et al. (1969)
3.1.–.–	Sphingomyelinase	Acts on sphingomyelin at pH 5.0. Brady (1966); Weinreb et al. (1968); Fowler (1969)
3.1.1.–	Phospholipase	Liver enzyme hydrolyzes the phospholipids—lecithin, lysolecithin, and phosphatidyl ethanolamine at pH 4.5. Van Deenen and de Haas (1966); Mellors and Tappel (1967); A. D. Smith and Winkler (1968); Stoffel and Trabert (1969)
3.1.1.–	Esterase	The esterase from liver catalyzes β-naphthyl acetate at pH 5.0. Shibko and Tappel (1964)
3.1.1.3	Triglyceride lipase	Liver and kidney lipase acts on glycerol trideoanoate at pH 4.2. Mahadevan and Tappel (1968a)
3.1.3.2	Acid phosphatase	The enzyme has an optimal range pH 5–6; acts on β-glycerol phosphate and most O-phosphoric esters. De Duve et al. (1955); Beaufay et al. (1959a,b); Schmidt (1961); Jackson and Black (1967); Helminen et al. (1968); Brightwell and Tappel (1968a,b); Paltemaa (1966); Lundquist and Öckerman (1969); Shamberger (1969); Dingle and Dott (1969); N. R. Miller and Rafferty (1969); D. Robinson and Willcox (1969); Ide and Fishman (1969)
3.1.3.4	Phosphatidate phosphatase	This enzyme is associated with lysosomes and microsomes. The lysosomal enzyme has an optimum at pH 6.4 and substrate specificity for phosphatidic acid. Wilgram and Kennedy (1963); Sedgwick and Hübscher (1965)
3.1.3.16	Phosphoprotein phosphatase	Phosphoprotein phosphatase hydrolyzing casein is partly lysosomal in rat and mouse liver and has a single pH optimum at 5.5. Paigen and Griffiths (1959); Roth et al. (1962)
3.1.4.1	Phosphodiesterase	The phosphodiesterase that hydrolyzes p-nitrophenyl-5-phosphothymidine most rapidly at pH 5.2, found in lysosomes has a K_m of 0.10 mM; is not affected by Mg^{2+}, Ca^{2+}, ethylenediaminetetraacetate, and dithiothreitol. Bernardi and Bernardi (1966); Brightwell and Tappel (1968a,b)
3.1.4.5–6	Acid deoxyribonuclease	Associated with lysosomes in various tissues; does not require Mg^{2+}. It acts on deoxyribonucleic acid at acid pH. De Lamirande et al. (1954); de Duve et al. (1955); Stevens and Reid (1956); Beaufay et al. (1957); Laskowski (1961); Novikoff (1961); Kurnick (1962); Bernardi and Griffe (1964); Bernardi (1966b); Helminen et al. (1968)
3.1.6.1	Arylsulfatases A and B	Distribution of arylsulfatase has been studied in various tissues. There are three distinct arylsulfatases of which A and B act on nitrocatecholsulfate at pH 4.9 and 5.9, respectively. Roy (1954, 1958, 1960); Dodgson et al. (1955); Viala and Gianetto (1955); Ugazio (1960); Novikoff (1961); Jackson and Black (1967); Helminen et al. (1968); Shamberger (1969)
3.2.1.–	Galactocerebrosidase	Like glucocerebrosidase, the enzyme has a single pH optimum at 6.0 and the enzyme is specific to galactocerebrosides substrates. Brady (1966); Weinreb et al. (1968)
3.2.1.–	O-Seryl-N-acetylgalactosamide glycosidase	Active at pH 4.4. Mahadevan and Tappel (1968b)
3.2.1.–	Glucocerebrosidase	Active at pH 6.0 on glucocerebrosides. Brady (1966); Weinreb et al. (1968)
3.2.1.–	Sialidase	Hydrolyzes sialic acid derivatives, glycoprotein, and glycolipid. Liver and kidney enzymes act on glycoprotein at pH 4.0 to 4.4. Rafelson et al. (1966a,b)
3.2.1.17	Lysozyme	Lysozyme, a lysosomal enzyme, contains 129 amino acid residues. It has a molecular weight of 14,400 ± 100. It is characterized by its ability to lyse suspensions of certain susceptible bacteria, notably Micrococcus lysodeikticus. The enzyme acts on bacterial cell wall mucopolysaccharides at pH 6.2. The backbone of this polysaccharide is composed of alternating residue of 2-acetamido-2-deoxy-D-glucose and its 3-O-lactyl derivative, N-acetyl-D-muramic acid (NAM) which are linked β-(1-4). Lysozyme depolymerizes this component by hydrolysis of the NAM-C-1 glycosidic bonds. The end product of the degradation is, therefore, the disaccharide. Chitin is also known to serve as a substrate but deacetylated chitin is not hydrolyzed

163

TABLE X (Continued)

3.2.1.17—Continued

		by the enzyme. Oligosaccharides derived from chitin do serve as substrate. Berger and Weiser (1957); Jolles (1960); Jeanloz et al. (1963); Gibian (1966); Sharon (1967); J. B. Howard and Glazer (1967); Hayashi et al. (1968); Kronman (1968); Davis et al. (1969); Ikeda and Hamaguchi (1969)
3.2.1.18	Neuraminidase	Hydrolyzes terminal α-2,6-links between N-acetylneuraminic acid and 2-acetamido-2-deoxy-D-galactose residues in various mucopolysaccharides. Taha and Carubelli (1967)
3.2.1.20	α-Glucosidase	Enzyme obtained from kidney acts on maltose at pH 5. It has substrate specificity for α-1,4-glucosidic linkages of polysaccharides especially glycogen. E. H. Fischer and Stein (1960); Novikoff (1961); Lejeune et al. (1963); Levvy and Conchie (1966); D. Fisher et al. (1967); D. Robinson et al. (1967); Lloyd (1969)
3.2.1.21	β-Glucosidase	Patel and Tappel (1969); Lloyd (1969)
3.2.1.23	β-Galactosidase	Liver lysosomal enzyme acts on nitrophenyl β-D-galactoside at pH 3.6. Sellinger et al. (1960); Wallenfels (1961); Furth and Robinson (1965); Stagni and de Bernard (1968); Shamberger (1969); Lundquist and Öckerman (1969); Alpers (1969)
3.2.1.24	α-Mannosidase	Like other glycosidases, the enzyme is associated with lysosomes. It has a substrate specificity for α-D-mannosides and related glycosides. De Duve (1960a,b); Sellinger et al. (1960); Conchie and Levvy (1960); Novikoff (1961); Lundquist and Öckerman (1969)
3.2.1.30	β-Acetylaminodeoxyglucosidase	This enzyme is predominantly lysosomal in liver and presumably associated with lysosomes in other tissues. Pugh et al. (1957); Sellinger et al. (1960); Conchie and Levvy (1960)
3.2.1.31	β-Glucuronidase	In rat liver, the major part of the activity is found in lysosomal fraction. Straus (1956, 1957); Levvy and Marsh (1959, 1960); Beaufay et al. (1959a,b); Caravaglios and Franzini (1959); Conchie and Levvy (1960); Fishman (1961); Paigen (1961); Novikoff (1961); Roth et al. (1962); Levvy and Conchie (1966); Paltemaa (1968); Stagni and de Bernard (1968); Dingle and Dott (1969); Schamberger (1969); Lundquist and Öckerman (1969); Ide and Fishman (1969)
3.2.1.35-36	Hyaluronidase	In liver this enzyme is found to be associated with lysosomes and acts on hyaluronate at pH 3.9. K. Meyer et al. (1960); Aronson and Davidson (1965, 1967a,b); Gibian (1966); Vaes (1967); Dott and Dingle (1968); Dingle and Dott (1969)
3.2.1.37	β-Xylosidase	When assayed in presence of p-nitrophenyl-β-D-xylopyranoside, this enzyme is highly active at pH 5.0. D. Robinson and Abrahams (1967); D. Fisher et al. (1967); Patel and Tappel (1969); D. Fisher and Kent (1969)
3.4.3.-	Peptidase	Liver peptidase hydrolyzes tyrosol glycine and other dipeptides. When assayed with tyrosyl glycine, activity was optimal at pH 7.8
3.4.4.-	Cathepsin A	Lysosomal distribution with pH optimum 5.4 when reacted with carbobenzoxy-L-glutamyl-L-tyrosine and pH 3.5 when assayed in presence of hemoglobin. Fruton (1960); Iodice et al. (1966); Iodice (1967).
3.4.4.-	Cathepsin B	Cathepsin B acts on benzoyl-L-argininamide at pH 5. Rademaker (1959); Fruton (1960); Bouma and Gruber (1964); Snellman (1969)
3.4.4.9	Cathepsin C	Associated with lysosomes but its presence in light mitochondria is also known; has pH optimum 5.0 when reacted with glycyl L-tyrosinamide and 5.3, when assayed with glycyl phenylalanine. Finkenstaedt (1957); Rademaker (1959); Fruton (1960); Planta et al. (1964); Bouma and Gruber (1954); Metrione et al. (1966); Voynick and Fruton (1968)
3.4.4.19	Clostridiopeptidase A	The lysosomal enzyme hydrolyzes collagen at pH 6.0. Mandl (1961); J. F. Woods and Nichols (1965); Anderson (1969)
3.4.4.23	Cathepsin D	Essentially lysosomal in liver and spleen. In other tissues, however, it has a diffuse distribution. Assays of this enzyme depend on its ability to degrade denatured hemoglobin at an optimum pH of 3.5 to 4.0. De Duve et al. (1955); Gianetto and de Duve (1955); Beaufay et al. (1959a,b); Planta et al. (1964); Iodice et al. (1966); Barrett (1967); Woessner (1968)

E. C. number	Enzyme (trivial name)	Comments and references
3.4.4.–	Cathepsin E	This lysosomal proteinase acts on albumin at pH 2.5. Lapresle and Webb (1962)
3.5.1.–	N-Acetyl-β-glucosamidase	Suematsu et al. (1970)
3.5.1.–	Aspartyl glucosylamine amidohydrolase	Liver and kidney hydrolyzes aspartyl glucosylamine at pH 7.6. Mahadevan and Tappel (1967)
3.6.1.1	Acid pyrophosphatase	Acid pyrophosphatase hydrolyzes adenosine triphosphate and other nucleoside triphosphates at pH 5 to 5.4. Brightwell and Tappel (1968a,b)
3.5.2.–	Arylamidase	Liver or kidney enzyme hydrolyzes 1-leucyl β-naphthylamide hydrochloride at pH 6.8 to 7.2. Mahadevan and Tappel (1967)

Particularly, acid phosphatases (Fig. 21) have been used by the electron microscopist as an identifiable entity for the localization of the lysosome. Acid phosphatase, acid ribonuclease, acid deoxyribonuclease, β-glu-

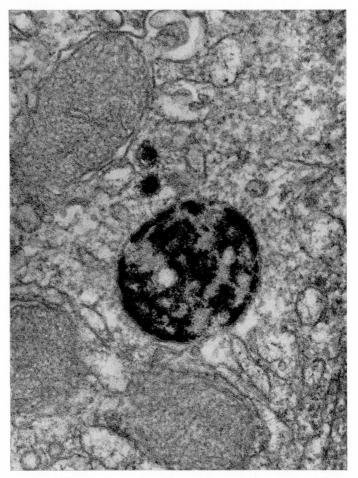

FIG. 21. Acid phosphatase activity of lysosome in rat vas deferens. (Photograph courtesy of Daniel S. Friend.)

curonidase, and cathepsin appear to be most uniformly identified with lysosome-like bodies (de Duve et al., 1955; de Duve, 1959). Other hydrolytic enzymes have also been reported in the lysosome fractions isolated by differential centrifugation from liver: arylsulfatase (Viala and Gianetto, 1955; Roy, 1958); β-galactosidase, β-n-acetylglucosamine hydrolase, α-mannosidase (Sellinger et al., 1960), collagenase (Frankland and

Wynn, 1962; Schaub, 1964), α-glucosidase (Lejeune *et al.*, 1963), hyaluronidase (Aronson and Davidson, 1965), and phosphatidate phosphatase (Wilgram and Kennedy, 1963; Sedgwick and Hübscher, 1965). Other enzymes that have been reported associated with the lysosomes from various tissues are phosphoprotein phosphatase, alkaline phosphatase, lipase, esterase, NADH$_2$ cytochrome c reductase, and ATPase. It must be recognized that the enzymes listed are those that appear to have been greatly enriched by lysosome particle isolation techniques. Since most authors have worked with admittedly impure preparations, the possibility that other enzymes are present in smaller amounts certainly remains distinct. In addition the identity of the listed enzymes as lysosomal on the basis of impure isolations cannot be considered certain. Most of these studies, however, have electron microscope histochemical substantiation.

The enzymological literature on the variety of acid hydrolases associated with lysosomes has centered upon characterization of these enzymes primarily with respect to their substrate specificities and their pH optima. Generally, the proteases have shown a broad range of activity; lysosomal nucleases hydrolyze both acidic and basic nucleic acids, and the glycosidases have a broad spectrum of activity in the hydrolysis of polysaccharides. The phosphatases are extremely diverse and appear to act upon most bioorganic phosphates of living cells. The pH optima, which for most phosphatases are around pH 5, presumably relate to the digestive function of these enzymes. Protein and peptide hydrolysis appears in nature to be a major function of the lysosomal enzymes. Eight distinct proteases and peptidases have been described (Iodice *et al.*, 1966; Coffey and de Duve, 1968).

Purified cathepsins isolated from lysozymes have been described (Iodice *et al.*, 1966; Metrione *et al.*, 1966; Barrett, 1967; Woessner, 1968; Iodice, 1967; Voynick and Fruton, 1968). The interaction of these proteases and peptidases in the degradation of glycoproteins by lysosomes has been reviewed by Tappel (1968). The presence in these same cellular inclusions of carbohydrolytic activities further results in an even more profound involvement in the catabolism of glycoproteins. β-Galactosidase, *N*-acetyl-β-glucosaminidase, and α-mannosidase, sialidase, β-aspartylglucosylamine amidohydrolase, and *O*-seryl-*N*-acetylgalactosamine glycosidase are examples of such activities. Lysosomal enzymes will degrade a wide variety of mucopolysaccharides (Hutterer, 1966). Hyaluronic acid, chondroitin 4-sulfate, chondroitin 6-sulfate, dermatan sulfate, heparin sulfate, heparin, and keratan sulfate are substrates for lysosomal enzymes. Mucopolysaccharide–protein complexes, e.g., chondroitin sulfate β-galactose–xylose-serine-peptide, are of so complex a structure that we should suspect that a number of lysosomal glycosidases and proteolytic enzymes are involved in their degradation—the glycosidases hydrolyzing the connecting sugars, and proteolytic enzymes hydrolyzing the peptide bonds. A variety of lipids serve as substrate for lysosomal enzymes; triglyceride lipase and phospholipases have been demonstrated as well.

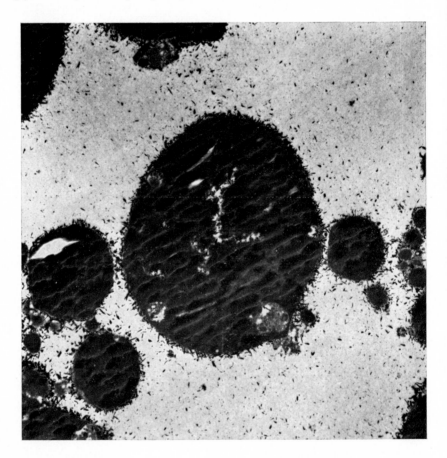

FIG. 22. Spherosomes isolated from castor endosperm. (Photograph courtesy of Lawrence Y. Yatsu.)

1. Spherosomes

Holcomb *et al.* (1967) have described sphaerosomes (Fig. 22) as being similar to small fat droplets. The bodies are vesicles that may represent buds from the endoplasmic reticulum. They contain a proteinaceous stroma which has acid phosphatase activity. Walek-Szernecka (1965) described a number of hydrolytic enzymes in spherosomes. Matila *et al.* (1965) have discussed hydrolytic enzymes such as acid phosphatases, acid ribonuclease, acid protease, and nonspecific esterases which are associated with these plant particles.

REFERENCES

Aas, M., and Bremer, J. (1968). *Biochim. Biophys. Acta.* **164,** 157.
Abe, K., Butcher, R. W., Nicholson, W. E., Baird, C. E., Liddle, R. A., and Liddle, G. W. (1969). *Endocrinology* **84,** 362.
Abood, L. G., and Gerard, R. W. (1954). *J. Cell. Comp. Physiol.* **43,** 379.
Abood, L. G., Gerard, R. W., Banks, J., and Tschirgi, R. D. (1952). *Amer. J. Physiol.* **168,** 728.
Abraham, A. D., Pora, E. A., and Hodosan, F. (1969). *Experientia* **25,** 820.
Adelstein, S. J., and Biggs, S. L. (1962). *Radiat. Res.* **16,** 422.
Agosin, M., Scaramelli, N., Gil, L., and Letelier, M. E. (1969). *Comp. Biochem. Physiol.* **29,** 785.
Agostoni, A., Vergani, C., and Villa, L. (1966). *Nature (London)* **209,** 1024.
Ahmed, Z., and King, E. J. (1959). *Biochim. Biophys. Acta* **34,** 313.
Aithal, H. N., and Ramasarma, T. (1969). *Biochem. J.* **115,** 77.
Akhtar, M., Rahimtula, A. D., and Wilton, D. C. (1969a). *Biochem. J.* **114,** 801.
Akhtar, M., Watkinson, I. A., Rahimtula, A. D., Wilton, D. C., and Munday, K. A. (1969b). *Biochem. J.* **111,** 757.
Aki, K. Takagi, T., Isemura, T., and Yamino, T. (1966). *Biochim. Biophys. Acta* **122,** 193.
Albers, R. W. (1967). *Annu. Rev. Biochem.* **36,** 727.
Albers, R. W., Fahn, S., and Koval, G. J. (1963). *Proc. Nat. Acad. Sci. U. S.* **50,** 474.
Albers, R. W., Koval, G. J., and Siegel, G. J. (1968). *Mol. Pharmacol.* **4,** 324.
Aldridge, W. N., and Johnson, M. K. (1959). *Biochem. J.* **73,** 270.
Alessio, L., Del Ninno, E., Pabis, A., and Dioguardi, N. (1967). *Ital. J. Biochem.* **16,** 161.
Allard, C., and Cantero, A. (1952). *Can. J. Med. Sci.* **30,** 295.
Allard, C., de Lamirande, G., Faria, H., and Cantero, A. (1954). *Can. J. Biochem. Physiol.* **32,** 383.
Allard, C., de Lamirande, G., Weber, G., and Cantero, A. (1956). *Can. J. Biochem. Physiol.* **34,** 170.
Allard, C., de Lamirande, G., and Cantero, A. (1957). *Cancer Res.* **17,** 862.
Allen, A. K., and Rosenberg, H. (1968). *Biochim. Biophys. Acta* **151,** 504.
Allen, J. M., and Beard, M. E. (1965). *Science* **149,** 1507.
Allende, J., Monro, R., and Lipmann, F. (1964). *Proc. Nat. Acad. Sci. U. S.* **51,** 1211.
Allfrey, V. G., and Mirsky, A. E. (1957). *Proc. Nat. Acad. Sci. U. S.* **43,** 589.
Allfrey, V. G., Mirsky, A. E., and Stern, H. (1955). *Advan. Enzymol.* **16,** 411.
Allmann, D. W., and Bachmann, E. (1967). *Methods Enzymol.* **10,** 438.
Allmann, D. W., Bachmann, E., and Green, D. E. (1966a). *Arch. Biochem. Biophys.* **115,** 165.
Allmann, D. W., Galzigna, L., McCaman, R. E., and Green, D. E. (1966b). *Arch. Biochem. Biophys.* **117,** 413.
Allmann, D. W., Harris, R. A., and Green, D. E. (1967). *Arch. Biochem. Biophys.* **122,** 766.
Allmann, D. W., Bachmann, E., Orme-Johnson, N., Tan, W. C., and Green, D. E. (1968). *Arch. Biochem. Biophys.* **125,** 981.

Alpers, D. H. (1969). *J. Biol. Chem.* **244,** 1238.
Amberson, W. R., Roisen, F. J., and Bauer, A. C. (1965). *J. Cell. Comp. Physiol.* **66,** 71.
Amons, R., Van Den Bergh, S. G., and Slater, E. C. (1968). *Biochim. Biophys. Acta* **162,** 452.
Anders, M. W. (1968). *Arch. Biochem. Biophys.* **126,** 269.
Anderson, A. J. (1969). *Biochemistry* **113,** 457.
Andreoli, T. E., Tieffenberg, M., and Tosteson, D. C. (1967). *J. Gen. Physiol.* **50,** 2527.
Anfinsen, C. B., and White, F. H. (1961). *In* "The Enzymes" (P. D. Boyer, H. Lardy, and K. Myrbäck, eds.), 2nd rev. ed., Vol. 5, p. 95. Academic Press, New York.
Appelmans, F., Wattiaux, R., and de Duve, C. (1955). *Biochem. J.* **59,** 438.
Arlinghouse, R., Schaffer, J., and Schweet, R. (1964). *Proc. Nat. Acad. Sci. U. S.* **51,** 1291.
Aronson, N. N., Jr., and Davidson, E. A. (1965). *J. Biol. Chem.* **240,** PC3222.
Aronson, N. N., Jr., and Davidson, E. A. (1967a). *J. Biol. Chem.* **242,** 441.
Aronson, N. N., Jr., and Davidson, E. A. (1967b). *J. Biol. Chem.* **242,** 437.
Ashworth, C. T., Luibel, F. J., and Stewart, S. C. (1963). *J. Cell Biol.* **17,** 1.
Atkinson, A., Hunt, S., and Lowe, A. G. (1968). *Biochim. Biophys. Acta* **167,** 469.
Avi-Dor, Y. (1960). *Biochem. J.* **76,** 370.
Avignon, J., and Steinberg, D. (1961). *J. Biol. Chem.* **236,** 2898.
Axelrod, J. (1955). *J. Biol. Chem.* **214,** 753.
Babad, H., Ben-Zvi, R., Bdolah, A., and Schramm, M. (1967). *Eur. J. Biochem.* **1,** 96.
Baccari, V., and Guerritore, A. (1955). *Arch. Sci. Biol. (Bologna)* **39,** 539.
Bachelard, H. S. (1967). *Biochem. J.* **104,** 286.
Bachmann, E., Allman, D. W., and Green, D. E. (1966). *Arch. Biochem. Biophys.* **115,** 153.
Bader, H., and Sen, A. K. (1966). *Biochim. Biophys. Acta* **118,** 116.
Baginski, E. S., Foa, P. O., and Zak, B. (1967). *Anal. Biochem.* **21,** 201.
Baginsky, M. L., and Hatefi, Y. (1969). *J. Biol. Chem.* **244,** 5313.
Baillie, M. J., and Morton, R. K. (1958). *Biochem. J.* **69,** 35.
Bakay, B., and Sorof, S. (1964). *Cancer Res.* **24,** 1814.
Baldwin, J. J., and Cornatzer, W. E. (1968). *Biochim. Biophys. Acta* **164,** 195.
Ball, J. H., and Kadis, B. (1965). *Arch. Biochem. Biophys.* **110,** 427.
Banay-Schwartz, M., Benziman, M., and Strecker, H. J. (1969). *Comp. Biochem. Physiol.* **28,** 177.
Barden, R. E., and Cleland, W. W. (1969). *J. Biol. Chem.* **244,** 3677.
Barker, P. J., Fincham, N. J., and Hardwick, D. C. (1968). *Biochem. J.* **110,** 739.
Barkley, D. S. (1969). *Science* **165,** 1133.
Barnett, R. E. (1970). *Fed. Proc., Fed. Amer. Soc. Exp. Biol.* **29,** 858.
Baron, E. J. (1966). *Biochim. Biophys. Acta* **116,** 425.
Barra, H. S., Cumar, F. A., and Kaputo, R. (1969). *J. Biol. Chem.* **244,** 6237.
Barrett, A. J. (1967). *Biochem. J.* **104,** 601.
Bar-Tana, J., and Shapiro, B. (1964). *Biochem. J.* **93,** 533.
Bartos, E., and Uziel, M. (1961). *J. Biol. Chem.* **236,** 1697.
Baskin, L. S., and Leslie, R. B. (1968). *Biochim. Biophys. Acta* **159,** 509.
Baskin, R. J., and Deamer, D. W. (1970). *J. Biol. Chem.* **245,** 1345.
Baudhuin, P., Muller, M., Poole, B., and de Duve, C. (1965). *Biochem. Biophys. Res. Commun.* **20,** 53.
Baudhuin, P., Hertoghe-Lefevre, E., and deDuve, C. (1969). *Biochem. Biophys. Res. Commun.* **35,** 548.

Baum, H., Rieske, J. S., Silman, I., and Lipton, S. H. (1967a). *Proc. Nat. Acad. Sci. U. S.* **57**, 798.

Baum, H., Silman, I., Rieske, J. S., and Lipton, S. H. (1967b). *J. Biochem.* (*Tokyo*) **242**, 4876.

Baxter, C. E., Van Reen, R., Pearson, P. B., and Rosenberg, C. (1958). *Biochim. Biophys. Acta* **27**, 584.

Beams, H. W., and Kessel, R. G. (1968). *Int. Rev. Cytol.* **23**, 209–276.

Beattie, D. S. (1968a). *Biochem. Biophys. Res. Commun.* **30**, 57.

Beattie, D. S. (1968b). *Biochem. Biophys. Res. Commun.* **31**, 901.

Beattie, D. S., and Basford, R. E. (1968). *J. Neurochem.* **15**, 325.

Beaufay, H., and de Duve, C. (1954). *Bull. Soc. Chim. Biol.* **36**, 1551.

Beaufay, H., Berleur, A. M., and Doyen, A. (1957). *Biochem. J.* **66**, 32P.

Beaufay, H., Bendall, D. S., Baudhuin, P., and de Duve, C. (1959a). *Biochem. J.* **73**, 623.

Beaufay, H., Bendall, D. S., Baudhuin, P., Wattiaux, R., and de Duve, C. (1959b). *Biochem. J.* **73**, 628.

Beechey, R. B. (1970). *Biochem. J.* **116**, 6p.

Bell, J. L., and Baron, D. N. (1964). *Proc. Biochem. Soc.* **90**, 8P.

Bellamy, D. (1959). *Biochem. J.* **72**, 165.

Belleau, B. (1960). *Adrenergic Mech., Ciba Found. Symp.* p. 223.

Bendall, D. S., and de Duve, C. (1960). *Biochem. J.* **74**, 444.

Benedetti, E. L., and Emmelot, P. (1965). *J. Cell Biol.* **26**, 299.

Benedetti, E. L., and Emmelot, P. (1968). *In* "*The Membranes*" (A. J. Dalton and F. Haguenau, eds.), pp. 33–120. Academic Press, New York.

Bengmark, S., Ekholm, R., and Olsson, R. (1967). *Acta Hepato-Splenol.* **14**, 80.

Benjamins, J. A., and Agranoff, B. W. (1969). *J. Neurochem.* **16**, 513.

Berger, L. R., and Weiser, R. S. (1957). *Biochim. Biophys. Acta* **26**, 517.

Berkes-Tomasevic, P., and Holzer, H. (1967). *Eur. J. Biochem.* **2**, 98.

Bernardi, G. (1966a). *In* "Procedures in Nucleic Acid Research" (G. L. Cantoni and D. R. Davies, eds.), p. 37. Harper, New York.

Bernardi, G. (1966b). *In* "Procedures in Nucleic Acid Research" (G. L. Cantoni and D. R. Davies, eds.), p. 102. Harper, New York.

Bernardi, G., and Bernardi, A. (1966). *In* "Procedures in Nucleic Acid Research" (G. L. Cantoni and D. R. Davies, eds.), p. 144. Harper, New York.

Bernardi, G., and Griffe, M. (1964). *Biochemistry* **3**, 1419.

Bernardi, G., Champagne, M., and Sadron, C. (1961). *Biochim. Biophys. Acta* **49**, 1.

Bernheim, M. L. (1969). *Arch. Biochem. Biophys.* **134**, 408.

Berthet, J. (1960). *Proc. Int. Congr. Biochem., 4th, 1958* Vol. 17, p. 107.

Berthet, J., and de Duve, C. (1951). *Biochemistry* **50**, 174.

Beutler, E., and Morrison, M. (1967). *J. Biol. Chem.* **242**, 5289.

Bewsher, P. D., and Ashmore, J. (1966). *Biochem. Biophys. Res. Commun.* **24**, 431.

Beyer, K. F., and Samuels, L. T. (1956). *J. Biol. Chem.* **219**, 69.

Bhargava, M. M., and Sreenivasan, A. (1968). *Biochem. J.* **108**, 619.

Bickerstaffe, R., and Annison, E. F. (1969). *Biochem. J.* **111**, 419.

Binkley, F. (1961). *J. Biol. Chem.* **236**, 1075.

Binkley, F., Davenport, J., and Eastall, F. (1959). *Biochem. Biophys. Res. Commun.* **1**, 206.

Birbeck, M. S. C., and Reid, E. (1956). *J. Biophys. Biochem. Cytol.* **2**, 609.

Birkmayer, G. D., Sebald, W., and Bücher, T. (1969). *Hoppe-Seyler's Z. Physiol. Chem.* **350**, 1159.

Birnstiel, M. L., Fleissner, E., and Borek, E. (1963). *Science* **142,** 1577.
Bishop, J. O., and Schweet, R. S. (1961). *Biochim. Biophys. Acta* **54,** 617.
Bishop, J. S., and Larner, J. (1969). *Biochim. Biophys. Acta* **171,** 374.
Bitensky, M. W., and Burstein, S. R. (1965). *Nature (London)* **208,** 1282.
Bjørnstad, P. (1966). *Biochim. Biophys. Acta* **116,** 500.
Blake, R. L., Blake, S. L., and Kun, E. (1967). *Biochim. Biophys. Acta* **148,** 293.
Blanchard, M., Green, D. E., Nocito, V., and Ratner, S. (1944). *J. Biol. Chem.* **155,** 421.
Blaschko, H. (1963). *In* "The Enzymes" (P. D. Boyer, H. Lardy, and K. Myrbäck, eds.), 2nd rev. ed., Vol. 8, p. 337. Academic Press, New York.
Blaschko, H., Hagen, P., and Welch, A. D. (1955). *J. Physiol. (London)* **129,** 27.
Blaschko, H., Hagen, J. M., and Hagen, P. (1957). *J. Physiol. (London)* **139,** 316.
Blecher, M. (1967). *Biochem. Biophys. Res. Commun.* **27,** 560.
Blecher, M., and White, A. (1960). *J. Biol. Chem.* **235,** 3404.
Blobel, G., and Potter, Van R. (1966). *Proc. Nat. Acad. Sci. U. S.* **55,** 1283.
Bloch, K. (1965). *Science* **150,** 19.
Bloom, B. M., and Goldman, I. M. (1966). *Advan. Drug Res.* **3,** 121.
Boell, E. J., and Weber, R. (1955). *Exp. Cell Res.* **9,** 559.
Bogdanski, D. F., Weissbach, H., and Udenfriend, S. (1957). *J. Neurochem.* **1,** 272.
Bogin, E., and Wallace, A. (1969). *Methods Enzymol.* **13,** 19.
Bogucka, K., and Wojtczak, L. (1966). *Biochim. Biophys. Acta* **122,** 381.
Bois, R., and Estabrook, R. W. (1969). *Arch. Biochem. Biophys.* **129,** 362.
Bojarski, T. B., and Wynne, A. M. (1957). *Proc. Can. Cancer Res. Conf.* **2,** 95.
Bonner, J. (1965a). *In* "Plant Biochemistry" (J. Bonner and J. E. Varner, eds.), 2nd ed., p. 3. Academic Press, New York.
Bonner, J. (1965b). *In* "Plant Biochemistry" (J. Bonner and J. E. Varner, eds.), 2nd ed., p. 38. Academic Press, New York.
Bont, W. S., Huizinga, F., Bloemendal, H., Van Weenen, M. F. and Bosch, L. (1965). *Arch. Biochem. Biophys.* **109,** 207.
Bonting, S. L., and Caravaggio, L. L. (1966). *Biochim. Biophys. Acta* **112,** 519.
Borst, P., and Peeters, E. M. (1961). *Biochim. Biophys. Acta* **54,** 188.
Böttger, I., Wieland, O., Brdiczka, D., and Pette, D. (1969). *Eur. J. Biochem.* **8,** 113.
Bouma, J. M. W., and Gruber, M. (1964). *Biochim. Biophys. Acta* **89,** 545.
Bowler, K., and Duncan, C. J. (1968a). *Comp. Biochem. Physiol.* **24,** 1043.
Bowler, K., and Duncan, C. J. (1968b). *Comp. Biochem. Physiol.* **24,** 223.
Boyd, J. W. (1961). *Biochem. J.* **81,** 434.
Boyer, P. D. (1965). *In* "Oxidases and Related Redox Systems" (T. E. King, H. S. Mason, and M. Morrison, eds.), Vol. 2, p. 994. Wiley, New York.
Boyer, P. D. (1967). *In* "Biological Oxidation" (T. P. Singer, ed.), p. 193. Wiley, New York.
Brachet, J., Deeroly-Briers, M., and Hoyez, J. (1958). *Bull. Soc. Chim. Biol.* **40,** 2039.
Brady, R. O. (1955). *J. Biol. Chem.* **217,** 213.
Brady, R. O. (1966). *N. Engl. J. Med.* **275,** 312.
Brand, L., and Mahler, H. R. (1959). *J. Biol. Chem.* **234,** 1615.
Brandt, E. E., and Finamore, F. J. (1963). *Biochim. Biophys. Acta* **68,** 618.
Brandt, I. K., Simmons, P., and Gutfinger, T. (1968). *Biochim. Biophys. Acta* **167,** 196.
Brdiczka, D., Pette, D., Brunner, A., and Miller, F. (1968). *Eur. J. Biochem.* **5,** 294.
Breckenridge, B. McL., and Johnston, R. E. (1969). *J. Histochem. Cytochem.* **17,** 505.
Bremer, J. (1955). *Acta Chem. Scand.* **9,** 268.

Brightwell, R., and Tappel, A. L. (1968a). *Arch. Biochem. Biophys.* **124**, 325.
Brightwell, R., and Tappel, A. L. (1968b). *Arch. Biochem. Biophys.* **124**, 333.
Brindley, D. N., and Hübscher, G. (1965). *Biochim. Biophys. Acta* **106**, 495.
Brindley, D. N., and Hübscher, G. (1966). *Biochim. Biophys. Acta* **125**, 92.
Brodbeck, J., and Ebner, K. E. (1966). *J. Biol. Chem.* **241**, 5526.
Brodie, B. B., Axelrod, J., Cooper, J. R., Gaudette, L., La Du, B. N., Mitoma, C., and Udenfriend, S. (1955). *Science* **121**, 603.
Brodie, B. B., Gillette, J. R., and La Du, B. N. (1958). *Annu. Rev. Biochem.* **27**, 427.
Brody, T. M., Wang, R. I. H., and Bain, J. A. (1952). *J. Biol. Chem.* **198**, 821.
Bronk, J. R., and Kielley, W. W. (1957). *Biochim. Biophys. Acta* **24**, 440.
Brosemer, R. W., and Rutter, W. J. (1961). *J. Biol. Chem.* **236**, 1253.
Brown, H. D. (1966a). *Proc. 2nd Int. Biophys. Congr., Abst.* 692.
Brown, H. D. (1966b). *Biochem. Pharmacol.* **15**, 2007.
Brown, H. D. (1966c). *In* "Membranes and Transport Phenomena" (F. Snell, ed.), p. 56. Biophys. Soc., St. Louis, Missouri.
Brown, H. D., and Altschul, A. M. (1964). *Biochem. Biophys Res. Commun.* **15**, 479.
Brown, H. D., Altschul, A. M., Evans, W. J., and Neucere, N. J. (1964a). *Plant Physiol.* **39**, Supp., lxi.
Brown, H. D., Evans, W. J., and Altschul, A. M. (1964b). *Life Sci.* **3**, 1487.
Brown, H. D., Neucere, N. J., Altschul, A. M., and Evans, W. J. (1965). *Life Sci.* **4**, 1439.
Brown, H. D., Chattopadhyay, S. K., and Patel, A. B. (1966). *Biochem. Biophys. Res. Commun.* **25**, 304.
Brown, H. D., Chattopadhyay, S. K., Patel, A., and Rigdon, R. H. (1967a). *Experientia* **23**, 522.
Brown, H. D., Chattopadhyay, S. K., and Patel, A. (1967b). *Enzymologia* **32**, 205.
Brown, H. D., Jackson, R. T., and Waitzman, M. B. (1967c). *Life Sci.* **6**, 1519.
Brown, H. D., Chattopadhyay, S. K., and Patel, A. (1967d). *Arch. Biochem Biophys.* **120**, 222.
Brown, H. D., Rigdon, R. H., Chattopadhyay, S. K., and Patel, A. (1968a). *Enzymol. Biol. Clin.* **9**, 433.
Brown, H. D., Chattopadhyay, S. K., and Patel, A. (1968b). *Metab. Clin. Exp.* **17**, 555.
Brown, H. D., Patel, A., Chattopadhyay, S. K., and Pennington, S. N. (1968c). *Enzymologia* **35**, 215.
Brown, H. D., Patel, A. B., Chattopadhyay, S. K., and Pennington, S. N. (1968d). *Enzymologia* **35**, 233.
Brown, H. D., Chattopadhyay, S. K., Patel, A. B., Pugh, R. P., Matthews, B., and Pennington, S. N. (1969a). *Missouri Med.* **66**, 418.
Brown, H. D., Chattopadhyay, S. K., Patel, A. B., Spjut, H. J., Spratt, J. S., Pugh, R. P., and Pennington, S. N. (1969b). *Brit. J. Cancer* **23**, 446.
Brown, H. D., Pennington, S. N., Chattopadhyay, S. K., Morris, H. P., and Spratt, J. S. (1969c). *In* "Carcinogen-Metabolism in Hepatoma" (S. Weinhouse, ed.), Temple Univ., Philadelphia, Pennsylvania.
Brown, H. D., Spratt, J. S., Jr., Morris, H. P., Chattopadhyay, S. K., and Pennington, S. N. (1969d). *Biophys. J.* **9**, A185.
Brown, H. D., Morris, H. P., Chattopadhyay, S. K., Patel, A. B., and Pennington, S. N. (1969e). *Experientia* **25**, 358.
Brown, J. L., and Johnston, J. M. (1964). *Biochim. Biophys. Acta* **84**, 264.
Bruni, A., and Racker, E. (1968). *J. Biol. Chem.* **243**, 962.
Brunner, G., and Bygrave, F. L. (1969). *Eur. J. Biochem.* **8**, 530.

174   HARRY D. BROWN AND SWARAJ K. CHATTOPADHYAY

Buchanan, W. E., and Schwartz, T. B. (1967). *Amer. J. Physiol.* **212**, 732.
Bucher, N. L. R., and McGarrahn, K. (1956). *J. Biol. Chem.* **222**, 1.
Bucher, N. L. R., Overath, P., and Lynen, F. (1960). *Biochim. Biophys. Acta* **40**, 491.
Burba, J. V. (1968). *Can. J. Biochem.* **46**, 367.
Burstein, S. (1967). *Biochim. Biophys. Acta* **146**, 529.
Burstein, S., Westort, C., and McDonald, C. F. (1967). *Endocrinology* **80**, 1120.
Busch, H. (1965). Quoted by Seibert and Humphrey. *Advan. Enzymol.* **27**, 268.
Busch, S., Chambon, P., Mandel, P., and Weill, J. D. (1962). *Biochem. Biophys. Res. Commun.* **7**, 255.
Butcher, R. W. (1966). *Pharmacol. Rev.* **18**, 237.
Butcher, R. W., and Sutherland, E. W. (1962). *J. Biol. Chem.* **237**, 1244.
Butcher, R. W., and Sutherland, E. W. (1967). *Ann. N. Y. Acad. Sci.* **139**, 849.
Butcher, R. W., Baird, C. E., and Sutherland, E. W. (1968). *J. Biol. Chem.* **243**, 1705.
Cammarano, P., Giudice, G., and Lukes, B. (1965). *Biochem. Biophys. Res. Commun.* **19**, 487.
Cammer, W., and Moore, C. L. (1969). *Arch. Biochem. Biophys.* **134**, 290.
Campbell, P. N., and Lawford, G. R. (1968). *In* "Structure and Function of the Endoplasmic Reticulum in Animal Cells" (F. C. Gran, ed.), p. 57. Academic Press, New York.
Campbell, P. N., Lowe, E., and Serck-Hanssen, G. (1967). *Biochem. J.* **103**, 280.
Caravaca, J., and Grisolia, S. (1960). *J. Biol. Chem.* **235**, 684.
Caravaglios, R., and Franzini, C. (1959). *Exp. Cell Res.* **17**, 22.
Carell, E. E., and Kahn, J. S. (1967). *Biochim. Biophys. Acta* **131**, 571.
Carey, N. J., and Greville, G. D. (1959). *Biochem. J.* **71**, 159.
Carruthers, C., and Baumler, A. (1962). *Arch. Biochem. Biophys.* **99**, 458.
Carruthers, C., Woernley, D. L., Baumler, A., and Davis, B. (1959). *Cancer Res.* **19**, 59.
Carter, J. R., Jr. (1968). *J. Lipid Res.* **9**, 748.
Castañeda, M., and Tyler, A. (1968). *Biochem. Biophys. Res. Commun.* **33**, 782.
Cerletti, P., and Zichella, L. (1960). *Clin. Chim. Acta* **5**, 748.
Cerletti, P., Fronticelli, C., and Zichella, L. (1960). *Clin. Chim. Acta* **5**, 439.
Cha, S. (1969). *Methods Enzymol.* **13**, 62–75.
Chambon, P., Weill, J. D., and Mandel, P. (1963). *Biochem. Biophys. Res. Commun.* **11**, 39.
Chance, B., and Schoener, B. (1964). *Biochem. Biophys. Res. Commun.* **17**, 416.
Chance, B., and Spencer, E. L., Jr. (1959). *Discuss. Faraday Soc.* **27**, 200.
Chance, B., Ernester, L., Garland, P. B., Lee, C. P., Light, P. A., Ohnishi, T., Ragan, E. I., and Wong, D. (1967). *Proc. Nat. Acad. Sci. U. S.* **57**, 1498.
Chang, Y.-Y. (1968). *Science* **160**, 57.
Chantrenne, H. (1947). *Biochim. Biophys. Acta* **1**, 437.
Chappel, J. B., and Perry, S. V. (1953). *Biochem. J.* **55**, 586.
Chassy, B. M., Love, L. L., and Krichevsky, M. I. (1969). *Proc. Nat. Acad. Sci. U. S.* **64**, 296.
Chatterjee, I. B., Ghosh, J. J., Ghosh, N. C., and Guha, B. C. (1959). *Naturwissenschaften* **46**, 580.
Chatterjee, I. B., Chatterjee, G. C., Ghosh, N. C., Ghosh, J. J., and Guha, B. C. (1960). *Biochem. J.* **76**, 279.
Chattopadhyay, S. K., Patel, A., Rigdon, R. H., and Brown, H. D. (1968). *Proc. 7th Int. Biochem. Congr., 1967* Vol. V, Art. J94, p. 970.
Chattopadhyay, S. K., Brown, H. D., and Patel, A. B. (1969). *Acta Biol. Med. Ger.* **22**, 1.

Chauveau, J., Moulé, Y., Rouiller, C., and Schneebeli, J. (1962). *J. Cell Biol.* **12,** 17.
Cheatum, S. G., Douville, A. W., and Warren, J. C. (1967). *Biochim. Biophys. Acta* **137,** 172.
Chen, R. F., and Plaut, G. W. E. (1963). *Biochemistry* **2,** 752.
Chignell, C. F. (1968). *Biochem. Pharmacol.* **17,** 1207.
Chignell, C. F., and Titus, E. (1968). *Biochim. Biophys. Acta* **159,** 345.
Christensen, H. N. (1964). *Mammalian Protein Metab.* **1,** 105.
Christie, G. S., and Judah, J. D. (1953). *Proc. Roy. Soc., Ser. B* **141,** 420.
Chytil, F. (1968). *J. Biol. Chem.* **243,** 893.
Chytil, F., and Skrivanova, J. (1963). *Biochim. Biophys. Acta* **67,** 164.
Cirillo, V. J., Andersen, O. F., Ham, E. A., and Gwatkin, R. B. L. (1969). *Exp. Cell Res.* **57,** 139.
Claude, A. (1943). *Biol. Symp.* **10,** 111.
Cleland, W. W., and Kennedy, E. P. (1960). *J. Biol. Chem.* **235,** 45.
Cleland, K. W., and Slater, E. C. (1953). *Biochem. J.* **53,** 547.
Coffey, J. W., and de Duve, C. (1968). *J. Biol. Chem.* **243,** 3255.
Cohen, G., and Monod, J. (1957). *Bacteriol. Rev.* **21,** 169.
Collins, D. C., Jirku, H., and Layne, D. S. (1968). *J. Biol. Chem.* **243,** 2928.
Collins, G. G. S., and Youdim, M. B. H. (1969). *Biochem. J.* **114,** 80P.
Colomb, M. G., Cheruy, A., and Vignais, P. V. (1969). *Biochemistry* **8,** 1926.
Conchie, J., and Levvy, G. A. (1960). *Biochem. J.* **73,** 12P.
Conney, A. H., Brown, R. R., Miller, J. A., and Miller, E. C. (1957). *Cancer Res.* **17,** 628.
Conway, T., and Lipmann, F. (1964). *Proc. Nat. Acad. Sci. U. S.* **52,** 1462.
Cooper, D. Y., Narasimhulu, S., Rosenthal, O., and Estabrook, R. W. (1965). *In* "Oxidases and Related Redox Systems" (T. E. King, H. S. Mason, and M. Morrison, eds.), Vol. 2, Part VII, p. 833. Wiley, New York.
Cooper, J. R., and Brodie, B. B. (1955). *J. Pharmacol. Exp. Ther.* **114,** 409.
Corman, L., Prescott, L. M., and Kaplan, N. O. (1967). *J. Biol. Chem.* **242,** 1383.
Cotzias, G. C., and Dole, V. P. (1952). *J. Biol. Chem.* **196,** 235.
Cox, G. F., and Davies, D. D. (1967). *Biochem. J.* **105,** 729.
Cox, G. F., and Davies, D. D. (1969). *Biochem. J.* **113,** 813.
Crane, R. K., and Sols, A. (1953). *J. Biol. Chem.* **203,** 273.
Craven, P. A., and Basford, R. E. (1969). *Biochemistry* **8,** 3520.
Craven, P. A., Goldblatt, P. J., and Basford, R. E. (1969). *Biochemistry* **8,** 3525.
Creange, J. E., and Roberts, S. (1965). *Biochem. Biophys. Res. Commun.* **19,** 73.
Creasey, W. A. (1962). *Biochim. Biophys. Acta* **64,** 559.
Cremer, J. E. (1958). *Biochem. J.* **68,** 685.
Crist, E. J., and Hulsmann, W. C. (1962). *Biochem. Biophys. Acta* **60,** 72.
Cryer, P. E., Jarett, L., and Kipnis, D. M. (1969). *Biochim. Biophys. Acta* **177,** 586.
Curtis, P. J., Burdon, M. C., and Smellie, R. M. S. (1966). *Biochem. J.* **98,** 813.
Dahlan, J. V., and Porter, J. W. (1968). *Arch. Biochem. Biophys.* **127,** 207.
Daikuhara, Y., Tsunemi, T., and Takeda, Y. (1968). *Biochim. Biophys. Acta* **158,** 51.
Dallman, P. R., Dallner, G., Bergstrand, A., and Ernster, L. (1969). *J. Cell Biol.* **41,** 357.
Dallner, G. (1963). *Acta Pathol. Microbiol. Scand.* Suppl. 166, 1.
Daly, J., Guroff, G., Udenfriend, S., and Witkop, B. (1968). *Biochem. Pharmacol.* **17,** 31.
Daly, J., Jerina, D., Farnsworth, J., and Guroff, G. (1969). *Arch. Biochem. Biophys.* **131,** 238.
Danielson, L., and Ernster, L. (1963). *Biochem. Biophys. Res. Commun.* **10,** 91.

Datta, P. K., and Shepard, T. H. (1959). *Arch. Biochem. Biophys.* **81,** 124.
Dauwalder, M., Kephart, J. E., and Whaley, W. G. (1966). *J. Cell Biol.* **31,** 25A.
Davies, D. D., and Kun, E. (1957). *Biochem. J.* **66,** 307.
Davies, R. C., Neuberger, A., and Wilson, B. M. (1969). *Biochim. Biophys. Acta* **178,** 294.
Davoren, P. R., and Sutherland, E. W. (1963). *J. Biol. Chem.* **238,** 3016.
Dawson, A. P., and Thorne, C. J. R. (1969). *Biochem. J.* **111,** 27.
Deamer, D. W., Utsumi, K., and Packer, L. (1967). *Arch. Biochem. Biophys.* **121,** 641.
Dean, P. D. G., Ortiz De Montellano, P. R., Bloch, K., and Corey, E. J. (1967). *J. Biol. Chem.* **242,** 3014.
de Baun, J. R., Rowley, J. Y., Miller, E. C., and Miller, J. A. (1968). *Proc. Soc. Exp. Biol. Med.* **129,** 268.
de Bernard, B., Stagni, N., Sottocasa, G. L., and Cremese, R. (1963). *Proc. 5th Int. Congr. Biochem., 1961* Vol. 29, p. 250.
de Crombrugghe, B., Varmus, H. E., and Perlman, R. L. (1970). *Biochem. Biophys. Res. Commun.* **38,** 894.
de Duve, C. (1959). *In* "Subcellular Particles" (T. Hayashi, ed.), p. 128. Ronald Press, New York.
de Duve, C. (1960a). *Bull. Soc. Chim. Biol.* **42,** 11.
de Duve, C. (1960b). *Nature (London)* **187,** 836.
de Duve, C., and Baudhuin, P. (1966). *Physiol. Rev.* **46,** 323.
de Duve, C., and Wattiaux, R. (1966). *Ann. Rev. Physiol.* **28,** 435.
de Duve, C., Pressman, B. C., Gianetto, R., Wattiaux, R., and Appelmans, F. (1955). *Biochem. J.* **60,** 604.
de Duve, C., Beaufay, H., Jacques, P., Rahman-Li, Y., Sellinger, O. Z., Wattiaux, R., and Wattiaux-De Coninck, S. (1960). *Biochim. Biophys. Acta* **40,** 186.
de Duve, C., Wattiaux, A., and Baudhuin, P. (1962). *Advan. Enzymol.* **24,** 291.
De Groot, L. J., and Carvalho, E. (1960). *J. Biol. Chem.* **235,** 1390.
de Lamirande, G., and Allard, C. (1957). *Proc. Can. Cancer Res. Conf.* **2,** 83.
de Lamirande, G., Allard, C., and Cantero, A. (1954). *Can. J. Biochem.* **32,** 35.
de Lamirande, G., Allard, C., and Cantero, A. (1958a). *Cancer Res.* **18,** 952.
de Lamirande, G., Allard, C., and Cantero, A. (1958b). *J. Biophys. Biochem. Cytol.* **4,** 373.
de Lamirande, G., Boileau, S., and Morais, R. (1966). *Can. J. Biochem.* **44,** 273.
de Lamirande, G., Morais, R., and Blackstein, M. (1967). *Arch. Biochem. Biophys.* **118,** 347.
Delbrück, A., Zebe, E., and Bücher, T. (1959a). *Biochem. Z.* **331,** 273.
Delbrück, A., Schimassek, H., Bartsch, K., and Bücher, T. (1959b). *Biochem. Z.* **331,** 297.
de Robertis, E., Rodriguez de Lores Arnaiz, G., Salganicoff, L., Pellegrino de Iraldi, A., and Zieber, L. M. (1963). *J. Neurochem.* **10,** 225.
de Robertis, E., Rodriguez de Lores Arnaiz, G., Alberici, M., Butcher, R. W., and Sutherland, E. W. (1967). *J. Biol. Chem.* **242,** 3487.
de Vincentiis, M., and Testa, M. (1959). *J. Histochem. Cytochem.* **7,** 393.
Dianzani, M. U. (1954a). *G. Biochem.* **3,** 29.
Dianzani, M. U. (1954b). *Biochim. Biophys. Acta* **14,** 514.
Dianzani, M. U. (1955). *Biochim. Biophys. Acta* **17,** 391.
Dickman, S. R. (1961). *In* "The Enzymes" (P. D. Boyer, H. Lardy, and K. Myrbäck, eds.), 2nd rev. ed., Vol. 5, p. 495. Academic Press, New York.
Dickman, S. R., and Speyer, J. F. (1954). *J. Biol. Chem.* **206,** 67.
Dingle, J. T., and Dott, H. M. (1969). *Biochem. J.* **111,** 35P.

Dingle, J. T., and Fell, H. B. (1969). "Lysosomes in Biology and Pathology." North-Holland Publ., Amsterdam.

di Prisco, G., Banay-Schwartz, M., and Strecker, H. J. (1968). *Biochem. Biophys. Res. Commun.* **33**, 606.

Dodgson, K. S., Spencer, B., and Thomas, J. (1954). *Biochem. J.* **56**, 177.

Dodgson, K. S., Spencer, B., and Thomas, J. (1955). *Biochem. J.* **59**, 29.

Domenech, C. E., and Blanco, A. (1967). *Biochem. Biophys. Res. Commun.* **28**, 209.

Dominas, H., Przelecka, A., Sarzala, M. G., and Taracha, M. (1963). *Folia Histochem. Cytochem.* **1**, 203.

Dott, H. M., and Dingle, J. T. (1968). *Exp. Cell Res.* **52**, 523.

Dounce, A. L., Kay, E. R. M., and Pate, S. (1953). *Fed. Proc., Fed. Amer. Soc. Exp. Biol.* **12**, 198.

Drabikowski, W., and Rafalowska, U. (1968). *Acta Biochim. Pol.* **15**, 45.

Dresden, M., and Hoagland, M. (1965). *Science* **149**, 647.

Duggan, P. F. (1968). *Life Sci.* **7**, 1265.

Dupourque, D., and Kun, E. (1968). *Eur. J. Biochem.* **6**, 151.

Dutton, G. J., and Storey, I. D. (1954). *Biochem. J.* **57**, 275.

Dyson, J. E. D., Anderson, W. B., and Nordlie, R. C. (1969). *J. Biol. Chem.* **244**, 560.

Eaton, R. H., and Moss, D. W. (1967). *Biochem. J.* **105**, 1307.

Eggerer, H., and Remberger, U. (1963). *Biochem. Z.* **337**, 202.

Eichel, H. J. (1957a). *Proc. Soc. Exp. Biol. Med.* **95**, 38.

Eichel, H. J. (1957b). *J. Biophys. Biochem. Cytol.* **3**, 397.

Eichel, H. J., and Bukovsky, J. (1961). *Nature (London)* **191**, 243.

Emanoil-Ravicovitch, R., and Herisson-Cavet, C. (1963). *Bull. Soc. Chim. Biol.* **45**, 989.

Emery, A. J., and Dounce, A. L. (1955). *J. Biophys. Biochem. Cytol.* **1**, 315.

Emmelot, P., and Bos, C. J. (1965). *Biochim. Biophys. Acta* **99**, 578.

Emmelot, P., and Bos, C. J. (1966). *Biochim. Biophys. Acta* **121**, 434.

Emmelot, P., and Bos, C. J. (1968). *Biochim. Biophys. Acta* **150**, 341.

Endahl, B. R., and Kochakian, C. D. (1956). *Amer. J. Physiol.* **185**, 250.

Endahl, G. L., Kochakian, C. D., and Hamm, D. (1960). *J. Biol. Chem.* **235**, 2792.

Englard, S., and Brieger, H. H. (1962). *Biochim. Biophys. Acta* **56**, 571.

Entman, M. L., Levey, G. S., and Epstein, S. E. (1969). *Biochem. Biophys. Res. Commun.* **35**, 728.

Epstein, E., and Rains, D. W. (1965). *Proc. Nat. Acad. Sci. U. S.* **53**, 1320.

Ernster, L., and Lee, C. P. (1964). *Annu. Rev. Biochem.* **33**, 729.

Ernster, L., and Jones, L. C. (1962). *J. Cell Biol.* **15**, 563.

Ernster, L., and Navazio, F. (1956). *Exp. Cell Res.* **11**, 483.

Ernster, L., Siekevitz, P., and Palade, G. E. (1962). *J. Cell Res.* **15**, 541.

Errera, M. (1949). *J. Biol. Chem.* **178**, 483.

Erwin, V. G., and Hellerman, L. (1968). *J. Biol. Chem.* **242**, 4230.

Essner, E., Novikoff, A. B., and Quintana, N. (1965). *J. Cell Biol.* **25**, 201.

Estabrook, R. W., and Sactor, B. (1958). *J. Biol. Chem.* **233**, 1014.

Estabrook, R. W., and Cohen, B. (1969). *In* "Microsomes and Drug Oxidations" (J. R. Gillette *et al.*, eds.), p. 95. Academic Press, New York.

Estabrook, R. W., and Nissley, S. P. (1963). *In* "Functionelle und morphologische Organisation der Zelle" (P. Karlson, ed.), p. 119. Springer, Berlin.

Estabrook, R. W., Hildebrandt, A., Pawar, S., Cohen, B., and Masters, B. S. S. (1969). *In* "Membrane Function Electron Transfer Oxygen" (W. J. Whelan and J. Schultz, eds.), p. 288. University of Miami.

Etemadi, A. H., Popjack, G., and Cornforth, J. W. (1969). *Biochem. J.* **111**, 445.

Ewald, W., Werbin, H., and Chaikoff, I. L. (1964). *Steroids* **3**, 505.

Exton, J. H., and Park, C. R. (1966). *Pharmacol. Rev.* **18**, 181.

Exton, J. H., and Park, C. R. (1968). *Advan. Enzyme Regul.* **6**, 391.

Exton, J. H., Jefferson, L. S., Butcher, R. W., and Park, C. R. (1966). *Amer. J. Med.* **40**, 709.

Fahien, L. A., and Strmecki, M. (1969a). *Arch. Biochem. Biophys.* **130**, 478.

Fahien, L. A., and Strmecki, M. (1969b). *Arch. Biochem. Biophys.* **130**, 456.

Fahien, L. A., Strmecki, M., and Smith, S. (1969). *Arch. Biochem. Biophys.* **130**, 449.

Faihmi, H. D., and Karnowsky, M. J. (1966). *J. Cell Biol.* **29**, 112.

Fanshier, D. W., Gottwald, L. K., and Kun, E. (1962). *J. Biol. Chem.* **239**, 3588.

Farstad, M., Bremer, J., and Norum, K. R. (1967). *Biochim. Biophys. Acta* **132**, 492.

Feldman, F., and Butler, L. G. (1969). *Biochem. Biophys. Res. Commun.* **36**, 119.

Fellman, J. H., Vanbellinghen, P. J., Jones, R. T., and Koler, R. D. (1969). *Biochemistry* **8**, 615.

Fenwick, M. L. (1958). *Nature (London)* **182**, 607.

Fernández-Morán, H. (1962). *Circulation* **26**, 1039.

Fernández-Morán, H. (1963). *Science* **140**, 381.

Fernley, H. N., and Walker, P. G. (1967). *Biochem. J.* **104**, 1011.

Fessenden, J. M. and Moldave, K. (1961). *Biochem. Biophys. Res. Commun.* **6**, 232.

Feuer, G., and Golberg, L. (1967). *Food Cosmet. Toxicol.* **5**, 665.

Feuer, G., Golberg, L., and Gibson, K. I. (1966). *Food Cosmet. Toxicol.* **4**, 283.

Fiala, E. S. (1967). *Experientia* **23**, 597.

Fillios, L. C., Yokono, O., Pronczuk, A., Gore, I., Satoh, T., and Koboayakawa, K. (1969). *J. Nutr.* **98**, 105.

Finkenstaedt, J. T. (1957). *Proc. Soc. Exp. Biol. Med.* **95**, 302.

Fischer, A. G., Schulz, A. R., and Oliner, L. (1968). *Endocrinology* **82**, 1098.

Fischer, E. H., and Stein, E. A. (1960). *In* "The Enzymes" (P. D. Boyer, H. Lardy, and K. Myrbäck, eds.), 2nd rev. ed., Vol. 4, p. 313. Academic Press, New York.

Fischer, F., Siebert, G., and Adloff, E. (1959). *Biochem. Z.* **332**, 131.

Fisher, C. J., and Stetten, M. R. (1966). *Biochim. Biophys. Acta* **121**, 102.

Fisher, D., and Kent, P. W. (1969). *Biochem. J.* **115**, 50P.

Fisher, D., Whitehouse, M. W., and Kent, P. W. (1967). *Nature (London)* **213**, 204.

Fishman, W. H. (1961). "Chemistry of Drug Metabolism." Thomas, Springfield, Illinois.

Fleischer, S., Brierley, G., Klouwen, H., and Slautterback, D. B. (1962). *J. Biol. Chem.* **237**, 3264.

Fleischer, B., Casu, P., and Fleischer, S. (1966). *Biochem. Biophys. Res. Commun.* **24**, 189.

Foster, D. W., and McWhorter, W. P. (1969). *J. Biol. Chem.* **244**, 260.

Foster, J. M. (1956). *J. Neurochem.* **1**, 84.

Fouts, J. R. (1961). *Biochem. Biophys. Res. Commun.* **6**, 373.

Fouts, J. R., Kamm, J. J., and Brodie, B. B. (1957). *J. Pharmacol. Exp. Ther.* **120**, 291.

Fouts, J. R., Rogers, L. A., and Gram, T. E. (1966). *Exp. Mol. Pathol.* **5**, 475.

Fowler, S. (1969). *Biochim. Biophys. Acta* **191**, 481.

Fox, C. F., and Kennedy, E. P. (1965). *Proc. Nat. Acad. Sci. U.S.* **54**, 891.

Frankland, D. M., and Wynn, C. H. (1962). *Biochem. J.* **84**, 20P.

Frantz, I. D., and Schroepfer, G. J. (1967). *Annu. Rev. Biochem.* **36**: 691.

Frantz, I. D., Davidson, A. G., Dulit, E., and Mobberley, M. L. (1959). *J. Biol. Chem.* **234**, 2290.

French, A. P., and Warren, J. C. (1967). *Biochem. J.* **105**, 233.

Frisell, W. R., Patwardhan, M. V., and Mackenzie, C. G. (1965). *J. Biol. Chem.* **240**, 1829.

Frontali, N., and Toschi, G. (1958). *Exp. Cell Res.* **15**, 446.

Fruton, J. S. (1960). *In* "The Enzymes" (P. D. Boyer, H. Lardy, and K. Myrbäck, eds.), 2nd rev. ed., Vol. 4, p. 233. Academic Press, New York.

Fujioka, M. (1969). *Biochim. Biophys. Acta* **185**, 228.

Fujita, M., Nagano, K., Mizuno, N., Tashima, Y., Nakao, T., and Nakao, M. (1968). *Biochem. J.* **106**, 113.

Fuller, R. W. (1968). *Arch. Int. Pharmacodyn. Ther.* **174**, 32.

Furth, A. J., and Robinson, D. (1965). *Biochem. J.* **97**, 59.

Futai, M., Miyato, S., and Mizuno, D. (1969). *J. Biol. Chem.* **244**, 4951.

Gainor, C., and Phillips, P. G. (1969). *Physiol. Plant.* **22**, 801.

Galzigna, L., Rossi, C. R., Sartorelli, L., and Gibson, D. M. (1967). *J. Biol. Chem.* **242**, 2111.

Ganguly, J. (1954). *Arch. Biochem. Biophys.* **52**, 186.

Ganguly, J., and Devel, H. J. (1953). *Nature (London)* **172**, 120.

Garland, P. B., Shepherd, D., Nicholls, D. G., Yates, D. W., and Annlight, P. (1969). *In* "Citric Acid Cycle" (J. M. Lowenstein, ed.), p. 163. Marcel Dekker, New York.

Gasior, E., and Moldave, K. (1965a). *J. Biol. Chem.* **240**, 3346.

Gasior, E., and Moldave, K. (1965b). *Biochim. Biophys. Acta* **95**, 679.

Gaull, G., and Villee, C. A. (1960). *Biochim. Biophys. Acta* **39**, 560.

Ghosh, N. C., Kar, N. C., and Chatterjee, I. (1963). *Nature (London)* **197**, 596.

Ghosh, S. K., and Ghosh, J. J. (1968). *J. Neurochem.* **15**, 1375.

Gianetto, R., and de Duve, C. (1955). *Biochem. J.* **59**, 433.

Gianetto, R., and Viala, R. (1955). *Science* **121**, 801.

Gibian, H. (1966). *In* "The Amino Sugars" (R. W. Jeanloz and E. A. Balazs, eds.), Vol. 2 B, p. 181. Academic Press, New York.

Gibson, K., and Harris, P. (1968). *Cardiovasc. Res.* **4**, 367.

Gibson, K., and Harris, P. (1969). *Biochem. Biophys. Res. Commun.* **35**, 75.

Gibson, K. D., Wilson, J. D., and Udenfriend, S. (1961). *J. Biol. Chem.* **236**, 673.

Gigon, P. L., Gram, T. E., and Gillette, J. R. (1968). *Biochem. Biophys. Res. Commun.* **31**, 558.

Gilbert, D., and Golberg, L. (1967). *Food Cosmet. Toxicol.* **5**, 481.

Gillette, J. R. (1963). *Progr. Drug. Res.* **6**, 13.

Gillette, J. R. (1966). *Advan. Pharmacol.* **4**, 219.

Gillette, J. R. (1969). *Ann. N. Y. Acad. Sci.* **160**, 558.

Gillette, J. R., and Gram, T. E. (1969). *In* "Microsomes and Drug Oxidation" (J. R. Gillette *et al.*, eds.), p. 133. Academic Press, New York.

Gillette, J. R., Brodie, B. B., and La Du, B. N. (1957). *J. Pharmacol. Exp. Ther.* **119**, 532.

Gillette, J. R., Conney, A. H., Cosmides, G. J., Estabrook, R. W., Fouts, J. R., and Mannering, G. J. (eds.) (1969). "Microsomes and Drug Oxidations." Academic Press, New York.

Ginsburg, V., and Hers, H. G. (1960). *Biochim. Biophys. Acta* **38**, 427.

Ginsburg, V., and Neufeld, E. F. (1969). *Annu. Rev. Biochem.* **38**, 371.

Giuditta, A., and Strecker, H. J. (1959). *J. Neurochem.* **5**, 50.

Glaser, L., and Brown, D. H. (1957a). *Biochim. Biophys. Acta* **23**, 449.

Glaser, L., and Brown, D. H. (1957b). *J. Biol. Chem.* **228**, 729.

Glock, G. E., and McLean, P. (1953). *Biochem. J.* **55**, 400.

Glynn, I. M. (1968). *Brit. Med. Bull.* **24,** 165.
Goffeau, A., Pedersen, P. L., and Lehninger, A. L. (1967). *J. Biol. Chem.* **242,** 1845.
Goffeau, A., Pedersen, P. L., and Lehninger, A. L. (1968). *J. Biol. Chem.* **243,** 1685.
Goldfarb, P. S. G., and Rodnight, R. (1969). *Biochem. J.* **111,** 22P.
Goldfischer, S. (1964). *J. Neuropathol. Exp. Neurol.* **23,** 36.
Goodlad, G. A., and Mills, G. T. (1957). *Biochem. J.* **66,** 346.
Goodwin, T. W. (ed.) (1966). "The Biochemistry of Chloroplasts." Academic Press, New York.
Gorkin, V. Z., and Akopyan, Zh. I. (1968). *Experientia* **24,** 1115.
Gotterer, G. S. (1964). Ph.D. Dissertation, Johns Hopkins University, Baltimore, Maryland.
Gotterer, G. S. (1967). *Biochemistry* **6,** 2139.
Goutier, R., and Goutier-Pirotte, M. (1955a). *Biochim. Biophys. Acta* **16,** 558.
Goutier, R., and Goutier-Pirotte, M. (1955b). *Biochim. Biophys. Acta* **16,** 361.
Goz, B. (1967). *Biochem. Pharmacol.* **16,** 593.
Graham, A. B., and Park, M. V. (1969). *Biochem. J.* **111,** 257.
Graham, A. B., and Wood, G. C. (1969). *Biochem. Biophys. Res. Commun.* **37,** 567.
Graham-Smith, D. G., Butcher, R. W., Ney, R. L., and Sutherland, E. W. (1967). *J. Biol. Chem.* **242,** 5535.
Gram, T. E., Guarino, A. M., Greene, F. E., Gigon, P. L., and Gillette, J. R. (1968a). *Biochem. Pharmacol.* **17,** 1769.
Gram, T. E., Hansen, A. R., and Fouts, J. R. (1968b). *Biochem. J.* **106,** 587.
Gray, E. G., and Whittaker, V. P. (1960). *J. Physiol. (London)* **153,** 35P.
Gray, G. D. (1966). *Arch. Biochem. Biophys.* **113,** 502.
Green, A. L., and Taylor, C. B. (1964). *Biochem. Biophys. Res. Commun.* **14,** 118.
Green, D. E. (1954). *In* "Chemical Pathways of Metabolism" (D. M. Greenberg, ed.), Vol. 1, p. 27. Academic Press, New York.
Green, D. E. (1967). *In* "Mitochondrial Structure and Compartmentation" (E. Quagliariello *et al.,* eds.), pp. 118, 121 Adriatica Editrice, Bari.
Green, D. E., and Baum, H. (1970). "Energy and the Mitochondrion." Academic Press, New York.
Green, D. E., and MacLennan, D. H. (1969). *BioScience* **19,** 213.
Green, D. E., and Perdue, J. F. (1966). *Ann. N. Y. Acad. Sci.* **137,** 667.
Green, D. E., and Tzagoloff, A. (1966). *Arch. Biochem. Biophys.* **116,** 293.
Green, D. E., Bachmann, E., Allmann, D. W., and Perdue, J. F. (1966). *Arch. Biochem. Biophys.* **115,** 172.
Green, D. E., Asai, J., Harris, R. A., and Penniston, J. T. (1968a). *Arch. Biochim. Biophys.* **125,** 684.
Green, D. E., Allman, D. W., Harris, R. A., and Tan, W. C. (1968b). *Biochem. Biophys. Res. Commun.* **31,** 368.
Greenbaum, A. L., Slater, T. F., and Wang, D. Y. (1960). *Nature (London)* **188,** 319.
Greenfield, P. C., and Boell, E. J. (1968). *J. Exp. Zool.* **168,** 491.
Greville, G. D. (1969). *In* "Citric Acid Cycle" (J. M. Lowenstein, ed.), p. 1. Marcel Dekker, New York.
Grillo, M. A., and Bedino, S. (1968). *Enzymologia* **35,** 1.
Grimm, F. C., and Doherty, D. V. (1961). *J. Biol. Chem.* **236,** 1980.
Groot, G. S. P., and Meyer, M. (1969). *Biochim. Biophys. Acta* **180,** 575.
Grisolia, S., and Cohen, P. P. (1953). *J. Biol. Chem.* **204,** 753.
Guarino, A. M., Gram, T. E., Gigon, P. L., Greene, F. E., and Gillette, J. R. (1969). *Mol. Pharmacol.* **5,** 131.
Guha, A., England, S., and Listowsky, I. (1968). *J. Biol. Chem.* **243,** 609.

Gul, B., and Dils, R. (1969). *Biochem. J.* **111**, 263.

Gurtoo, H. L., Campbell, T. C., Webb, R. E., and Plowman, K. M. (1968). *Biochem. Biophys. Res. Commun.* **31**, 588.

Güttler, F. (1967). *Protides Biol. Fluids, Proc. Colloq.* **15**, 167.

Güttler, F., and Clausen, J. (1967). *Enzymol. Biol. Clin.* **8**, 456.

Hackenbrock, C. R. (1966). *J. Cell Biol.* **30**, 269.

Hackenbrock, C. R. (1968). *J. Cell Biol.* **37**, 345.

Hadley, M. E., and Goldman, J. M. (1969). *Brit. J. Pharmacol.* **37**, 650.

Hafkenscheid, J. C. M., and Bonting, S. L. (1968). *Biochim. Biophys. Acta* **151**, 204.

Hagen, P. (1955). *J. Physiol. (London)* **129**, 50.

Hall, D. O., and Whatley, F. R. (1967). *In* "Enzyme Cytology" (D. B. Roodyn, ed.), p. 181. Academic Press, New York.

Hall, J. L. (1969). *Planta* **85**, 105.

Hall, P. F., and Koritz, S. B. (1965). *Biochemistry* **4**, 1037.

Hallaway, M. (1965). *Biol. Rev. Cambridge Phil. Soc.* **40**, 188.

Hallaway, M. (1968). *In* "Plant Cell Organelles" (J. B. Pridham, ed.), p. 1. Academic Press, New York.

Hansen, J. N., Dinovo. E. C., and Boyer, P. D. (1969). *J. Biol. Chem.* **244**, 6270.

Hanson, T. L., and Fromm, H. J. (1967). *J. Biol. Chem.* **242**, 501.

Hanson, T. L., and Nordlie, R. (1970). *Biochim. Biophys. Acta* **198**, 66.

Harbury, A. J., and Foley, K. A. (1958). *Proc. Nat. Acad. Sci. U. S.* **44**, 662.

Harlan, W. R., and Wakil, S. J. (1962). *Biochem. Biophys. Res. Commun.* **8**, 131.

Harlan, W. R., and Wakil, S. J. (1963). *J. Biochem.* **238**, 3216.

Harold, F. M., and Baarda, J. R. (1969). *J. Biol. Chem.* **244**, 2261.

Harris, J. B., and Alonso, D. (1965). *Fed. Proc., Fed. Amer. Soc. Exp. Biol.* **24**, 1368.

Harris, R. A., Penniston, J. T., Asai, J., and Green, D. E. (1968). *Proc. Nat. Acad. Sci. U. S.* **59**, 830.

Hartman, B., Kloepter, H., and Yasunobu, K. (1969). *Fed. Proc., Fed. Amer. Soc. Exp. Biol.* **28**, 857.

Hatefi, Y., and Stempel, K. E. (1969). *J. Biol. Chem.* **244**, 2350.

Hatefi, Y., Stempfel, K. E., and Hanstein, W. G. (1969). *J. Biol. Chem.* **244**, 2358.

Haugaard, N., and Hess, M. E. (1965). *Pharmacol. Rev.* **17**, 27.

Hawkins, J. (1952). *Biochem. J.* **50**, 577.

Hayaishi, O. (1969). *Annu. Rev. Biochem.* **38**, 21.

Hayase, K., and Tappel, A. L. (1969). *J. Biol. Chem.* **244**, 2269.

Hayashi, K., Kugimiya, M., and Funatsu, M. (1968). *J. Biochem. (Tokyo)* **63**, 93.

Haynes, R. C., Sutherland, E. W., and Rall, T. W. (1960). *Recent Progr. Horm. Res.* **16**, 121.

Hebb, C. O., and Whittaker, V. P. (1958). *J. Physiol. (London)* **142**, 187.

Hecht, L. I., and Novikoff, A. B. (1952). *Cancer Res.* **12**, 269.

Hechter, O., Yoshinaga, K.. Halkerston, D. D. K., and Birchall, K. (1967). *Arch. Biochem. Biophys.* **122**, 449.

Hegyvary, C. G., and Post, R. L. (1969). *In* "The Molecular Basis of Membrane Function" (D. C. Tosteson, ed.), p. 519. Prentice-Hall, Englewood Cliffs, New Jersey.

Heinz, E. (1967). *Annu. Rev. Physiol.* **29**, 21.

Heller, L., and Bargoni, N. (1950). *Ark. Kemi* **1**, 447.

Hellerman, L., and Erwin, V. G. (1968). *J. Biol. Chem.* **243**, 5234.

Helminen, H. J., Ericcson, J. L. E., and Orrenius, S. (1968). *J. Ultrastruct. Res.* **25**, 240.

Henderson, N. S. (1966). *Arch. Biochem. Biophys.* **117**, 28.

Henderson, N. S. (1968). *Ann. N. Y. Acad. Sci.* **151,** 429.

Henley, D. D., Lindahl, T., and Fresco, J. R. (1966). *Proc. Nat. Acad. Sci. U. S.* **55,** 191.

Heppel, L. A., Ortiz, P., and Ochoa, S. (1956). *Science* **123,** 415.

Hernandez, A., and Crane, R. K. (1966). *Arch. Biochem. Biophys.* **113,** 223.

Hernandez, P. H., Gillette, J. R., and Mazel, P. (1967a). *Biochem. Pharmacol.* **16,** 1859.

Hernandez, P. H., Mazel, P., and Gillette, J. R. (1967b). *Biochem. Pharmacol.* **16,** 1877.

Hers, H. G., and de Duve, C. (1950). *Bull. Soc. Chim. Biol.* **32,** 20.

Hers, H. G., Berthet, J., Berthet, L., and de Duve, C. (1951). *Bull. Soc. Chim. Biol.* **33,** 21.

Heymann, E., Krisch, K., and Pahlich, E. (1969). *Hoppe-Seyler's Z. Physiol. Chem.* **350,** 1177.

Higashi, T., and Peters, T. (1963a). *J. Biol. Chem.* **238,** 3945.

Higashi, T., and Peters, T. (1963b). *J. Biol. Chem.* **238,** 3952.

Higashi, T., Maruyama, E., Otani, T., and Sakamoto, Y. (1965). *J. Biochem.* (*Tokyo*) **57,** 793.

Higgins, J. A., and Barrnett, R. J. (1970). *J. Cell Sci.* **6,** 29.

Hill, R. J., and Bradshaw, R. A. (1969). *Methods Enzymol.* **13,** 91.

Hird, F. J. R., and Rowsell, E. V. (1950). *Nature* (*London*) **166,** 517.

Hirsch-Hoffmann, A. M., Holzel, F., and Maass, H. (1964). *Naturwissenschaften* **51,** 414.

Hoffman, D. G., Worth, H. M., and Anderson, R. C. (1968). *Toxicol. Appl. Pharmacol.* **12,** 464.

Hogeboom, G. H. (1949). *J. Biol. Chem.* **177,** 847.

Hogeboom, G. H. (1950). *J. Nat. Cancer Inst.* **10,** 983.

Hogeboom, G. H., and Schneider, W. C. (1950). *J. Biol. Chem.* **186,** 417.

Hogeboom, G. H., and Schneider, W. C. (1953). *J. Biol. Chem.* **204,** 233.

Hogeboom, G. H., Schneider, W. C., and Palade, G. E. (1948). *J. Biol. Chem.* **172,** 619.

Hogeboom, G. H., Schneider, W. C., and Striebich, M. J. (1952). *J. Biol. Chem.* **196,** 111.

Hogeboom, G. H., Schneider, W. C., and Striebich, M. J. (1953). *Cancer Res.* **13,** 617.

Hogeboom, G. H., Kuff, E. L. and Schneider, W. C. (1957). *Int. Rev. Cytol.* **6,** 425.

Hokin, L. E., (1955). *Biochim. Biophys. Acta* **18,** 379.

Hokin, L. E., and Reasa, D. (1964). *Biochim. Biophys. Acta* **90,** 176.

Hokin, L. E., and Yoda, A. (1964). *Biochemistry* **52,** 454.

Hokin, L. E., and Yoda, A. (1965). *Biochim. Biophys. Acta* **97,** 594.

Holcomb, G. E., Hildebrandt, A. C., and Evert, R. F. (1967). *Amer. J. Bot.* **54,** 1204.

Holden, H. F., and Lemberg, R. (1939). *Aust. J. Exp. Biol. Med. Sci.* **17,** 133.

Holland, D. L., and Perry, S. V. (1969). *Biochem. J.* **114,** 161.

Holley, R. W., Apgar, J., Everett, G. A., Madison, J. T., Marquisee, M., Merrill, S. H., Penswick, J. R., and Zamir, A. (1965). *Science* **147,** 1462.

Holter, H. (1954). *Proc. Roy. Soc. Ser. B* **142,** 140.

Holter, H. (1955). *In* "Symposium on Fine Structure of Cells," p. 71. Noordhoff, Gröningen.

Honova, E., Drahota, Z., and Hahn, P. (1967). *Experientia* **23,** 632.

Hook, R. H., and Vestling, C. S. (1962). *Biochim. Biophys. Acta* **65,** 358.

Hoppel, C., and Cooper, C. (1968). *Biochem. J.* **107,** 367.

Horgan, D. J., and Singer, T. P. (1968). *J. Biol. Chem.* **243**, 834.

Hosie, R. J. A. (1965). *Biochem. J.* **96**, 404.

Howard, C. F. (1970). *J. Biol. Chem.* **245**, 462.

Howard, J. B., and Glazer, A. N. (1967). *J. Biol. Chem.* **242**, 5715.

Howell, R. R., Loeb, J. N., and Tomkins, G. M. (1964). *Proc. Nat. Acad. Sci. U.S.* **52**, 1241.

Hruban, Z., and Recheigl, M., Jr. (1969). "Microbodies and Related Particles," pp. 126–201. Academic Press, New York.

Hübscher, G. (1962). *Biochim. Biophys. Acta* **57**, 555.

Hübscher, G., Clark, D., Webb, M. E., and Sherratt, H. S. A. (1963). *In* "Biochemical Problems of Lipids" (A. C. Frazer, ed.), p. 201. Elsevier, Amsterdam.

Hübscher, G., Smith, M. E., and Gurr, M. I. (1964). *In* "Metabolism and Physiological Significance of Lipids" (R. M. C. Dawson and D. N. Rhodes, eds.), p. 229. Wiley, New York.

Hulsmann, W. C. (1960). *Biochim. Biophys. Acta* **45**, 623.

Hulsmann, W. C. (1962). *Biochim. Biophys. Acta* **58**, 417.

Hultin, T. (1957). *Exp. Cell Res.* **13**, 47.

Hultin, T. (1959). *Exp. Cell Res.* **18**, 112.

Hurlock, B., and Talalay, P. (1959). *Arch. Biochem. Biophys.* **80**, 468.

Hussein, K. A., and Kochakian, C. D. (1968). *Acta Endocrinol. (Copenhagen)* **59**, 459.

Hutterer, F. (1966). *Biochim. Biophys. Acta* **115**, 312.

Ichikawa, Y., and Yamano, T. (1969). *J. Biochem. (Tokyo)* **66**, 351.

Ide, H., and Fishman, W. H. (1969). *Histochemie* **20**, 300.

Igaue, I., Gomes, B., and Yasunobu, K. T. (1967). *Biochem. Biophys. Res. Commun.* **29**, 562.

Ikeda, I., and Hamaguchi, K. (1969). *J. Biochem. (Tokyo)* **66**, 513.

Ikemoto, N., Sreter, F. A., Nakamura, A., and Gergely, J. (1968). *J. Ultrastruct. Res.* **23**, 216.

Imai, Y., and Sato, R. (1966a). *Biochem. Biophys. Res. Commun.* **23**, 5.

Imai, Y., and Sato, R. (1966b). *Biochem. Biophys. Res. Commun.* **22**, 620.

Imai, Y., and Sato, R. (1968a). *J. Biochem. (Tokyo)* **64**, 147.

Imai, Y., and Sato, R. (1968b). *J. Biochem. (Tokyo)* **63**, 380.

Inano, H., Machino, A., and Tamaoki, B. (1969). *Steroids* **13**, 357.

Inturrisi, C. E. (1969). *Biochim. Biophys. Acta* **173**, 569.

Iodice, A. A. (1967). *Arch. Biochem. Biophys.* **121**, 241.

Iodice, A. A., Leong, V., and Weinstock, I. M. (1966). *Arch. Biochem. Biophys.* **117**, 477.

Irias, J. J., Olmsted, M. R., and Utter, M. F. (1969). *Biochemistry* **8**, 5136.

Isherwood, F. A., Mapson, L. W., and Chen, Y. T. (1960). *Biochem. J.* **76**, 157.

Ishida, M., and Mizushima, S. (1969a). *J. Biochem. (Tokyo)* **66**, 33.

Ishida, M., and Mizushima, S. (1969b). *J. Biochem. (Tokyo)* **66**, 133.

Ishikawa, E., Oliver, R. M., and Reed, L. J. (1966). *Proc. Nat. Acad. Sci. U. S.* **56**, 534.

Israel, Y. (1969). *In* "The Molecular Basis of Membrane Function" (D. C. Tosteson, ed.), p. 529. Prentice-Hall, Englewood Cliffs, New Jersey.

Israel, Y., and Salazar, I. (1967). *Arch. Biochem. Biophys.* **122**, 310.

Israel, Y., and Titus, E. (1967). *Biochim. Biophys. Acta* **139**, 450.

Israel, Y., Kalant, H., and Laufer, I. (1965). *Biochem. Pharmacol.* **14**, 1803.

Jackson, C., and Black, R. E. (1967). *Biol. Bull.* **132**, 1.

Jacobson, K. B., and Kaplan, N. O. (1957a). *J. Biol. Chem.* **226**, 427.

Jacobson, K. B., and Kaplan, N. O. (1957b). *J. Biophys. Biochem. Cytol.* **3**, 31.

Jacques, P. (1968). *Biochem. J.* **111**, 25P.

Jaffe, H., Fuji, K., Sengupta, M., Guerin, H., and Epstein, S. S. (1968). *Life Sci.* **7**, 1051.

Jakobsson, S. V., and Dallner, G. (1968). *Biochim. Biophys. Acta* **165**, 380.

James, L. K., and Augenstein, L. G. (1967). *Advan. Enzymol.* **28**, 1.

Jarnefelt, J. (1964a). *Biochem. Biophys. Res. Commun.* **17**, 330.

Jarnefelt, J. (1964b). *Abstr. 6th Int. Congr. Biochem.*, 1964 p. 613.

Jarrott, B., and Iversen, L. L. (1968). *Biochem. Pharmacol.* **17**, 1619.

Jean, D. H., and Bader, H. (1967). *Biochem. Biophys. Res. Commun.* **27**, 650.

Jeanloz, R. W., Sharon, N., and Flowers, H. M. (1963). *Biochem. Biophys. Res. Commun.* **13**, 20.

Jefcoate, C. R. E., and Gaylor, J. L. (1969). *J. Amer. Chem. Soc.* **91**, 4610.

Jerina, D. M., Daly, J. W., Witkop, B., Zaltzman-Nirenberg, P., and Udenfriend, S. (1970). *Biochemistry* **9**, 147.

Johnson, M. K. (1960). *Biochem. J.* **77**, 610.

Johnston, J. M. (1959). *J. Biol. Chem.* **234**, 1965.

Johnston, J. M., Rao, G. A., Lowe, P. A., and Schwarz, B. E. (1967). *Lipids* **2**, 14.

Jòlles, P. (1960). *In* "The Enzymes" (P. D. Boyer, H. Lardy, and K. Myrbäck, eds.), 2nd rev. ed., Vol. 4, p. 431. Academic Press, New York.

Jones, M. S., and Jones, O. T. G. (1968). *Biochem. Biophys. Res. Commun.* **31**, 977.

Jones, P. D., and Wakil, S. J. (1967). *J. Biol. Chem.* **242**, 5267.

Jordan, W. K., and March, R. (1956). *J. Histochem. Cytochem.* **4**, 301.

Jorgensen, P. L., and Skou, J. C. (1969). *Biochem. Biophys. Res. Commun.* **37**, 39.

Josefsson, L. I., and Lagerstedt, S. (1962). *Methods Biochem. Anal.* **9**, 39.

Joshi, V. C., Kurup, C. K. R., and Ramasarma, T. (1969). *Biochem. J.* **111**, 297.

Jost, J. P., Hsie, A. W., and Rickenberg, H. V. (1969). *Biochem. Biophys. Res. Commun.* **34**, 748.

Jost, J. P., Hsie, A. W., Hughes, D., and Ryan, L. (1970). *J. Biol. Chem.* **245**, 351.

Jurtshuk, P., Jr., Sekuzu, I., and Green, D. E. (1963). *J. Biol. Chem.* **238**, 3595.

Kaback, H. R. (1969). *In* "The Molecular Basis of Membrane Function" (D. C. Tosteson, ed.), p. 421. Prentice-Hall, Englewood Cliffs, New Jersey.

Kabat, E. A. (1941). *Science* **93**, 43.

Kafer, E., and Pollack, J. K. (1961). *Exp. Cell Res.* **22**, 120.

Kagawa, Y. (1969). *J. Biochem. (Tokyo)* **65**, 925.

Kagawa, Y., and Racker, E. (1966a). *J. Biol. Chem.* **241**, 2467.

Kagawa, Y., and Racker, E. (1966b). *J. Biol. Chem.* **241**, 2475.

Kahlenberg, A., Dulak, N. C., Dixon, J. F., Galsworthy, P. R., and Hokin, L. E. (1969). *Arch. Biochem. Biophys.* **131**, 253.

Kalina, M., Weavers, B., and Pearse, A. G. E. (1969). *Nature (London)* **221**, 479.

Kallman, F. G. (1951). *J. Cell Comp. Physiol.* **38**, 137.

Kamin, H., Masters, B. S. S., Gibson, Q. H., and Williams, C. H. (1965). *Fed. Proc., Fed. Amer. Soc. Exp. Biol.* **24**, 1164.

Kamin, H., Masters, B. S. S., and Gibson, Q. H. (1966). *In* "Flavins and Flavoproteins" (E. C. Slater, ed.), pp. 306–324. Elsevier, Amsterdam.

Kandutsch, A. A. (1967). *Steroids* **10**, 31.

Kandutsch, A. A., and Saucier, S. E. (1969). *Arch. Biochem. Biophys.* **135**, 201.

Kanfer, J., Burns, J. J., and Ashwell, G. (1959). *Biochim. Biophys. Acta* **31**, 556.

Kaplan, N. M. (1965). *J. Clin. Invest.* **44**, 2029.

Kaplan, N. O. (1968). *Ann. N. Y. Acad. Sci.* **151**, 382.

Kaplan, N. O., Colowick, S. P., and Neufeld, E. F. (1953). *J. Biol. Chem.* **205**, 1.

Kar, N. C., Chatterjee, I. B., Ghosh, N. C., and Guha, B. C. (1962). *Biochem. J.* **84,** 16.

Karaboyas, G. C., and Koritz, S. B. (1965). *Biochemistry* **4,** 462.

Kasbekar, D. K., Lavate, W. V., Rege, D. V., and Sreenivasan, A. (1959). *Biochem. J.* **72,** 374.

Kato, R., and Takahashi, A. (1968). *Mol. Pharmacol.* **4,** 109.

Kato, R., and Takanaka, A. (1968). *Jap. J. Pharmacol.* **18,** 381.

Kato, R., Takanaka, A., and Takayanagi, M. (1968). *Jap. J. Pharmacol.* **18,** 482.

Kato, R., Takayanagi, M., and Oshima, T. (1969). *Jap. J. Pharmacol.* **19,** 53.

Kawai, K. (1960). *Biochim. Biophys. Acta* **44,** 202.

Keech, D. B., and Utter, M. F. (1962). *J. Biol. Chem.* **238,** 2609.

Keller, P. J., and Cohen, E. (1961). *J. Biol. Chem.* **236,** 1407.

Kennedy, E. P. (1961). *Fed. Proc., Fed. Amer. Soc. Exp. Biol.* **20,** 934.

Kennedy, E. P., and Lehninger, A. L. (1948). *J. Biol. Chem.* **172,** 847.

Kennedy, K. G., and Nayler, W. G. (1965). *Biochim. Biophys. Acta* **110,** 174.

Kensler, C. J., and Langemann, H. (1951). *J. Biol. Chem.* **192,** 551.

Kepes, A. (1969). *In* "The Molecular Basis of Membrane Function" (D. C. Tosteson, ed.), p. 353. Prentice-Hall, Englewood Cliffs, New Jersey.

Kepner, G. R., and Macey, R. I. (1968). *Biochem. Biophys. Res. Commun.* **30,** 582.

Kepner, G. R., and Macey, R. I. (1969). *Biochim. Biophys. Acta* **183,** 241.

Khairallah, E. A., and Pitot, H. C. (1967). *Biochem. Biophys. Res. Commun.* **29,** 269.

Kielley, R. K., and Schneider, W. C. (1950). *J. Biol. Chem.* **185,** 869.

Kielley, W. W., and Kielley, R. K. (1953). *J. Biol. Chem.* **200,** 213.

Kiessling, K. H., and Tilander, K. (1960). *Biochim. Biophys. Acta* **43,** 335.

Kikuchi, G., Ramirez, J., and Barron, E. S. G. (1959). *Biochim. Biophys. Acta* **36,** 335.

Kimhi, Y., and Littauer, U. Z. (1967). *Biochemistry* **6,** 2066.

Kimmich, G. A., and Rasmussen, H. (1969). *J. Biol. Chem.* **244,** 190.

King, C. M., and Gutmann, H. R. (1964). *Cancer Res.* **24,** 770.

King, T. E., Kuboyama, M., and Takemorei, S. (1965). *In* "Oxidases and Related Redox Systems" (T. E. King, H. S. Mason, and M. Morrison, eds.), Vol. 2, p. 707. Wiley, New York.

King, T. E., Bayley, P. M., and Mackler, B. (1969). *J. Biol. Chem.* **244,** 1890.

Kirchner, J., Watson, J. G., and Chaykin, S. (1966). *J. Biol. Chem.* **241,** 953.

Kitada, J. (1959). *Dobutsugaku Zasshi* **68,** 245.

Kitto, G. B., and Kaplan, N. O. (1966). *Biochemistry* **5,** 3966.

Kitto, G. B., Wassarman, P. M., Michjeda, J., and Kaplan, N. O. (1966). *Biochem. Biophys. Res. Commun.* **22,** 75.

Kitto, G. B., Stolzebach, F. E., and Kaplan, N. O. (1970). *Biochem. Biophys. Res. Commun.* **38,** 31.

Klainer, L. M., Chi, Y. M., Friedberg, S. L., Rall, T. W., and Sutherland, E. W. (1962). *J. Biol. Chem.* **237,** 1239.

Klingenberg, M., and Pette, D. (1962). *Biochem. Biophys. Res. Commun.* **7,** 430.

Klingenberg, M., and Schollmeyer, P. (1963). *Proc. 5th Int. Congr. Biochem., 1961* Vol. V, p. 46.

Klingenberg, M., and Slenczka, W. (1959). *Biochem. J.* **331,** 486.

Koerner, D. R. (1969). *Biochim. Biophys. Acta* **176,** 377.

Kohen, E., Seibert, G., and Kohen, C. (1964). *Histochemie* **3,** 477.

Koide, S. S., and Torres, M. T. (1965). *Biochim. Biophys. Acta* **105,** 115.

Kondo, Y. (1966). *J. Phys. Soc. Jap.* **28,** 108.

Korner, A., and Tarver, H. (1958). *J. Gen. Physiol.* **41,** 219.

Koskow, D. P., and Rose, I. A. (1968). *J. Biol. Chem.* **243,** 3623.

Krakow, J. S., Coutsogeorgopoulos, C., and Canellakis, E. S. (1962). *Biochim. Biophys. Acta* **55,** 639.

Kramer, R., Muller, M., and Salvenmoser, F. (1968). *Biochim. Biophys. Asta* **162,** 289.

Kratz, F. (1968). *Biochim. Biophys. Acta* **165,** 176.

Krause, R. F., and Powell, L. T. (1953). *Arch. Biochem. Biophys.* **44,** 57.

Krebs, E. G., DeLange, R. J., Kemp, R. G., and Riley, W. D. (1966). *Pharmacol. Rev.* **18,** 163.

Krebs, H. A., Benett, D. A. H., de Gasquet, P., Gascoyne, T., and Yoshida, T. (1963). *Biochem. J.* **86,** 22.

Kretchmer, N., and Dickerman, H. W. (1954). *J. Exp. Med.* **99,** 629.

Krisch, K. (1963). *Biochem. Z.* **337,** 531.

Krishnamurthy, S., Seshadri Sastri, P., and Ganguly, J. (1958). *Arch. Biochem. Biophys.* **75,** 6.

Kronman, M. J. (1968). *Biochem. Biophys. Res. Commun.* **33,** 535.

Kruskemper, H. L., Corchielli, E., and Ringold, H. G. (1964). *Steroids* **3,** 295.

Kuff, E. L. (1954). *J. Biol. Chem.* **207,** 361.

Kuff, E. L., and Schneider, W. C. (1954). *J. Biol. Chem.* **206,** 677.

Kuff, E. L., Hogeboom, G. H., and Dalton, A. J. (1956). *J. Biophys. Biochem. Cytol.* **2,** 33.

Kun, E., and Fanshier, D. W. (1959). *Biochim. Biophys. Acta* **32,** 338.

Kun, E., Talalay, P., and Williams-Ashman, H. G. (1951). *Cancer Res.* **11,** 855.

Kundig, W., Ghosh, S., and Roseman, S. (1964). *Proc. Nat. Acad. Sci. U.S.* **52,** 1067.

Kuo, J. F., and Greengard, P. (1969). *J. Biol. Chem.* **244,** 3417.

Kurnick, N. B. (1962). *Methods Biochem. Anal.* **9,** 1.

Kurokawa, M., Sakamoto, T., and Kato, M. (1965). *Biochem. J.* **97,** 833.

La Bella, F. S., and Brown, J. H. U. (1958). *J. Biophys. Biochem. Cytol.* **4,** 833.

Lands, W. E., and Merkl, I. (1963). *J. Biol. Chem.* **238,** 898.

Lang, K., and Lang, H. (1958). *Biochem. Z.* **329,** 577.

Langan, T. A. (1969). *J. Biol. Chem.* **244,** 5763.

Langdon, R. G. (1957). *J. Biol. Chem.* **226,** 615.

Langmuir, I., and Schaefer, V. J. (1938). *J. Amer. Chem. Soc.* **60,** 1351.

Lapresle, C., and Webb, T. (1962). *Biochem. J.* **84,** 455.

Lardy, H. A., and Ferguson, S. M. (1969). *Ann. Rev. Biochem.* **38,** 991.

Lardy, H. A., and Wellman, H. (1953). *J. Biol. Chem.* **201,** 357.

Lardy, H. A., Johnson, D., and McMurray, W. C. (1958). *Arch. Biochem. Biophys.* **78,** 587.

Larner, J. (1966). *Trans. N. Y. Acad. Sci.* [2] **29,** 192.

Laskowski, M. (1961). *In* "The Enzymes" (P. D. Boyer, H. Lardy, and K. Myrbäck, eds.), 2nd rev. ed., Vol. 5, p. 123. Academic Press, New York.

Lawford, G. R., Langford, P., and Schachter, H. (1966). *J. Biol. Chem.* **241,** 1835.

Lazarus, S. S., and Barden, H. (1962). *J. Histochem. Cytochem.* **10,** 368.

Lee, S. H. (1969). *Histochemie* **19,** 99.

Lee, S. H., and Torack, R. M. (1968). *J. Cell Biol.* **39,** 725.

Lee, Y.-P., and Hsu, H. H.-T. (1969). *Endocrinology* **85,** 251.

Leech, R. M. (1968). *In* "Plant Cell Organelles" (J. B. Pridham, ed.), p. 137. Academic Press, New York.

Lehninger, A. L. (1965). "The Mitochondrion," p. 112. Benjamin, New York.

Lehninger, A. L., Sudduth, H. C., and Wise, J. B. (1960). *J. Biol. Chem.* **235,** 2450.

Lejeune, N., Thines-Sempoux, D., and Hers, H. G. (1963). *Biochem. J.* **86,** 16.

LeJohn, H. B. (1968). *Biochem. Biophys. Res. Commun.* **32,** 278.

LeJohn, H. B., McCrea, B. E., Suzuki, I., and Jackson, S. (1969). *J. Biol. Chem.* **244,** 2484.

Leloir, L. F., and Goldemberg, S. H. (1960). *J. Biol. Chem.* **235,** 919.

Lenaz, G. (1969). *Acta Vitaminol. Enzymol.* **23,** 169.

Lengyel, P., and Söll, D. (1969). *Bacteriol. Rev.* **33,** 264.

Leslie, I. (1961). *Nature (London)* **189,** 260.

Leuthardt, F., and Müller, A. F. (1948). *Experientia* **4,** 478.

Leuthardt, F., and Nielsen, H. (1951). *Helv. Chim. Acta* **34,** 1618.

Leuthardt, F., Müller, A. F., and Nielsen, H. (1949). *Helv. Chim. Acta* **32,** 744.

Leventer, L. L., Buchanan, J. L., Ross, J. E., and Tapley, D. F. (1965). *Biochim. Biophys. Acta* **110,** 428.

Levin, W., Alvares, A., Jacobson, M., and Kuntzman, R. (1969). *Biochem. Pharmacol.* **18,** 883.

Levvy, G. A., and Conchie, J. (1966). *In* "Glucuronic Acid: Free and Combined Chemistry, Biochemistry, Pharmacology, and Medicine" (D. J. Dutton, ed.), p. 301. Academic Press, New York.

Levvy, G. A., and Marsh, C. A. (1959). *Advan. Carbohyd. Chem.* **14,** 381.

Levvy, G. A., and Marsh, C. A. (1960). *In* "The Enzymes" (P. D. Boyer, H. Lardy, and K. Myrbäck, eds.), 2nd rev. ed., Vol. 4, p. 397. Academic Press, New York.

Levy, M., Toury, R., and Andre, J. (1967). *Biochim. Biophys. Acta* **135,** 599.

Lieberman, M., and Baker, J. E. (1965). *Annu. Rev. Plant Physiol.* **16,** 343.

Likar, I. N., and Rosenkrantz, H. (1964). *Lab. Invest.* **13,** 246.

Lin, J.-K., Miller, J. A., and Miller, E. C. (1967). *Biochem. Biophys. Res. Commun.* **28,** 1040.

Lindahl, T., Adams, A., and Fresco, J. R. (1966). *Proc. Nat. Acad. Sci. U. S.* **55,** 941.

Linn, T. C. (1967a). *J. Biol. Chem.* **242,** 990.

Linn, T. C. (1967b). *J. Biol. Chem.* **242,** 981.

Lipner, H. J., and Barker, S. B. (1953). *Endocrinology* **52,** 367.

Lloyd, J. B. (1969). *Biochem. J.* **115,** 52P.

Lotlikar, P. D., Enomoto, M., Miller, J. A., and Miller, E. C. (1967). *Proc. Soc. Exp. Biol. Med.* **125,** 341.

Lowe, C. U., and Lehninger, A. L. (1955). *J. Biophys. Biochem. Cytol.* **1,** 89.

Lowenstein, J. M. (ed.) (1969). "Citric Acid Cycle." Marcel Dekker, New York.

Lowenstein, J. M., and Smith, S. R. (1962). *Biochim. Biophys. Acta* **56,** 385.

Lu, A. Y. H., Jun, K. W., and Coon, M. J. (1969). *J. Biol. Chem.* **244,** 3714.

Ludewig, S., and Chanutin, A. (1950). *Arch. Biochem.* **29,** 441.

Lundquist, A., and Öckerman, P. A. (1969). *Enzymol. Biol. Clin.* **10,** 300.

Lygre, D. G., and Nordlie, R. C. (1969). *Biochim. Biophys. Acta* **178,** 389.

Lynen, F. (1961). *Fed. Proc., Fed. Amer. Soc. Exp. Biol.* **20,** 941.

Lynen, F. (1962). *Proc. Robert A. Welch Found. Conf. Chem. Res.* **5,** 293.

Lynen, F. (1967). *In* "Organizational Biosynthesis" (H. J. Vogel, J. O. Lampen, and V. Bryson, eds.), p. 243. Academic Press, New York.

McArdle, B., Thompson, R. H. S., and Webster, G. R. (1960). *J. Neurochem.* **5,** 135.

McCann, W. P. (1957). *J. Biol. Chem.* **226,** 15.

McComb, R. B., and Yushok, W. D. (1959). *Biochim. Biophys. Acta* **34,** 515.

McEwen, C. M., Jr., Sasaki, G., and Lenz, W. R., Jr. (1968). *J. Biol. Chem.* **243,** 5217.

McEwen, C. M., Jr., Sasaki, G., and Jones, D. C. (1969a). *Biochemistry* **8,** 3952.

McEwen, C. M., Jr., Sasaki, G., and Jones, D. C. (1969b). *Biochemistry* **8,** 3963.

Machinist, J. M., Orme-Johnson, W. H., and Ziegler, D. M. (1966). *Biochemistry* **5,** 2939.

Machinist, J. M., Dehner, E. W., and Ziegler, D. M. (1968). *Arch. Biochem. Biophys.*
    125, 858.
McKay, R., Druyan, R., Getz, G. S., and Rabinowitz, M. (1969). *Biochem. J.* **114**, 455.
McKnight, R. C., and Hunter, F. E., Jr. (1966). *J. Biol. Chem.* **241**, 2757.
McLagan, N. F., and Reid, D. (1957). *Ciba Found. Colloq. Endocrinol.* [*Proc.*] **10**,
    190.
MacLennan, D. H., and Asai, J. (1968). *Biochem. Biophys. Res. Commun.* **33**, 441.
MacLennan, D. H., and Tzagoloff, A. (1968). *Biochemistry* **7**, 1603.
McLeod, R. M., Farkas, W., Fridovitch, I., and Handler, P. (1961). *J. Biol. Chem.*
    236, 1841.
McSham, W. H., Rozich, R., and Meyer, R. K. (1953). *Endocrinology* **52**, 215.
Mahadevan, S., and Tappel, A. L. (1967). *J. Biol. Chem.* **242**, 2369.
Mahadevan, S., and Tappel, A. L. (1968a). *J. Biol. Chem.* **243**, 2849.
Mahadevan, S., and Tappel, A. L. (1968b). *Arch. Biochem. Biophys.* **128**, 129.
Mahesh, V. B., and Ulrich, F. (1960). *J. Biol. Chem.* **235**, 356.
Mahler, H. R. (1963). *In* "The Enzymes" (P. D. Boyer, H. Lardy, and K. Myrbäck,
    eds.), 2nd ed., Vol. 8, pp. 285–296. Academic Press, New York.
Mahler, H. R., Hübscher, G., and Baum, H. (1955). *J. Biol. Chem.* **216**, 625.
Mahler, H. R., Wittenberger, M. H., and Brand, L. (1958). *J. Biol. Chem.* **233**, 770.
Malamud, D. (1969). *Biochem. Biophys. Res. Commun.* **35**, 754.
Mandel, P., Revel, M., Weill, J. D., Busch, S., and Chambon, P. (1962). *Biochem. J.*
    84, 88P.
Mandl, I. (1961). *Advan. Enzymol.* **23**, 174.
Mann, K. G., and Vestling, C. S. (1969). *Biochemistry* **8**, 1105.
Mansour, T. E. (1966). *Pharmacol. Rev.* **18**, 173.
Margolis, S. A., and Baum, H. (1966). *Arch. Biochem. Biophys.* **114**, 445.
Margreth, A., Catani, C., and Schiaffino, S. (1967). *Biochem. J.* **102**, 35c.
Marinetti, G. V., Ray, T. K., and Tomasi, V. (1969). *Biochem. Biophys. Res. Commun.* **36**, 185.
Marker, C. L., and Hunter, R. L. (1959). *J. Histochem. Cytochem.* **7**, 42.
Marsh, J. M., Butcher, R. W., Savard, K., and Sutherland, E. W. (1966). *J. Biol.
    Chem.* **241**, 5436.
Martin, S. J., England, H., Turkington, V., and Leslie, I. (1963). *Biochem. J.* **89**, 327.
Martonosi, A. (1967). *Biochem. Biophys. Res. Commun.* **26**, 753.
Martonosi, A. (1968). *J. Biol. Chem.* **243**, 71.
Martonosi, A., Donley, J., and Halpin, R. A. (1968). *J. Biol. Chem.* **243**, 61.
Marunouchi, T. (1969). *J. Biochem.* (*Tokyo*) **66**, 113.
Marunouchi, T., and Mori, T. (1967). *J. Biochem.* (*Tokyo*) **62**, 401.
Masko, H., Munk, K., Homan, J. D. H., Bouman, J., and Matthÿsen, R. (1956). *Z.
    Naturforsch. B* **11**, 407.
Mason, H. S. (1965). *Annu. Rev. Biochem.* **34**, 595.
Mason, H. S., North, J. C., and Vaneste, M. (1965). *Fed. Proc., Fed. Amer. Soc. Exp.
    Biol.* **24**, 1172.
Mason, T. L., and Hooper, A. B. (1969). *Develop. Biol.* **20**, 472.
Massaro, D., Weiss, H., and Simon, M. R. (1970). *Amer. Rev. Resp. Dis.* **101**, 198.
Massey, V., and Curti, B. (1966). *J. Biol. Chem.* **241**, 3417.
Masters, B. S. S., Billimoria, M. H., Kamin, H., and Gibson, Q. H. (1965a). *J. Biol.
    Chem.* **240**, 4081.
Masters, B. S. S., Kamin, H., Gibson, Q. H., and Williams, C. H., Jr. (1965b). *J. Biol.
    Chem.* **240**, 921.

Matila, P., Balz, J. P., and Semadeni, E. (1965). Z. Naturforsch. B **20,** 693.

Matsui, H., and Schwartz, A. (1966). Biochim. Biophys. Acta **128,** 380.

Maver, M. E., and Greco, A. E. (1951). J. Nat. Cancer Inst. **12,** 37.

Maxwell, E. S., and Ashwell, G. (1953). Arch. Biochem. Biophys. **43,** 389.

May, L., Miyazaki, M., and Grenell, R. G. (1959). J. Neurochem. **4,** 269.

Medzihradsky, F., Kline, M. H., and Hokin, L. (1967). Arch. Biochem. Biophys. **121,** 311.

Meister, A., Wellner, D., and Scott, S. J. (1960). J. Nat. Cancer Inst. **24,** 31.

Melani, F., Farnararo, M., and Sgaragli, G. (1968). Experientia **24,** 114.

Mellors, A., and Tappel, A. L. (1967). J. Lipid Res. **8,** 479.

Mendicino, J., Beaudreau, C., and Bhattacharyva, R. N. (1966). Arch. Biochem. Biophys. **116,** 436.

Menon, I. A., and Haberman, H. F. (1970). Arch. Biochem. Biophys. **137,** 231.

Metrione, R. M., Neves, A. G., and Fruton, J. S. (1966). Biochemistry **5,** 1597.

Meyer, K., Hoffman, P., and Linker, A. (1960). In "The Enzymes" (P. D. Boyer, H. Lardy, and K. Myrbäck, eds.), 2nd rev. ed., Vol. 4, p. 447. Academic Press, New York.

Meyer, R. K., and Clifton, K. H. (1956). Arch. Biochem. Biophys. **62,** 198.

Meyer, R. R., and Simpson, M. V. (1968). Proc. Nat. Acad. Sci. U. S. **61,** 130.

Meyer, R. R., and Simpson, M. V. (1969). Biochem. Biophys. Res. Commun. **34,** 238.

Michuda, C. M., and Martinez-Carrion, M. (1969). Biochemistry **8,** 1095.

Miller, E. C., and Miller, J. A. (1967). Proc. Soc. Exp. Biol. Med. **124,** 915.

Miller, J. E., and Litwack, G. (1969). Arch. Biochem. Biophys. **134,** 149.

Miller, L. A., and Goldfeder, A. (1961). Exp. Cell Res. **23,** 311.

Miller, N. R., and Rafferty, N. S. (1969). J. Morphol. **129,** 359.

Mills, R. R., and Cochran, D. G. (1967). Comp. Biochem. Physiol. **20,** 919.

Milsom, D. W., Rose, F. A., and Dodgson, K. S. (1968). Biochem. J. **109,** 40P.

Mirsky, A. E., and Osawa, S. (1961). In "The Cell" (J. Brachet and A. E. Mirsky, eds.), Vol. 2, p. 677. Academic Press, New York.

Mirsky, A. E., Osawa, S., and Allfrey, V. G. (1956). Cold Spring Harbor Symp. Quart. Biol. **21,** 49.

Mitchell, P. (1961). Nature (London) **191,** 144.

Mitchell, P. (1963). In "Cell Interface Reactions" (H. D. Brown, ed.), p. 33. Scholar's Library, New York.

Mitchell, P. (1968). "Chemiosmotic Coupling and Energy Transduction." Glynn Res. Ltd., Bodmin.

Mitchell, P. (1969). In "The Molecular Basis of Membrane Function" (D. C. Tosteson, ed.), p. 483. Prentice-Hall, Englewood Cliffs, New Jersey.

Mitchell, P., and Moyle, J. (1967). In "Biochemistry of Mitochondria" (E. C. Slater, Z. Kaniuga, and L. Wojtczak, eds.), p. 53. Academic Press, New York.

Mitchell, P., and Moyle, J. (1969). Eur. J. Biochem. **7,** 471.

Mitoma, C., Posner, H. S., Reitz, H. C., and Udenfriend, S. (1956). Arch. Biochem. Biophys. **61,** 431.

Mitzmegg, P., Heim, F., and Meythaler, B. (1970). Life Sci. **9,** 121.

Miyake, Y., Aki, K., Shashimoto, S., and Yamano, T. (1965). Biochim. Biophys. Acta **105,** 86.

Miyamoto, E., Kuo, J. F., and Greengard, P. (1969). Science **165,** 63.

Mizutani, A. (1968). J. Histochem. Cytochem. **16,** 172.

Mizutani, A., and Fujita, H. (1969). J. Electronmicrosc. **18,** 17.

Moldave, K. (1965). Annu. Rev. Biochem. **34,** 419.

Mollenhauer, H. H., and Morré, D. J. (1966). *Annu. Rev. Plant Physiol.* **17,** 27.

Moore, C. L. (1968). *Arch. Biochem. Biophys.* **128,** 734.

Moore, J. T., and Gaylor, J. L. (1969). *J. Biol. Chem.* **244,** 6334.

Moorjani, S., and Lemonde, A. (1967). *Can. J. Biochem.* **45,** 1393.

Morais, R., and de Lamirande, G. (1965). *Biochim. Biophys. Acta* **95,** 40.

Morais, R., Blackstein, M., and de Lamirande, G. (1967). *Arch. Biochem. Biophys.* **121,** 711.

Mordoh, J., Krisman, C. R., Parodi, A. J., and Leloir, L. F. (1968). *Arch. Biochem. Biophys.* **127,** 193.

Morino, Y., and Watanabe, T. (1969). *Biochemistry* **8,** 3412.

Morino, Y., Itoh, H., and Wada, H. (1963). *Biochem. Biophys. Res. Commun.* **13,** 348.

Morton, R. K. (1954a). *Biochem. J.* **57,** 231.

Morton, R. K. (1954b). *Biochem. J.* **57,** 595.

Moses, J. L., Rosenthal, A. S., Beaver, D. L., and Schuffman, S. S. (1966). *J. Histochem. Cytochem.* **14,** 702.

Moss, D. W., Eaton, R. H., Smith, J. K., and Whitby, L. G. (1967). *Biochem. J.* **102,** 53.

Mudd, J. B., Golde, L. M. G., and Van Deenen, L. L. M. (1969). *Biochim. Biophys. Acta* **176,** 547.

Mueller, G. C., and Miller, J. A. (1949). *J. Biol. Chem.* **180,** 1125.

Mukherjee, S., and Bhose, A. (1968). *Biochim. Biophys. Acta* **164,** 357.

Müller, A. F., and Leuthardt, F. (1950a). *Helv. Chim. Acta* **33,** 262.

Müller, A. F., and Leuthardt, F. (1950b). *Helv. Chim. Acta* **33,** 268.

Munoz, E., Nachbar, M. S., Schor, M. T., and Salton, M. R. J. (1968a). *Biochem. Biophys. Res. Commun.* **32,** 539.

Munoz, E., Freer, J. H., Ellar, D. J., and Salton, M. R. J. (1968b). *Biochim. Biophys. Acta* **150,** 531.

Murad, F., Chi, Y. M., Rall, T. W., and Sutherland, E. R. (1962). *J. Biol. Chem.* **237,** 1233.

Myers, D. K. (1962). *Can. J. Biochem. Physiol.* **40,** 619.

Myers, D. K., and Slater, E. C. (1957). *Biochem. J.* **67,** 558.

Myers, L. T., and Worthen, H. G. (1961). *Fed. Proc., Fed. Amer. Soc. Exp. Biol.* **20,** 218.

Nachbaur, J., and Vignais, P. M. (1968). *Biochem. Biophys. Res. Commun.* **33,** 315.

Nagano, K., Kanazawa, T., Mizuno, N., Tashima, Y., Nakao, T., and Nakao, M. (1965). *Biochem. Biophys. Res. Commun.* **19,** 759.

Nakamaru, Y. (1968). *J. Biochem. (Toyko)* **63,** 626.

Nakamaru, Y., and Konishi, K. (1968). *Biochim. Biophys. Acta* **159,** 206.

Nakamaru, Y., Kosakai, M., and Konishi, K. (1967). *Arch. Biochem. Biophys.* **120,** 15.

Nakano, Y., Fujioka, M., and Wada, H. (1968). *Biochim. Biophys. Acta* **159,** 19.

Nakao, T., Nagano, K., Adachi, K., and Nakao, M. (1963). *Biochem. Biophys. Res. Commun.* **13,** 444.

Nakao, T., Tashima, Y., Nagano, K., and Nakao, M. (1965). *Biochem. Biophys. Res. Commun.* **19,** 755.

Narasimhulu, S., Cooper, D. Y., and Rosenthal, O. (1965). *Life Sci.* **4,** 2101.

Nathan, P., and Aprison, M. H. (1955). *Fed. Proc., Fed. Amer. Soc. Exp. Biol.* **14,** 106.

Nebert, D. W., and Gelboin, H. V. (1968a). *J. Biol. Chem.* **243,** 6242.

Nebert, D. W., and Gelboin, H. V. (1968b). *J. Biol. Chem.* **243,** 6250.

Nebert, D. W., and Gelboin, H. V. (1969). *Arch. Biochem. Biophys.* **134,** 76.

Nebert, D. W., and Gelboin, H. V. (1970). *J. Biol. Chem.* **245**, 160.
Nelson, L. (1954). *Biochim. Biophys. Acta* **14**, 312.
Nelson, L. (1955). *Biochim. Biophys. Acta* **16**, 494.
Neville, A. M., and Engel, L. L. (1968). *Endocrinology* **83**, 864.
Neville, A. M., Orr, J. C., and Engel, L. L. (1969). *J. Endocrinol.* **43**, 599.
Newsholme, E. A., Rolleston, F. S., and Taylor, K. (1968). *Biochem. J.* **106**, 193.
Nichoalds, G. E., and Oxender, D. L. (1970). *Fed. Proc., Fed. Amer. Soc. Exp. Biol.* **29**, 341.
Nicholls, D. G., and Garland, P. B. (1969). *Biochem. J.* **114**, 215.
Nichols, D. W., and James, A. T. (1968). *In* "Plant Cell Organelles" (J. B. Pridham, ed.), p. 163. Academic Press, New York.
Nielsen, H., and Leuthardt, F. (1949). *Helv. Physiol. Acta* **7**, C53.
Nirenberg, M. W. (1964). *Methods Enzymol.* **6**, 17.
Nishibayashi, H., and Sato, R. (1968). *J. Biochem. (Tokyo)* **63**, 766.
Nishida, G., and Labbe, R. F. (1959). *Biochim. Biophys. Acta* **31**, 519.
Nisselbaum, J. S., and Bodansky, O. (1963). *Fed. Proc., Fed. Amer. Soc. Exp. Biol.* **22**, 241.
Nordlie, R. C., and Johns, P. T. (1968). *Biochemistry* **7**, 1473.
Nordlie, R. C., and Lygre, D. G. (1966). *J. Biol. Chem.* **241**, 3136.
Nordlie, R. C., and Snoke, R. E. (1967). *Biochim. Biophys. Acta* **148**, 222.
Nordlie, R. C., Hanson, T. L., and Johns, P. T. (1967). *J. Biol. Chem.* **242**, 4144.
Nordlie, R. C., Arion, W. J., Hanson, T. L., Gilsdorf, J. R., Horne, R. N., Soodsma, J. F., Anderson, W. B., and Lygre, D. G. (1968). *J. Biol. Chem.* **243**, 1140.
Noronha, J. M., and Sreenivasan, A. (1960). *Biochim. Biophys. Acta* **44**, 64.
Norum, K. R. (1966). *Acta Physiol. Scand.* **66**, 172.
Norum, K. R., Farstad, M., and Bremer, J. (1966a). *Biochem. Biophys. Res. Commun.* **24**, 488.
Norum, K. R., Farstad, M., and Bremer, J. (1966b). *J. Biochem. Biophys. Res. Commun.* **24**, 797.
Noveles, R. R., and Davis, W. J. (1967). *Endocrinology* **81**, 283.
Novikoff, A. B. (1957). *Cancer Res.* **17**, 1010.
Novikoff, A. B. (1960). "Cell Physiology of Neoplasia," p. 219. Univ. of Texas Press, Austin, Texas.
Novikoff, A. B. (1961). *In* "The Cell" (J. Brachet and A. E. Mirsky, eds.), Vol. 2, p. 423. Academic Press, New York.
Novikoff, A. B., and Goldfischer, S. (1961). *Proc. Nat. Acad. Sci. U. S.* **47**, 802.
Novikoff, A. B., and Heus, M. (1963). *J. Biol. Chem.* **238**, 710.
Novikoff, A. B., Podber, E., and Ryan, J. (1950). *Fed. Proc., Fed. Amer. Soc. Exp. Biol.* **9**, 210.
Novikoff, A. B., Hecht, L. I., Podber, E., and Ryan, J. (1952). *J. Biol. Chem.* **194**, 153.
Novikoff, A. B., Podber, E., and Ryan, J. (1953a). *J. Biol. Chem.* **203**, 665.
Novikoff, A. B., Podber, E., Ryan, J., and Noe, E. (1953b). *J. Histochem. Cytochem.* **1**, 27.
Novikoff, A. B., Hausman, D. H., and Podber, E. (1958). *J. Histochem. Cytochem.* **6**, 61.
Novikoff, A. B., Hirsh, R., and Quintana, N. (1965). *J. Cell Biol.* **27**, 73A.
O'Donovan, D. J., and Lotspeich, W. D. (1966). *Nature (London)* **212**, 930.
Ogawa, H., Sawada, M., and Kawada, M. (1966). *J. Biochem. (Tokyo)* **59**, 126.
Okada, Y., and Okunuki, K. (1969). *J. Biochem. (Tokyo)* **65**, 581.

Okamato, H., Yamamato, S., Nozaki, M., and Hayaishi, O. (1967). *Biochem. Biophys. Res. Commun.* **26**, 309.

Okuyama, H., and Lands, W. E. (1969). *J. Biol. Chem.* **244**, 6514.

Oldham, S. B., Bell, J. J., and Harding, B. W. (1968). *Arch. Biochem. Biophys.* **123**, 496.

Oleinick, N. L., and Koritz, S. B. (1966a). *Biochemistry* **5**, 715.

Oleinick, N. L., and Koritz, S. B. (1966b). *Biochemistry* **5**, 3400.

Olson, J. A., Jr., Lindberg, M., and Bloch, K. (1957). *J. Biol. Chem.* **226**, 941.

Omachi, A., Barnum, C. P., and Glick, D. (1948). *Proc. Soc. Exp. Biol. Med.* **67**, 133.

Omura, T., and Sato, R. (1962). *J. Biol. Chem.* **237**, 1375.

Orloff, J., and Handler, J. (1967). *Amer. J. Med.* **42**, 757.

Orme-Johnson, W. H., and Ziegler, D. M. (1965). *Biochem. Biophys. Res. Commun.* **21**, 78.

Orrenius, S. (1965). *J. Cell Biol.* **26**, 725.

Orrenius, S., and Thor, H. (1969). *Eur. J. Biochem.* **9**, 415.

Osterhout, W. J. V. (1933). *Ergeb. Physiol. Exp. Pharmakol.* **35**, 967.

Oxender, D. L. (1968). *J. Biol. Chem.* **243**, 5921.

Paigen, K. (1954). *J. Biol. Chem.* **206**, 945.

Paigen, K. (1961). *Exp. Cell Res.* **25**, 246.

Paigen, K., and Griffiths, S. K. (1959). *J. Biol. Chem.* **234**, 299.

Palade, G. E. (1951). *Arch. Biochem.* **30**, 144.

Palade, G. E. (1952). *Anat. Rec.* **114**, 427.

Palade, G. E., and Siekevitz, P. (1956). *J. Biophys. Biochem. Cytol.* **2**, 171.

Paltemaa, S. (1968). *Acta Rheumatol. Scand.* **14**, 161.

Papa, S., Tager, J. M., Franciavilla, A., De Hann, E. J., and Quagliariello, E. (1967). *Biochim. Biophys. Acta* **131**, 14.

Papa, S., Alifano, A., Tager, J. M., and Quagliariello, E. (1968). *Biochim. Biophys. Acta* **153**, 303.

Parodi, A. J., Mordoh, J., Chrisman, C. R., and Leloir, L. F. (1969). *Arch. Biochem. Biophys.* **132**, 111.

Parsons, D. F. (1963). *Science* **140**, 985.

Parsons, D. F. (1965). *Int. Rev. Exp. Pathol.* **4**, 1.

Parsons, D. F., and Williams, G. R. (1967). *Methods Enzymol.* **10**, 443.

Parsons, D. F., Williams, G. R., and Chance, B. (1966). *Ann. N. Y. Acad. Sci.* **137**, 643.

Parsons, D. F., Williams, G. R., Thomson, W., Wilson, D. F., and Chance, B. (1967). *In* "Mitochondrial Structure and Compartmentation" (E. Quagliariello *et al.*, eds.), p. 29. Adriatica Editrice, Bari.

Parvin, R. (1969). *Methods Enzymol.* **13**, 16.

Pastan, I. (1966). *Biochem. Biophys. Res. Commun.* **25**, 14.

Patel, V., and Tappel, A. L. (1969). *Biochim. Biophys. Acta* **191**, 86.

Patton, G. W., and Nishimura, E. T. (1967). *Cancer Res.* **27**, 117.

Paul, J., and Fottrel, P. F. (1961). *Ann. N. Y. Acad. Sci.* **94**, 668.

Pedersen, P. L. (1968). *J. Biol. Chem.* **243**, 4305.

Penefsky, H. S., Pullman, M. E., Datta, A., and Racker, E. (1960). *J. Biol. Chem.* **235**, 3330.

Penniall, R., and Holbrook, J. P. (1968). *Biochim. Biophys. Acta* **151**, 700.

Pennington, S. N., Brown, H. D., Patel, A. B., and Chattopadhyay, S. K. (1970a). *Biophys. J.* **9**, A184.

Pennington, S. N., Chattopadhyay, S. K., and Brown, H. D. (1970b). *Quart. J. Stud. Alc., Part A* **31**, 13.

Penniston, J. T., Harris, R. A., Asai, J., and Green, D. E. (1968). *Proc. Nat. Acad. Sci. U. S.* **59**, 624.

Perdue, J. F., and Sneider, J. (1970). *Biochim. Biophys. Acta* **196**, 125.

Perry, S. V. (1952). *Biochim. Biophys. Acta* **8**, 499.

Person, P., Zipper, H., and Felton, J. H. (1969a). *Arch. Biochem. Biophys.* **131**, 459.

Person, P., Felton, J. H., O'Connell, D. J., Zipper, H., and Philpott, D. E. (1969b). *Arch. Biochem. Biophys.* **131**, 470.

Peters, T. (1962). *J. Biol. Chem.* **237**, 1186.

Petrack, B., Greengard, P., Craston, A., and Kalinsky, H. J. (1963). *Biochem. Biophys. Res. Commun.* **13**, 472.

Pette, D., and Brdiczka, D. (1966). *In* "Regulation of Metabolic Processes in Mitochondria" (J. M. Tager *et al.*, eds.), p. 28. Elsevier, Amsterdam.

Pette, D., Klingenberg, M., and Bucher, T. (1962). *Biochem. Biophys. Res. Commun.* **7**, 425.

Phillips, A. H., and Langdon, R. G. (1962). *J. Biol. Chem.* **237**, 2652.

Planta, R. J., Gorter, J., and Gruber, M. (1964). *Biochim. Biophys. Acta* **89**, 511.

Plaut, G. W. E. (1969). *Methods Enzymol.* **12**, 34.

Pokrovskii, A. A., and Karovnikore, K. A. (1969). *Vop. Med. Khim.* **15**, 382.

Posner, H., Mitoma, C., and Udenfriend, S. (1961). *Arch. Biochem. Biophys.* **94**, 269.

Possmayer, F., Scherphof, G. L., Dubbelman, T. M. A. R., Van Golde, L. M. G., and van Deenen, L. L. M. (1969). *Biochim. Biophys. Acta* **176**, 95.

Post, R. L., and Sen, A. K. (1967a). *Methods Enzymol.* 762.

Post, R. L., and Sen, A. K. (1967b). *Methods Enzymol.* 773.

Post, R. L., Sen, A. K., and Rosenthal, A. S. (1965). *J. Biol. Chem.* **240**, 1437.

Post, R. L., Kume, S., Tobin, T., Orcutt, B., and Sen, A. K. (1969). *J. Gen. Psysiol.* **54**, 3065.

Potter, L. T., and Glover, V. A. S. (1968). *J. Biol. Chem.* **243**, 3864.

Potter, V. R., and Reif, A. E. (1952). *J. Biol. Chem.* **194**, 287.

Potter, V. R., Siekevitz, P., and Simonson, H. C. (1953). *J. Biol. Chem.* **205**, 893.

Prabhu, V. G. (1969). *J. Amer. Osteopath. Ass.* **68**, 1058.

Pridham, J. B., ed. (1968). "Plant Cell Organelles." Academic Press, New York.

Pryor, J., and Berthet, J. (1960). *Biochim. Biophys. Acta* **43**, 556.

Pugh, D., Leaback, D. H., and Walker, P. G. (1957). *Biochem. J.* **65**, 16P.

Pullman, M. E., and Schatz, G. (1967). *Annu. Rev. Biochem.* **36**, 539.

Pullman, M. E., Penefsky, H., and Racker, E. (1958). *Arch. Biochem. Biophys.* **76**, 227.

Pullman, M. E., Penefsky, H., Datta, A., and Racker, E. (1960). *J. Biol. Chem.* **235**, 2322.

Pulsinelli, W. A., and Eik-Nes, K. B. (1970). *Fed. Proc., Fed. Amer. Soc. Exp. Biol.* **29**, 918.

Quagliariello, E., Papa, S., Slater, E. C., and Tager. J. M. (eds.) (1967). "Mitochondrial Structure and Compartmentation." Adriatica Editrice, Bari.

Quigley, J. P., and Gotterer, G. S. (1969). *Biochim. Biophys. Acta* **173**, 456.

Raacke, I., and Fiala, J. (1964). *Proc. Nat. Acad. Sci. U. S.* **52**, 1283.

Rabinowitz, M., Desalles, L., Meisler, J., and Lorand, L. (1965). *Biochim. Biophys. Acta* **97**, 29.

Racker, E. (1963). *Biochem. Biophys. Res. Commun.* **10**, 435.

Racker, E. (1965). "Mechanisms in Bioenergetics," p. 14. Academic Press, New York.

Racker, E. (1967). *Fed. Proc., Fed. Amer. Soc. Exp. Biol.* **26**, 1335.

Racker, E., and Horstman, L. L. (1967). *J. Biol. Chem.* **242**, 2547.

Raczynska-Bojanowska, K., and Gasiorowska, I. (1963). *Acta Biochim. Pol.* **10**, 117.

194   HARRY D. BROWN AND SWARAJ K. CHATTOPADHYAY

Rademaker, W. J. (1959). "De localisatie van enige proteasen in de levercel." Kemink en zoon, Utrecht.
Rafelson, M. E., Jr., Gold, S., and Priede, I. (1966a). *Methods Enzymol.* **8,** 677.
Rafelson, M. E., Jr., Schneir, M., and Wilson, V. W., Jr. (1966b). *In* "The Amino Sugars" (R. W. Jeanloz and E. A. Balazs, eds.), Vol. 2B, p. 171. Academic Press, New York.
Rao, G. A., and Johnston, J. M. (1966). *Biochim. Biophys. Acta* **125,** 465.
Rawson, M. D., and Pincus, J. H. (1968). *Biochem. Pharmacol.* **17,** 573.
Razzell, W. E. (1961). *J. Biol. Chem.* **236,** 3028.
Rechardt, L., and Kokko, A. (1967). *Histochemie* **10,** 278.
Rechcigl, M. Jr., and Price, V. E. (1968). *Progr. Exp. Tumor Res.* **10,** 112.
Rechcigl, M. Jr., Hruban, Z., and Morris, H. P. (1969). *Enzymol. Biol. Clin.* **10,** 161.
Recknagel, R. O. (1950). *J. Cell. Comp. Physiol.* **35,** 111.
Reddi, T. G., and Nath, M. C. (1969). *Can. J. Biochem.* **47,** 297.
Redman, C. M., Siekevitz, P., and Palade, G. E. (1966). *J. Biol. Chem.* **241,** 1150.
Reichard, P. (1954). *Acta Chem. Scand.* **8,** 795.
Reid, E. (1961a). *Biochim. Biophys. Acta* **49,** 218.
Reid, E. (1961b). *In* "Biochemists' Handbook" (C. Long, ed.), p. 814. Spon, London.
Reid, E., O'Neal, M. A., and Lewin, I. (1956). *Biochem. J.* **64,** 730.
Reid, E., El Aaser, A. A., Turner, M. K., and Siebert, G. (1964). *Hoppe-Seyler's Z. Physiol. Chem.* **339,** 135.
Reinauer, H., and Junger, E. (1969). *Hoppe Seyler's Z. Physiol. Chem.* **350,** 1161.
Rendi, R., and Uhr, M. L. (1964). *Biochim. Biophys. Acta* **89,** 520.
Rendina, G., and Singer, T. P. (1959). *J. Biol. Chem.* **234,** 1605.
Rendon, A., and Waksman, A. (1969). *Biochem. Biophys. Res. Commun.* **35,** 324.
Rhodin, J. (1954). "Correlation of Ultrastructural Organization and Function in Normal and Experimentally Changed Proximal Convoluted Tubule Cells of the Mouse Kidney." Stockholm Karolinska Institutet, Stockholm, Aktiebologet, Godvil.
Rigdon, R. H., Brown, H. D., Chattopadhyay, S. K., and Patel, A. (1968). *Arch. Pathol.* **85,** 208.
Rizack, M. A. (1964). *J. Biol. Chem.* **239,** 392.
Robbins, K. C. (1968). *Arch. Biochem. Biophys.* **123,** 531.
Robertson, H. E., and Boyer, P. D. (1955). *J. Biol. Chem.* **214,** 295.
Robinson, D., and Abrahams, H. E. (1967). *Biochim. Biophys. Acta* **132,** 212.
Robinson, D., and Willcox, P. (1969). *Biochim. Biophys. Acta* **191,** 183.
Robinson, D., Price, R. G., and Dance, N. (1967). *Biochem. J.* **102,** 525.
Robinson, G. A., Butcher, R. W., and Sutherland, E. W. (1967). *Ann. N. Y. Acad. Sci.* **139,** 703.
Robinson, G. A., Butcher, R. W., and Sutherland, E. W. (1968). *Annu. Rev. Biochem.* **37,** 149.
Robinson, J. C., Keay, L., Molinari, R., and Sizer, I. W. (1962). *J. Biol. Chem.* **237,** 2001.
Robinson, J. D. (1967). *Biochemistry* **6,** 3250.
Robinson, J. D., Lowinger, J., and Bettinger, B. (1968). *Biochem. Pharmacol.* **17,** 1113.
Roche, J., and Ricaud, P. (1955). *C. R. Soc. Biol.* **149,** 1364.
Rodbell, M. (1967). *Biochem. J.* **105,** 2P.
Roodyn, D. B. (1959). *Int. Rev. Cytol.* **8,** 279.
Roodyn, D. B. (1963). *Biochem. Soc. Symp.* **23,** 20.
Rosenthal, O., and Vars, H. M. (1954). *Proc. Soc. Exp. Biol. Med.* **86,** 555.

Rosenthal, O., Novack, B. G., and Brodie, J. W. (1952). *Fed. Proc., Fed. Amer. Soc. Exp. Biol.* **11,** 276.

Rosenthal, O., Gottlieb, B., Gorry, J. D., and Vars, H. M. (1956). *J. Biol. Chem.* **223,** 469.

Rosenthal, O., Thind, S. K., and Conger, N. (1960). *Abstr. 138th Meet. Amer. Chem. Soc.* p. 10C.

Roth, J. S., Bukovsky, J., and Eichel, H. J. (1962). *Radiat. Res.* **16,** 27.

Rothfield, L., and Finkelstein, A. (1968). *Annu. Rev. Biochem.* **37,** 463.

Rothfield, L., Takeshita, M., Pearlman, M., and Horne, R. W. (1966). *Fed. Proc., Fed. Amer. Soc. Exp. Biol.* **25,** 1495.

Roy, A. B. (1954). *Biochim. Biophys. Acta* **14,** 149.

Roy, A. B. (1958). *Biochem. J.* **68,** 519.

Roy, A. B. (1960). *Biochem. J.* **77,** 380.

Rubin, B. B., and Katz, A. M. (1967). *Science* **158,** 1189.

Rudney, H. (1957). *J. Biol. Chem.* **227,** 363.

Rudolph, N., and Betheil, J. J. (1970). *J. Nutr.* **100,** 21.

Rutter, W. J., and Lardy, H. A. (1958). *J. Biol. Chem.* **233,** 374.

Ryan, W. L., and Heidrick, M. L. (1968). *Science* **162,** 1484.

Sachs, G., Rose, J. D., and Hirschowitz, B. I. (1967). *Arch. Biochem. Biophys.* **119,** 277.

Sacktor, B. (1953). *J. Gen. Physiol.* **36,** 371.

Sacktor, B. (1955). *J. Biophys. Biochem. Cytol.* **1,** 29.

Salganicoff, L., and Koeppe, R. E. (1968). *J. Biol. Chem.* **243,** 3416.

Saltman, P. (1953). *J. Biol. Chem.* **200,** 145.

Samson, F. E., Jr., Balfour, W. M., and Jacobs, R. J. (1961). *Fed. Proc., Fed. Amer. Soc. Exp. Biol.* **20,** 340.

Sanadi, D. R. (1965). *Annu. Rev. Biochem.* **34,** 21.

Sanadi, D. R., Fluharty, A. L., and Andreoli, T. E. (1962). *Biochem. Biophys. Res. Commun.* **8,** 200.

Sánchez de Jiménez, E., and Cleland, W. W. (1969). *Biochim. Biophys. Acta* **176,** 685.

Sano, S., and Granick, S. (1961). *J. Biol. Chem.* **236,** 1173.

Sargeant, J. R., and Campbell, P. N. (1965). *Biochem. J.* **96,** 134.

Sato, R., Nishibayashi, H., and Ito, A. (1969). *In* "Microsomes and Drug Oxidations" (J. R. Gillette *et al.,* eds.), p. 111. Academic Press, New York.

Sauer, L. A., Martin, A. P., and Stotz, E. (1960). *Cancer Res.* **20,** 251.

Scallen, T. J., Dean, W. J., and Schuster, M. W. (1968). *J. Biol. Chem.* **243,** 5202.

Scevola, M. E., and de Barbieri, A. (1957). *Bull. Soc. Chim. Biol.* **39,** 1305.

Schatzmann, H. J. (1962). *Nature (London)* **196,** 677.

Schaub, M. C. (1964). *Helv. Physiol. Pharmacol. Acta* **22,** 271.

Schechter, A. N., and Epstein, C. J. (1968). *Science* **159,** 997.

Schein, A. H., and Young, E. (1952). *Exp. Cell Res.* **3,** 383.

Schein, A. H., Podber, E., and Novikoff, A. B. (1951). *J. Biol. Chem.* **190,** 331.

Schenkman, J. B., Frey, I., Remmer, H., and Estabrook, R. W. (1967). *Mol. Pharmacol.* **3,** 516.

Schick, L., and Butler, L. G. (1969). *J. Cell Biol.* **42,** 235.

Schmidt, G. (1961). *In* "The Enzymes" (P. D. Boyer, H. Lardy, and K. Myrbäck, eds.), 2nd rev. ed., Vol. 5, p. 37. Academic Press, New York.

Schnaitman, C., and Greenawalt, J. W. (1968). *J. Cell Biol.* **38,** 158.

Schnaitman, C. A., and Pedersen, P. L. (1968). *Biochem. Biophys. Res. Commun.* **30,** 428.

Schnaitman, C., Erwin, V. G., and Greenawalt, J. W. (1967). *J. Cell Biol.* **32,** 719.
Schneider, W. C. (1946). *Cancer Res.* **6,** 685.
Schneider, W. C. (1948). *J. Biol. Chem.* **176,** 259.
Schneider, W. C. (1959). *Advan. Enzymol.* **21,** 1.
Schneider, W. C. (1963). *J. Biol. Chem.* **238,** 3572.
Schneider, W. C., and Hogeboom, G. H. (1950a). *J. Biol. Chem.* **183,** 123.
Schneider, W. C., and Hogeboom, G. H. (1950b). *J. Nat. Cancer Inst.* **10,** 969.
Schneider, W. C., Claude, A., and Hogeboom, G. H. (1948). *J. Biol. Chem.* **172,** 451.
Schneider, W. C., Hogeboom, G. H., and Ross, H. E. (1950). *J. Nat. Cancer Inst.* **10,** 977.
Scholte, H. R. (1969). *Biochim. Biophys. Acta* **178,** 137.
Schonbrod, R. D., Khan, M. A. Q., Terriere, L. C., and Plapp, F. W., Jr. (1968). *Life Sci.* **7,** 681.
Schoner, W., Beusch, R., and Kramer, R. (1967). *Eur. J. Biochem.* **1,** 334.
Schotz, M. C., Rice, L. L., and Alfin-Slater, R. B. (1954). *J. Biol. Chem.* **207,** 665.
Schramm, M. (1967). *Annu. Rev. Biochem.* **36,** 307.
Schrier, S. L., Giberman, E., and Katchalski, E. (1969). *Biochim. Biophys. Acta* **183,** 397.
Schultz, R. L., and Meyer, R. K. (1958). *J. Cell. Comp. Physiol.* **52,** 1.
Schultz, S. G. (1969). *In* "The Molecular Basis of Membrane Function" (D. C. Tosteson, ed.), p. 401. Prentice-Hall, Englewood Cliffs, New Jersey.
Schulze, W., and Wollenberger, A. (1965). *Histochemie* **5,** 417.
Schwartz, A. (1962). *Biochem. Biophys. Res. Commun.* **9,** 301.
Schwartz, A. (1965). *Biochim. Biophys. Acta* **100,** 202.
Schwartz, A., and Laseter, A. H. (1964). *Biochem. Pharmacol.* **13,** 921.
Schweet, R., and Heintz, R. (1966). *Annu. Rev. Biochem.* **35,** 723.
Seal, U. S., and Gutmann, H. R. (1959). *J. Biol. Chem.* **234,** 648.
Sedgwick, B., and Hübscher, G. (1965). *Biochim. Biophys. Acta* **106,** 63.
Segal, H. L., and Washko, M. E. (1959). *J. Biol. Chem.* **234,** 1937.
Sekuzu, I., Jurtshuk, P., Jr., and Green, D. E. (1961). *Biochem. Biophys. Res. Commun.* **6,** 71.
Sekuzu, I., Jurtshuk, P., Jr., and Green, D. E. (1963). *J. Biol. Chem.* **238,** 975.
Sellinger, O. Z., Beaufay, H., Jacques, P., Doyen, A., and de Duve, C. (1960). *Biochem. J.* **74,** 450.
Selwyn, M. J. (1967). *Biochem. J.* **105,** 279.
Selwyn, M. J. (1968). *Nature (London)* **219,** 490.
Sen, A. K., Tobin, T., and Post, R. L. (1969). *J. Biol. Chem.* **244,** 6596.
Senior, J. R., and Isselbacher, K. J. (1961). *Fed. Proc., Fed. Amer. Soc. Exp. Biol.* **20,** 245.
Senior, J. R., and Isselbacher, K. J. (1962). *J. Biol. Chem.* **237,** 1454.
Seshadri Sastri, P., Jayaraman, J., and Ramasarma, T. (1961). *Nature (London)* **189,** 577.
Shack, J. (1943). *J. Nat. Cancer Inst.* **3,** 389.
Shamberger, R. J. (1969). *Biochem. J.* **111,** 375.
Shanthaveerappa, T. R., and Bourne, G. H. (1965). *Cellule* **65,** 201.
Shapiro, B. (1967). *Ann. Rev. Biochem.* **36,** 247.
Sharon, N. (1967). *Proc. Roy. Soc., Ser. B* **167,** 402.
Shepherd, D., and Garland, P. B. (1969a). *Biochem. J.* **114,** 597.
Shepherd, D., and Garland, P. B. (1969b). *Methods Enzymol.* **13,** 11.
Shepherd, J. A., and Kalnitski, G. (1951). *J. Biol. Chem.* **192,** 1.

Shepherd, J. A., and Kalnitsky, G. (1954). *J. Biol. Chem.* **207,** 605.
Shepherd, J. A., Li, Y. W., Mason, E. E., and Ziffren, S. E. (1955). *J. Biol. Chem.* **213,** 405.
Shibko, S., and Tappel, A. L. (1964). *Arch. Biochem. Biophys.* **106,** 259.
Shiga, T., and Piette, L. H. (1964). *Photochem. Photobiol.* **3,** 213.
Shiga, T., Layani, M., and Douzou, P. (1968). *In* "Flavins and Flavoproteins" (K. Yagi, ed.), p. 140. Univ. of Tokyo Press, Tokyo.
Shikita, M., Ogiso, T., and Tamaoki, B.-I. (1965). *Biochim. Biophys. Acta* **105,** 516.
Shull, K. H. (1959). *Nature (London)* **183,** 259.
Siebert, G. (1961). *Biochem. Z.* **334,** 369.
Siebert, G. (1968). *Comp. Biochem.* **23,** 1.
Siebert, G., and Hannover, R. (1963). *Biochem. Z.* **339,** 162.
Siebert, G., and Humphrey, G. B. (1965). *Advan. Enzymol.* **27,** 239.
Siebert, G., Bassler, K. H., Hannover, R., Adloff, E., and Bayer, R. (1961). *Biochem. Z.* **334,** 388.
Siegel, G. J., and Albers, R. W. (1967). *J. Biol. Chem.* **242,** 4972.
Siegel, L., and England, S. (1960). *Biochem. Biophys. Res. Commun.* **3,** 253.
Siegel, L., and England, S. (1961). *Biochim. Biophys. Acta* **54,** 67.
Siegel, L., and England, S. (1962). *Biochim. Biophys. Acta* **64,** 101.
Siekevitz, P. (1963). *Annu. Rev. Physiol.* **25,** 15.
Siekevitz, P., and Palade, G. E. (1966). *J. Cell Biol.* **30,** 519.
Siekevitz, P., and Potter, V. R. (1953). *J. Biol. Chem.* **201,** 1.
Siekevitz, P., Low, H., Ernster, L., and Linberg, O. (1958). *Biochim. Biophys. Acta* **29,** 378.
Siggins, G. R., Hoffer, B. J., and Bloom, F. E. (1969). *Science* **165,** 1018.
Silbert, C. K., and Martin, D. B. (1968). *Biochem. Biophys. Res. Commun.* **31,** 818.
Silverstein, E., and Sulebele, G. (1969a). *Biochemistry* **8,** 2543.
Silverstein, E., and Sulebele, G. (1969b). *Biochim. Biophys. Acta* **185,** 297.
Silverstein, E., and Sulebele, G. (1970). *Biochemistry* **9,** 274.
Simpson, E. R., and Estabrook, R. W. (1968). *Arch. Biochem. Biophys.* **126,** 977.
Simpson, E. R., Cooper, D. Y., and Estabrook, R. W. (1969). *Recent Progr. Horm. Res.* **25,** 523.
Singer, T. P., and Kearney, E. B. (1955). Quoted by Singer and Kearney. *In* "The Enzymes" (P. D. Boyer, H. Lardy, and K. Myrbäck, eds.), Vol. 7, p. 383. Academic Press, New York, 1963.
Sjöstrand, F. S. (1962). *Ciba Found. Symp. Pancreas* pp. 1–22.
Skou, J. C. (1957). *Biochim. Biophys. Acta* **23,** 394.
Skou, J. C. (1963). *Biochem. Biophys. Res. Commun.* **10,** 79.
Skou, J. C. (1965). *Physiol. Rev.* **45,** 596.
Skou, J. C. (1967). *Protoplasma* **63,** 303.
Skou, J. C. (1969). *In* "The Molecular Basis of Membrane Function" (D. C. Tosteson, ed.), p. 455. Prentice-Hall, Englewood Cliffs, New Jersey.
Slater, E. C. (1966). *Compr. Biochem.* **14,** 327.
Slater, E. C. (1967a). *Euro. J. Biochem.* **1,** 317.
Slater, E. C. (1967b). *In* "Biochemistry of Mitochondria" (E. C. Slater, Z. Kaniuga, and L. Wojtczak, eds.), p. 1. Academic Press, New York.
Slater, T. F. (1959). *Nature (London)* **183,** 1679.
Slater, T. F., and Planterose, D. N. (1960). *Biochem. J.* **74,** 584.
Smallman, B. N., and Wolfe, L. S. (1956). *J. Cell. Comp. Physiol.* **48,** 197.
Smith, A. D., and Winkler, H. (1968). *Biochem. J.* **108,** 867.

Smith, D. S. (1963). *J. Cell Biol.* **19**, 115.
Smith, E. L., and Hill, R. L. (1960). *In* "The Enzymes" (P. D. Boyer, H. Lardy, and K. Myrbäck, eds.), 2nd rev. ed., Vol. 4, pp. 37–62. Academic Press, New York.
Smith, J. C., Foldes, V., and Foldes, F. F. (1960). *Fed. Proc., Fed. Amer. Soc. Exp. Biol.* **19**, 260.
Smith, M. E., and Hübscher, G. (1966). *Biochem. J.* **101**, 308.
Smith, M. E., Sedgwick, B., Brindley, D. N., and Hübscher, G. (1967). *Eur. J. Biochem.* **3**, 70.
Snellman, O. (1969). *Biochem. J.* **114**, 673.
Sobel, H. J. (1962). *Anat. Rec.* **143**, 389.
Solomon, J. B. (1959). *Develop. Biol.* **1**, 182.
Somogyi, J. (1964). *Biochim. Biophys. Acta* **92**, 615.
Sone, N., Furuya, E., and Hagihara, B. (1969). *J. Biochem. (Tokyo)* **65**, 935.
Song, C. S., and Bodansky, O. (1967). *J. Biol. Chem.* **242**, 694.
Song, C. S., Nisselbaum, J. S., Tandler, B., and Bodansky, O. (1968). *Biochim. Biophys. Acta* **150**, 300.
Soodsma, J. F., and Nordlie, R. C. (1966). *Biochim. Biophys. Acta* **122**, 510.
Soodsma, J. F., Legler, B., and Nordlie, R. C. (1967). *J. Biol. Chem.* **242**, 1955.
Sottocasa, G. L. (1967). *Biochem. J.* **105**, 1P.
Sottocasa, G. L., and Sandri, G. (1969). *In* "Enzymes and Isoenzymes" (D. Shuger, ed.), p. 211. Academic Press, New York.
Sottocasa, G. L., Kuylenstierna, B., Ernster, L., and Bergstrand, A. (1967a). *J. Cell Biol.* **32**, 415.
Sottocasa, G. L., Kuylenstierna, B., Ernster, L., and Bergstrand, A. (1967b). *Methods Enzymol.* **10**, 448.
Sourkes, T. L. (1968). *Advan. Pharmacol.* **6F**, 61–69.
Spiegel, H. E., and Wainio, W. W. (1969). *J. Pharmacol. Exp. Ther.* **165**, 23.
Spiro, M. J., and Ball, E. G. (1961a). *J. Biol. Chem.* **236**, 225.
Spiro, M. J., and Ball, E. G. (1961b). *J. Biol. Chem.* **236**, 231.
Srere, P. A. (1969). *Methods Enzymol.* **13**, 3.
Srivastava, S. K., and Beutler, E. (1969). *J. Biol. Chem.* **244**, 6377.
Stagni, N., and de Bernard, B. (1968). *Biochem. Biophys. Acta* **170**, 129.
Stahl, W. L. (1967). *Arch. Biochem. Biophys.* **120**, 230.
Stahl, W. L., Sattin, A., and McIlwain, H. (1966). *Biochem. J.* **99**, 404.
Stanbury, J. B. (1957). *J. Biol. Chem.* **228**, 801.
Stanbury, J. B., and Morris, M. L. (1958). *J. Biol. Chem.* **233**, 106.
Stark, G., and Jung, W. (1956). *Arch. Gynaekol.* **187**, 398.
Stein, A. M., Kirkman, S. K., and Stein, J. H. (1967). *Biochemistry* **6**, 3197.
Stein, Y., and Shapiro, B. (1957). *Biochim. Biophys. Acta* **24**, 197.
Stein, Y., and Shapiro, B. (1958). *Biochim. Biophys. Acta* **30**, 271.
Stein, Y., Tietz, A., and Shapiro, B. (1957). *Biochim. Biophys. Acta* **26**, 286.
Stern, H., and Mirsky, A. E. (1953). *J. Gen. Physiol.* **36**, 181.
Stetten, M. R., and Burnett, F. F. (1966). *Biochim. Biophys. Acta* **128**, 344.
Stetten, M. R., and Burnett. F. F. (1967). *Biochim. Biophys. Acta* **139**, 138.
Stevens, B. M., and Reid, E. (1956). *Biochem. J.* **64**, 735.
Stevenson, I. H., and Turnbull, M. J. (1968). *Biochem. Pharmacol.* **17**, 2297.
Stirpe, F., and Aldridge, W. N. (1961). *Biochem. J.* **80**, 481.
Stjarne, L., Roth, R. H., and Giarman, N. J. (1968). *Biochem. Pharmacol.* **17**, 2008.
Stoeckenius, W. (1963). *J. Cell Biol.* **17**, 443.
Stoffel, W., and Trabert, U. (1969). *Hoppe-Seyler's Z. Physiol. Chem.* **350**, 836.

Stone, D. B., and Mansour, T. E. (1967). *Mol. Pharmacol.* **3,** 177.
Straus, W. (1954). *J. Biol. Chem.* **207,** 745.
Straus, W. (1956). *J. Biophys. Biochem. Cytol.* **2,** 513.
Straus, W. (1957). *J. Biophys. Biochem. Cytol.* **3,** 933.
Strecker, H. J., and Mela, P. (1955). *Biochim. Biophys. Acta* **17,** 580.
Strittmatter, C. F., and Ball, E. G. (1954). *J. Cell. Comp. Physiol.* **43,** 57.
Strittmatter, C. F., and Umberger, F. T. (1969). *Biochim. Biophys. Acta* **180,** 18.
Strittmatter, P. (1958). *J. Biol. Chem.* **233,** 748.
Strittmatter, P. (1959). *J. Biol. Chem.* **234,** 2661.
Strittmatter, P., and Velick, S. F. (1956). *J. Biol. Chem.* **221,** 253.
Strominger, J. L., Maxwell, E. S., Axelrod, J., and Kalckar, H. M. (1957). *J. Biol. Chem.* **224,** 79.
Struck, E., Ashmore, J., and Wieland, O. (1966). *Enzymol. Biol. Clin.* **7,** 38.
Suematsu, T., Iwabori, N., and Koizumi, T. (1970). *Biochim. Biophys. Acta* **201,** 378.
Sulebele, G., and Silverstein, E. (1970). *Biochemistry* **9,** 283.
Sulimovici, S., and Boyd, G. S. (1968). *Eur. J. Biochem.* **3,** 332.
Sulimovici, S., and Boyd, G. S. (1969). *Eur. J. Biochem.* **7,** 549.
Sullivan, R. J., Miller, O. N., and Sellinger, O. Z. (1968). *J. Neurochem.* **15,** 115.
Sun, F. F., and Jacobs, E. E. (1967). *Biochim. Biophys. Acta* **143,** 639.
Sung, S. C., and Williams, J. N., Jr. (1952). *J. Biol. Chem.* **197,** 175.
Sussman, K. E., Vaughan, G. D. (1967). *Diabetes* **16,** 449.
Sutherland, E. W., and Rall, T. W. (1960). *Pharmacol. Rev.* **12,** 265.
Sutherland, E. W., and Robinson, G. A. (1966). *Pharmacol. Rev.* **18,** 145.
Sutherland, E. W., Rall, T. W., and Menon, T. (1962). *J. Biol. Chem.* **237,** 1220.
Sutherland, E. W., Øye, I., and Butcher, R. W. (1965). *Recent Progr. Horm. Res.* **21,** 623.
Suzuki, K., Mano, Y., and Shimazono, N. (1960). *J. Biochem. (Tokyo)* **47,** 846.
Swanson, M. A. (1950a). *J. Biol. Chem.* **184,** 647.
Swanson, M. A. (1950b). *Fed. Proc., Fed. Amer. Soc. Exp. Biol.* **9,** 236.
Swanson, P. D., Bradford, H. F., and McIlwain, H. (1964). *Biochem. J.* **92,** 235.
Swanson, P. D. (1967). *J. Neurochem.* **14,** 343.
Sweat, M. L. (1951). *J. Amer. Chem. Soc.* **73,** 4056.
Sweat, M. L., and Lipscomb, M. D. (1955). *J. Amer. Chem. Soc.* **77,** 5185.
Swick, R. W., Barnstein, P. L., and Stange, J. L. (1965a). *J. Biol. Chem.* **240,** 3334.
Swick, R. W., Barnstein, P. L., and Stange, J. L. (1965b). *J. Biol. Chem.* **240,** 3341.
Szego, C. M. (1965). *Fed. Proc., Fed. Amer. Soc. Exp. Biol.* **24,** 1343.
Tager, J. M., Papa, S., Quagliariello, E., and Slater, E. C. (eds.) (1966). "Regulation of Metabolic Processes in Mitochondria." Elsevier, Amsterdam.
Taha, B. H., and Carubelli, R. (1967). *Arch. Biochem. Biophys.* **119,** 55.
Takagi, T., Aki, K., Isemura, T., and Yamino, T. (1966). *Biochem. Biophys. Res. Commun.* **24,** 501.
Takeuchi, Y., Yoshida, K., and Sato, S. (1969). *Plant Cell. Physiol.* **10,** 733.
Tanaka, R., and Abood, L. G. (1964). *Arch. Biochem. Biophys.* **108,** 47.
Tanaka, R., and Sakamoto, T. (1969). *Biochim. Biophys. Acta* **193,** 384.
Tanaka, R., and Strickland, K. P. (1965). *Arch. Biochem. Biophys.* **111,** 583.
Tappel, A. L. (1968). *Comp. Biochem.* **23,** 77.
Tashiro, Y. (1957). *Acta Sch. Med. Univ. Kioto* **34,** 238.
Tata, J. R. (1958). *Biochim. Biophys. Acta* **28,** 95.
Tata, J. R. (1964). *Biochem. J.* **90,** 284.
Taylor, C. B. (1963). *Biochem. Pharmacol.* **12,** 539.

Tchen, T. T., and Bloch, K. (1957). *J. Biol. Chem.* **226**, 921.
Teichgräber, P., and Biesold, D. (1968). *J. Neurochem.* **15**, 979.
Tenhunen, R., Marver, H. S., and Schmid, R. (1969). *J. Biol. Chem.* **244**, 6388.
Thines-Sempoux, D., Amar-Costesec, A., Beaufay, H., and Berthet, J. (1969). *J. Cell Biol.* **43**, 189.
Thomas, M., and Aldridge, W. N. (1966). *Biochem. J.* **98**, 94.
Thompson, M. F., and Bachelard, H. S. (1969). *Biochem. J.* **111**, 18P.
Thomson, J. F., and Klipfel, F. J. (1957). *Arch. Biochem. Biophys.* **70**, 224.
Thomson, J. F., and Klipfel, F. J. (1958). *Exp. Cell. Res.* **14**, 612.
Thomson, J. F., and Mikuta, E. T. (1954). *Arch. Biochem. Biophys.* **51**, 487.
Thorell, B., and Yamada, E. (1959). *Biochim. Biophys. Acta* **31**, 104.
Thorne, C. J. R. (1960). *Biochim. Biophys. Acta* **42**, 175.
Tice, L. W., and Engel, A. G. (1966). *J. Cell Biol.* **31**, 489.
Tipton, K. F., and Dawson, A. P. (1968). *Biochem. J.* **108**, 95.
Tobin, T., and Sen, A. K. (1970). *Biochim. Biophys. Acta* **198**, 120.
Toda, G., Hashimoto, T., Asakura, T., and Minakami, S. (1967). *Biochim. Biophys. Acta* **135**, 570.
Toschi, G. (1959). *Exp. Cell Res.* **16**, 232.
Towle, D. W., and Copenhaver, J. H., Jr. (1970). *Biochim. Biophys. Acta* **203**, 124.
Trotter, J. L., and Burton, R. M. (1969). *J. Neurochem.* **16**, 805.
Tryfiates, G. P., and Litwack, G. (1964). *Biochemistry* **3**, 1483.
Tsou, K. C., Goodwin, C. W., Seamond, B., and Lynn, D. (1968). *J. Histochem. Cytochem.* **16**, 487.
Tsuboi, K. K. (1952). *Biochim. Biophys. Acta* **8**, 173.
Tsuchiyama, M., and Osaka, D. (1965). *Igaku Zasshi* **11**, 4299.
Tsukada, K., and Lieberman, I. (1965). *Biochem. Biophys. Res. Commun.* **19**, 702.
Tsukada, K., Mariyama, T., Doi, O., and Lieberman, I. (1968). *J. Biol. Chem.* **243**, 1152.
Tubbs, P. K., and Garland, P. B. (1968). *Brit. Med. Bull.* **24**, 158.
Turtle, J. R., Littleton, G. K., and Kipnis, D. M. (1967). *Nature (London)* **213**, 727.
Tzur, R., and Shapiro, B. (1964). *J. Lipid Res.* **5**, 542.
Uesugi, S., Kahlenberg, A., Medzihradsky, F., and Hokin, L. E. (1969). *Arch. Biochem. Biophys.* **130**, 156.
Ugazio, G. (1960). *G. Biochem.* **9**, 96.
Ulrich, F. (1963). *Biochem. J.* **88**, 193.
Ulrich, F. (1964). *J. Biol. Chem.* **239**, 3532.
Ulrich, F. (1965). *Biochim. Biophys. Acta* **105**, 460.
Underhay, E., Holt, S. J., Beaufay, H., and de Duve, C. (1956). *J. Biophys. Biochem. Cytol.* **2**, 635.
Unemoto, T., Takahashi, F., and Hayashi, M. (1969). *Biochim. Biophys. Acta* **185**, 134.
Vaes, G. (1967). *Biochem. J.* **103**, 802.
Valette, G., Cohen, Y., and Burkard, W. (1954). *C. R. Soc. Biol.* **148**, 1762.
Vallee, B. L., and Coleman, J. E. (1965). *Comp. Biochem.* **12**, 231.
Vallejos, R. H., Van den Bergh, S. G., and Slater, E. C. (1968). *Biochim. Biophys. Acta* **153**, 509.
Van Deenen, L. L. M., and de Haas, G. H. (1966). *Annu. Rev. Biochem.* **35**, 157.
Van den Bosch, H., Van Golde, L. M. G., Eibl, H., and Van Deenen, L. L. M. (1967). *Biochim. Biophys. Acta* **144**, 613.
Van den Bosch, H., Van Golde, L. M. G., Slotboom, A. J., and Van Deenen, L. L. M. (1968). *Biochim. Biophys. Acta* **152**, 694.

Van den Bosch, H., Slotboom, A. J., and Van Deenen, L. L. M. (1969). *Biochim. Biophys. Acta* **176**, 632.
Van Lancker, J. L. (1960). *Biochim. Biophys. Acta* **45**, 63.
Van Lancker, J. L., and Holtzer, R. L. (1959). *J. Biol. Chem.* **234**, 2359.
Van Tol, A., and Hülsmann, W. C. (1969). *Biochim. Biophys. Acta* **189**, 342.
Vary, M. J., Edwards, C. L., and Stewart, P. R. (1969). *Arch. Biochem. Biophys.* **130**, 235.
Vaughan, M., and Murad, F. (1969). *Biochemistry* **8**, 3092.
Veldsema-Currie, R. D., and Slater, E. C. (1968). *Biochim. Biophys. Acta* **162**, 310.
Vessell, E. S. (1968). *Ann. N. Y. Acad. Sci.* **151**, Art. 1, 1–689.
Viala, R., and Gianetto, R. (1955). *Can. J. Biochem. Physiol.* **33**, 839.
Vigers, G. A., and Ziegler, F. D. (1968). *Biochem. Biophys. Res. Commun.* **30**, 83.
Villee, C. A., and Spencer, J. M. (1960). *J. Biol. Chem.* **235**, 3615.
Vogel, H. J., and Vogel, R. H. (1967). *Annu. Rev. Biochem.* **36**, 519.
Vogell, W., Bishai, F. R., Bucher, T., Klingenberg, M., Pette, D., and Zebe, E. (1959-1960). *Biochem. Z.* **332**, 81.
Voynick, I. M., and Fruton, J. S. (1968). *Biochemistry* **7**, 40.
Wada, F., Shimakawa, H., Takasugi, M., Kotake, T., Sakamoto, Y. (1968). *J. Biochem. (Tokyo)* **64**, 109.
Waite, M., and Van Deenen, L. L. M. (1967). *Biochim. Biophys. Acta* **137**, 498.
Waite, M., Scherphof, G. L., Boshouwers, F. M. G., and Van Deenen, L. L. M. (1969). *J. Lipid Res.* **10**, 411.
Wakid, N., and Kerr, S. E. (1955). *J. Histochem. Cytochem.* **3**, 75.
Wakid, N. W., and Needham, D. M. (1960). *Biochem. J.* **76**, 95.
Waksman, A., and Bloch, M. (1968). *J. Neurochem.* **15**, 99.
Waksman, A., and Rendon, A. (1968). *Arch. Biochem. Biophys.* **123**, 201.
Waldschmidt, M. (1967). *Strahlentherapie* **132**, 463.
Walek-Szernecka, A. (1965). *Acta Soc. Bot. Pol.* **34**, 573.
Walker, P. G. (1952). *Biochem. J.* **51**, 223.
Wallenfels, K. (1961). *Advan. Carbohyd. Chem.* **16**, 239.
Walsh, D. A., Perkins, J. P., and Krebs, E. G. (1968). *J. Biol. Chem.* **243**, 3763.
Watson, M. L., and Siekevitz, P. (1956). *J. Biophys. Biochem. Cytol.* **2**, 639.
Wattiaux-De Coninck, S., and Wattiaux, R. (1969). *Biochim. Biophys. Acta* **183**, 118.
Wattiaux, R., Baudhuin, P., Berleur, A. M., and de Duve, C. (1956). *Biochem. J.* **63**, 608.
Weber, R. (1955). *In* "Symposium on Fine Structure of Cells," p. 60. Noordhoff, Gröningen.
Webster, L. T., Jr., Gerowin, L. D., and Rakita, L. (1965). *J. Biol. Chem.* **240**, 29.
Weier, T. E., Stocking, C. R., and Schumway, L. K. (1967). *Brookhaven Symp. Biol.* **19**, 353.
Weinbach, E. C., and Garbus, J. (1966). *J. Biol. Chem.* **241**, 169.
Weinbach, E. C., Garbus, J., and Sheffield, H. G. (1967). *Exp. Cell Res.* **46**, 129.
Weiner, J. H., Berger, E. A., Hamilton, M. N., and Heppel, L. A. (1970). *Fed. Proc., Fed. Amer. Soc. Exp. Biol.* **29**, 341.
Weiner, N. (1960). *J. Neurochem.* **6**, 79.
Weinreb, N. J., Brady, R. O., and Tappel, A. L. (1968). *Biochim. Biophys. Acta* **159**, 141.
Weiss, B. (1953). *J. Biol. Chem.* **205**, 193.
Weiss, S. B., and Kennedy, E. P. (1956). *J. Amer. Chem. Soc.* **78**, 3550.
Weissman, G. (1965). *N. Eng. J. Med.* **273**, 1084 and 1143 (article in two parts).

202   HARRY D. BROWN AND SWARAJ K. CHATTOPADHYAY

Weitzman, P. J. D. (1969). *Method Enzymol.* **13,** 22.
Wettstein, F. O., Staehelin, T., and Noll, H. (1963). *Nature (London)* **197,** 430.
Whereat, A. F., Hull, F. E., Orishimo, M. W., and Rabinowitz, J. L. (1967). *J. Biol. Chem.* **242,** 4013.
Whereat, A. F., Orishimo, M. W., Nelson, J., and Phillips, S. J. (1969). *J. Biol. Chem.* **244,** 6498.
Whittaker, V. P. (1959). *Biochem. J.* **72,** 694.
Whittaker, V. P. (1966). *In* "Regulation of Metabolic Processes in Mitochondria" (J. M. Tager, S. Papa, E. Quagliariello, and E. C. Salter, eds.), Vol. 7, p. 1. Elsevier (BBA Library), Amsterdam.
Wicks, W. D. (1969). *J. Biol. Chem.* **244,** 3941.
Wieland, T., Pfleiderer, G., Haupt, I., and Wöerner, W. (1959–1960). *Biochem. Z.* **332,** 1.
Wilgram, G. F., and Kennedy, E. P. (1963). *J. Biol. Chem.* **238,** 2615.
Williams, C. H., Jr., and Kamin, H. (1962). *J. Biol. Chem.* **237,** 587.
Williams, C. H., Jr., Gibbs, R. H., and Kamin, H. (1959). *Biochim. Biophys. Acta* **32,** 568.
Williams, J. N., Jr. (1952a). *J. Biol. Chem.* **194,** 139.
Williams, J. N., Jr. (1952b). *J. Biol. Chem.* **195,** 37.
Wilson, J. E. (1967). *Biochem. Biophys. Res. Commun.* **28,** 123.
Wilson, J. T., and Fouts, J. R. (1967). *Biochem. Pharmacol.* **16,** 215.
Winkelman, J., and Lehninger, A. L. (1958). *J. Biol. Chem.* **233,** 794.
Wintersberger, E. (1968). *In* "Round Table Discussion on Biochemical Aspects of the Biogenesis of Mitochondria" (E. C. Slater, *et al.*, eds.), p. 189. Adriatica Editrice, Bari.
Wintersberger, V., and Wintersberger, E. (1970). *Eur. J. Biochem.* **13,** 20.
Woernley, D. L., Carruthers, C., Ligla, K. T., and Baumler, A. (1959). *Arch. Biochem. Biophys.* **84,** 157.
Woessner, J. R., Jr. (1968). *Abstr. 7th Int. Congr. Biochem., 1967* Vol. IV, p. 798.
Wood, T. (1963). *Biochem. J.* **89,** 210.
Wood, T. and Swanson, P. D. (1964). *J. Neurochem.* **11,** 301.
Woods, J. F., and Nichols, G. (1965). *J. Cell Biol.* **26,** 747.
Woods, M. W. (1954). *Proc. Soc. Exp. Biol. Med.* **87,** 71.
Wong-Leung, Y. L., and Kenney, A. J. (1968). *Biochem. J.* **110,** 5P.
Wong-Leung, Y. L., George, S. G., Aparicio, S. G. R., and Kenney, A. J. (1968). *Biochem. J.* **110,** 5P.
Wu, C. (1961). *Fed. Proc., Fed. Amer. Soc. Exp. Biol.* **20,** 218.
Yago, N., and Ichii, S. (1969). *J. Biochem. (Tokyo)* **65,** 215.
Yamamoto, S., and Bloch, K. (1969). *Biochem. J.* **113,** 19P.
Yamamoto, S., Lin, K., and Bloch, K. (1969). *Proc. Nat. Acad. Sci. U.S.* **63,** 110.
Yamamoto, Y. (1969). *Plant Physiol.* **44:** 262.
Yamashita, S., and Racker, E. (1968). *J. Biol. Chem.* **243,** 2446.
Yamashita, S., and Racker, E. (1969). *J. Biol. Chem.* **244,** 1220.
Yamazaki, E., and Slingerland, D. W. (1959). *Endocrinology* **64,** 126.
Yasunobu, K. T., Igaue, I., and Gomes, B. (1968). *Advan. Pharmacol.* **6A,** 43.
Yoshida, H., Izumi, F., and Nagai, K. (1966). *Biochim. Biophys. Acta* **120,** 183.
Yoshida, H., Nagai, K., Kamei, M., and Nakagawa, Y. (1968). *Biochim. Biophys. Acta* **150,** 162.
Youdim, M. B. H., and Sandler, M. (1968). *Eur. J. Pharmacol.* **4,** 105.
Zachau, H. G., Dutting, D., and Feldmann, H. (1966). *Angew. Chem.* **78,** 392.

Zaheer, N., Tewari, K. K., and Krishnan, P. S. (1967). *Arch. Biochem. Biophys.* **120,** 22.

Zahler, W. L., and Cleland, W. W. (1969). *Biochim. Biophys. Acta* **176,** 699.

Zalkin, H., Tappel, A. L., Desai, I., Caldwell, K., and Peterson, D. W. (1961). *Fed. Proc., Fed. Amer. Soc. Exp. Biol.* **20,** 303.

Zebe, E. (1959–1960). *Biochem. Z.* **332,** 328.

Zebe, E., Delbrück, A., and Bücher, T. (1959). *Biochem. Z.* **331,** 254.

Ziegenbein, R. (1966). *Nature (London)* **212,** 935.

Ziegler, D. M., and Melchior, J. B. (1956). *J. Biol. Chem.* **222,** 721.

Ziegler, D. M., and Pettit, F. H. (1966). *Biochemistry* **5,** 2932.

Zieve, P. D., and Greenough, W. B., III. (1969). *Biochem. Biophys. Res. Commun.* **35,** 462.

Zile, M., and Lardy, H. A. (1959). *Arch. Biochem. Biophys.* **82,** 411.

Zuckerman, N. G., and Hagerman, D. D. (1969). *Arch. Biochem. Biophys.* **135,** 410.

CHAPTER IV

Lipid–Protein Interface Chemistry

H. T. TIEN AND LAYLIN K. JAMES, JR.

I. Introduction

The word "interface" denotes heterogeneity and a large area of contact in two dimensions. Traditionally, the study of phenomena associated with fluid–fluid (e.g., air–water and oil–water) interfaces has been mainly the province of the colloid and surface chemists, although a small number of scientists whose primary training was in medicine and physiology were also closely involved in the development of colloid and surface chemistry in the first half of this century. The interest of life scientists in surface and colloid chemistry can be explained by the fact that living organisms via manifestation of cellular membranes possess enormous interfacial areas. Therefore, an adequate understanding of the living system and life itself in molecular terms necessitates the acquisition of knowledge dealing with interfaces. At the biological interface, flows and exchanges take place across ubiquitous cell membranes the major constituents of which are lipids and proteins, organized in a micellar structure of the lamellar form separating two aqueous environments. The lamellar micelle as a basic structural element of the biological membrane has been well recognized in the past. At present, the lamellar micelle, in isolated form, constitutes the matrix of two experimental membrane models for a variety of biological membranes.

To understand the functions and structure of biological membranes in physicochemical terms, their various constituents, mainly lipids and proteins, must be isolated and studied. Studies of lipid–lipid and lipid–protein interactions, either in model systems or biological membranes, are the subject matter of lipid–protein interface chemistry. This chapter attempts to present a unified approach toward the understanding of the structure and functions of biological membranes by reviewing the literature and recent experimental work. Covered in the first section are the topics of monolayers, micelles, lipoproteins, bimolecular lipid membranes (BLM), and microvesicles (liposomes). These topics are followed by a section on lipid–protein interactions in biological membranes and a brief review of the physical techniques employed in the current work.

II. Lipid–Lipid and Lipid–Protein Interactions

A. MONOLAYERS

Both lipids and proteins are capable of forming macromolecular layers at interfaces and both have been extensively studied separately, particularly at the air–water interface. Many types of polar lipids such as di- and triglycerides, long-chain fatty acids and alcohols, or sterol esters have been investigated and found to be typical monolayer formers, being insoluble at the air–water interface and frequently showing characteristic phase changes as the area available per molecule is varied (Harkins and Boyd, 1941; Cadenhead and Phillips, 1968). Information about the area available per molecule in the monolayer and the amount of material spread at the surface permits calculation of the cross-sectional area per molecule, e.g., 19 to 21 $Å^2$ for fatty acids or their common ionic derivatives. Other properties of the monolayer which can be studied are the surface potential, surface viscosity (Davies and Rideal, 1963), and material transport (Blank, 1968; Miller, 1968).

In simplest terms, it is considered that the polar or ionic groups of lipids in monolayers extend into the aqueous layer and that the hydrocarbon chains extend into the air phase (or oil phase at the oil–water interface) in order to produce minimum interference with the hydrogen bonding between water molecules and, hence, produce the lowest potential energy for the interfacial system.

Proteins also can spread under suitable conditions at an interface to produce a layer one molecule in thickness, although the process is rather complex seeming to involve large conformational changes in the protein which are frequently irreversible (James and Augenstein, 1966). The type of orientation of proteins at interfaces is quite indefinite because of dispersion of hydrophobic and hydrophilic groups along the polypeptide backbone. Energetically, one would expect that as many hydrophilic groups as possible would be accommodated in the aqueous phase by extensive protein unfolding and/or refolding and that similarly, the maximum possible number of hydrophobic groups would tend to escape from contact with water and enter the air or organic phase. Before reviewing the literature on lipid–protein interactions in monolayers, it may be helpful to consider briefly the nature and behavior of lipid in bulk phase alone.

Small (1968) has proposed a detailed classification of lipids based primarily upon their physical properties in bulk aqueous systems but also considering their interfacial properties. The major classifications are "nonpolar" and "polar" lipids. The second class is divided into various categories such as "insoluble, nonswelling, amphiphilic lipids," e.g., lecithins, sphingomyelins, monoglycerides; "soluble amphiphiles" capable of forming liquid crystals, e.g., many classic detergents, lysolecithin; "soluble amphiphiles" not forming liquid crystals, e.g., some bile salts, saponins. The interfacial properties and micelle-forming abilities of the various categories of polar lipids are also considered.

Illustrations of interactions of lipids of the various classes with one another and with water are provided by means of ternary and quaternary phase diagrams. Small (1968) believes that the arrangement of lipids in membranes, cellular organelles, tissues, and lipoproteins can be indicated by such studies. Ideally, perhaps, phase studies involving lipid, water, and protein would be informative. However, such studies involving macromolecular materials which are often impure and not well defined would be difficult and imprecise. Phase studies on such systems, if they were possible to carry out with any degree of precision, should answer important questions about the interactions and physical arrangements in a variety of biological systems.

Although there exists a considerable body of information about lipid monolayers and some information about insoluble protein monolayers when present separately, the physical state of an interface at which lipids and proteins are simultaneously present is not well defined. A number of susbtances have been spread together to form mixed films (for references, see Adamson, 1967). The behavior found ranged from apparent ideal solution behavior to apparent compound formation and mutual immiscibility. With lipids and proteins the technique commonly employed is one of penetration in which protein is injected beneath a preexisting lipid monolayer. Penetration is said to occur if polar and nonpolar parts of the injected protein interact strongly with the corresponding regions of the lipid monolayer (Gurd, 1960). This penetration is thought to result in an expansion of the monolayer maintained at constant film pressure or in an increase in pressure if the film area is held constant. Surface potential may also change which is considered evidence of penetration. Cholesterol and gliadin formed such a mixed film from which the gliadin could be displaced on compression (Schulman and Rideal, 1937).

Studies of the interaction of protein injected beneath a lipid monolayer were once thought to indicate that a layer of spread protein was formed beneath the monolayer followed often by a second layer of unspread, globular protein molecules beneath the first (Eley and Hedge, 1956). It was believed that the nonpolar protein side chains penetrated into the lipid monolayer interacting with the hydrocarbon chains of the lipid. It has been pointed out that such results can be explained, perhaps with better justification, by considering that whole molecules of injected protein at low film pressure are dissolving in the film, denaturing, and exerting their own film pressure independently in the mixed film (Dawson, 1968). Recent studies have tended to refute the idea that the hydrophobic moiety of protein penetrates between lipid hypdocarbon chains (Colacicco and Rapport, 1968). The following evidence is cited: pronase or trypsin injected into the subphase beneath other protein in equilibrium with a lipid monolayer produced no fall of surface pressure, although protein hydrolysis in the subphase was extensive; phospholipase A injected beneath leci· thin monolayer at equilibrium with globulin can attack lecithin; the protein does not seem to be covering lecithin polar groups and rendering them inaccessible to the lipolytic enzyme.

Studies of adsorption of proteins onto charged lipid monolayers (Matalon and Schulman, 1949) are believed to indicate that primary association is by way of charged groups on lipid and protein (Dawson, 1968). Albumin or hemoglobin injected beneath a negatively charged monolayer of cardiolipin was adsorbed if the bulk pH was below the protein isoelectric point so that it was positively charged. This electrostatic binding was essentially a reversible phenomenon. If the film was held at a constant pressure, injection of oppositely charged protein caused expansion of the film and probably extensive penetration of protein into the surface where it might well be held by other forces in addition to electrostatic ones. This penetration was not reversible since it would involve marked unfolding of the protein. Such results, of course, do not prove complex formation but merely indicate interaction between charged groups followed by polar or other interaction (Dawson, 1968). Interaction between ribonuclease and phospholipid monolayers has been recently reported by Khaiat and Miller (1969).

The activity of the enzyme catalase has been measured following exposure to hydrocarbon–water emulsions stabilized by various agents (Fraser et al., 1955). Adsorption of the enzyme was said to parallel its ability to penetrate monolayers of the stabilizing agents. It was concluded that catalase interacted with polar groups of the stabilizers, and partial unfolding of the enzyme without complete loss of activity was postulated. Trypsin showed a similar variety of behavior at the various interfaces. In the absence of stabilizer the enzyme was denatured.

Dawson (1968) has given a detailed summary of phospholipase reactions at the phospholipid-water interface. In these systems the zeta potential of the interface along with the charge on the enzyme is highly important. Here, of course, the enzyme must be adsorbed at the interface in such a way that its conformation does not change sufficiently to destroy the enzymic activity. Proteins frequently undergo marked conformational changes at interfaces sufficient to cause decreased solubility, loss of biological activity, and denaturation (James and Augenstein, 1966). However, where the energy of the interface is low or the area available per protein molecule is small, adsorption may not be followed by denaturation. By means of deuterium exchange and infrared spectroscopy and electron diffraction, indications have been found that the alpha helix may be the stable configuration in polypetide monolayers at air–water and, perhaps, oil–water interfaces (Malcolm, 1968). It is suggested that unfolding at interfaces may represent simply a loss of tertiary protein structure; hence specific interactions between protein and lipids should be considered, perhaps, in light of the possible stability of alpha helices at lipid–water interfaces (Lucy, 1968).

Studies have been made of the stability of air bubbles formed beneath stearic acid monolayers with poly-L-lysine present in the subphase (Shah, 1969). Effects of varying the pH of the subphase upon bubble stability and surface potential suggest that the helical conformation of the polypeptide is retained at the surface. Using the pendant drop, Ghosh and Bull (1962) have studied absorbed films of bovine serum albumin at air–water and oil–water interfaces.

The possibility of demonstrating complex formation between lipid and protein, lipid and lipid, or other combinations at interfaces is still somewhat disputed. Studies of a mixed film formed by spreading protein into a film of cholesterol indicate a linear relationship between the surface pressure of the mixed monolayer and the mole fraction of amino acid residues in the film (Vilallonga et al., 1967). This is considered to indicate a nonspecific, two-dimensional dissolution process in which the hydrocarbon chains of the amino acid residues and cholesterol are involved. The ΔG (mixing) for this process and for the surface dissolution of cetyl alcohol and cholesterol is practically the same as for an ideal system, indicating that no specific interactions occur between components of the surface solution.

Radiotracer studies have refuted earlier reports of complex formation in mixed films of lipids and sodium alkyl sulfate (Matsubara, 1965). The detergent is found to adsorb and exert its film pressure independently of the lipid film.

Colacicco (1969) has discussed the problem of specific lipid–protein interactions in monolayers and concluded that there are several cases of bona fide interactions of a specific nature. These include the lipid–hapten antibody interaction, the interaction between mitochondrial structural protein and phospholipid, and the interaction between the apoprotein from high-density serum lipoprotein and its lipid. The author concluded that proteins which are found associated with lipid in nature seem to have unique surface activity and form the mixed films spontaneously when injected. In his summary of studies of molecular interactions in monolayers, Lucy (1968) pointed out that such systems, useful as they are for considering lipid–protein interactions, are not very satisfactory as models for biological membranes.

James (1958) has studied adsorption of soluble lipoproteins at the aqueous solution—organic interface. These interfacial tension measurements indicate that centrifugally isolated low-density lipoproteins from human blood serum or from hen's egg yolk are considerably more effective in reducing the tension at the aqueous–heptane interface than protein alone. The rates of adsorption and levels of interfacial tension characteristic of various lipoproteins depend upon their lipid composition and are markedly affected by presaturation of the lipoprotein solutions with heptane. Under the influence of surface forces, the lipoprotein particles adsorb, undergo physical changes, and form thick surface or interfacial films having characteristic viscoelastic properties (James, 1958, 1968).

B. MICELLES

Another context in which lipid–protein interfacial interactions can be considered arises from the well-known capacity of water-insoluble, amphiphatic lipid molecules to aggregate into organized arrays of micelles in aqueous media (McBain, 1950). Such micellar or liquid crystalline systems can contain spherical aggregates, rodlike aggregates of special interest in the case of phospholipid dispersions, or hexagonal phases or lamellar phases (Luzzati, 1968). In these micellar arrangements, the polar portions of the lipids are placed on the exterior in contact with the water and the nonpolar hydrocarbon parts are buried in the interior of the particles, out of contact with water or other solvent. Although, in the case of detergent micelles, equilibrium seems to exist between free molecules and the micelles, no such equilibrium has been found for natural phospholipids (Green and Fleischer, 1964).

A great deal of information exists about such micelles (Shinoda *et al.*, 1963). Here discussion is limited to protein interactions with such lipid aggregates. Basic proteins and peptides such as cytochrome c of animal origin, protamine, polylysine, histone, and ribonuclease form stable complexes with micelles of acidic phospholipids (Green and Fleischer, 1964). Every protein with an isoelectric point greater than pH 9 has been found capable of forming a complex with such micelles. These micelles can be formed of a single acidic lipid or phosphatide, such as cardiolipin or phosphatidyl inositol, or may be made up of more than one acidic phospholipid or even of both zwitterionic and acidic lipids.

The interactions seem to involve titration of the positively charged groups of the protein with a corresponding number of negative charges in the micelle. Evidence that points to a simple electrostatic or saltlike interaction includes suppression of interaction and dissociation of the complex by high salt concentrations, inability of proteins with neutral or acidic isoelectric points to interact with acidic micelles, and disappearance of the capacity to bind phospholipid when a protein's positively charged groups are acylated (Dawson, 1968). Purified acidic phospholipids form complexes with cytochrome c which are isooctane-insoluble. These complexes can further react with lecithin or phosphatidylethanolamine which are zwitterionic and, therefore, much less acidic. Such further complexes are now soluble in isooctane. Hydrophobic or apolar binding may explain the complexing with lecithin or phosphatidyl ethanolamine.

C. Soluble Lipoproteins

Another group of complexes will be discussed in an attempt to gain further information about how lipids and proteins interact. These are the soluble lipoproteins in which considerable amounts of hydrophobic lipid can be maintained in a fairly stable dispersed form in an aqueous environment. A general discussion on lipoproteins has been published in a monograph by the Faraday Society (1949).

This group of substances, here loosely termed the "soluble lipoproteins," is found in mammalian blood serum, hen's egg yolk, and a variety of other sources. With one exception there does not seem to be any reports that lipid and protein are found covalently bound together in these lipoproteins (Fisher and Gurin, 1964). After extensive extraction with organic solvents, small quantities of long-chain fatty acids remain. Following enzymatic digestion of protein and partial purification of resulting peptides, fatty acids are found associated with a peptide fraction which contains organic phosphate. Fatty acids are firmly bound, presumably covalently.

Commonly those lipoproteins of lowest density are isolated by centrifugation in a medium sufficiently dense to cause them to float and may, thus, be classified on the basis of their negative sedimentation or flotation rates or S_f values (Searcy and Bergquist, 1962).

The low-density lipoproteins of blood serum tend to float during the usual centrifugal isolation procedures and have been studied extensively as to lipid and protein content and possible relation to disease. As the S_f value increases, protein, cholesterol, and phospholipid content all decrease whereas triglycerides increase. Oncley (1963) has summarized the state of knowledge about the structure of these lipoprotein complexes up to about 1960.

Briefly, the low-density lipoproteins were considered compact particles with exterior surfaces made up of protein and polar lipids and interior surfaces consisting of neutral lipid. Most of the lipids were found to be labile and readily exchangeable; yet solubility, titration, and electrophoretic behavior made it seem unlikely that all lipid was on the outside shell of the lipoprotein structure. Most models put the protein and phospholipid on the exterior and cover the outer surface of the particle with a layer of water of hydration. This was thought to be necessary to explain difficulties in extracting lipid from the lipoprotein with ether, unless the complex had been first frozen, treated with detergent, or a polar solvent. Oncley (1963) pointed out that the importance of water was being questioned in more recent work. He indicated that recent data put the amount of water in the structure at about 0.1 gm/gm lipoprotein rather than the earlier value of 0.6 gm/gm.

The protein components or apoproteins remaining after careful extraction of lipid have been obtained in water–soluble form. The molecular weights of these proteins are in reasonably good agreement with estimations based upon analyses of terminal amino acids (Finean, 1967).

Recent studies by Scanu et al. (1969) indicate that the delipidated low-density apolipoproteins showed spectral properties markedly differing from the native protein. Ultraviolet spectroscopy suggested that tyrosine and tryptophan, which had been shielded in the lipoprotein, became sus-

IV. LIPID–PROTEIN INTERFACE CHEMISTRY

ceptible to solvent perturbation after lipid was removed. The occurrence of β structure in both human β-lipoprotein and the apolipoprotein is indicated in infrared spectroscopy and ultraviolet circular dichroism (Gotto et al., 1968). The apoprotein probably has some increase in disordered structure, suggesting a structural role for lipid and an important conformational instability of the protein once lipid is removed (Scanu et al., 1969).

Cook and Martin (1962) report that lipoproteins from different sources have a wide and nearly continuous range of lipid content but consider that there is good evidence for a transition in composition in the range where protein, phosphilipid, and neutral lipid are present in approximately 1:1:1 proportions. Lipoproteins with protein content less than 33% have neutral lipid–phospholipid ratios ranging from 1:1 to 10:1, but this ratio does not vary much from 1:1 in those lipoproteins with protein content greater than 33%. The protein content of the former type decreases with increasing particle size and is apparently sufficient only to cover partially the surface of the particle with an extended monolayer. Hence, this type of particle must have a core of neutral lipid with much of the phospholipid present in the surface. Such particles are said to resemble microemulsions or micelles. The high-protein type of lipoprotein is considered to have a more definite type of molecular structure, perhaps resembling a protein in structural integrity. The lipid content seems to be independent of molecular weight, and a high-protein lipoprotein from egg yolk, for example, undergoes a pH-dependent, reversible dissociation into two subunits (Cook and Martin, 1962).

An analogy to the micellar hypothesis of lipoprotein structure has been found in the first protein for which the structure has been completely determined, myoglobin (Kendrew, 1962). Evidence indicates that all internal portions of myoglobin are arranged to permit association of hydrophobic side chains (Richards, 1963).

Exchange of unesterified cholesterol between human low-density lipoprotein and rat red blood cell ghosts has been investigated (Bruckdorfer and Green, 1967). The process is pH dependent and not particularly dependent on temperature or ionic strength. However, compounds such as urea, alcohols, acetone, dimethylsulfoxide, or tetraalkyl ammonium salts greatly accelerate the exchange. All of these are considered to have potent effects on the local structure of water, weakening hydrophobic bonding and, perhaps, making nonpolar lipids in the exchanging systems more exposed to the aqueous medium and more readily exchangeable on collision. It is believed that the lipoprotein structure must be one in which protein, phospholipid, and nonpolar lipid are all directly exposed to the aqueous medium. Extensive coverage of protein by lipid is considered to be unlikely. The membrane is said to be a mosaic with patches of lipid interspersed with protein molecules.

High-resolution proton, nuclear magnetic resonance (NMR) spectra of low- and high-density lipoproteins from human serum are very similar to those of aqueous dispersions of lipid from the lipoproteins. Line widths of absorptions of hydrocarbon protons are not increased in the lipoprotein spectra, whereas a polar binding by serum albumin of lysolecithin causes marked line broadening and a chemical shift. The results are said to be consistent with a micellar structure for the lipoproteins rather than indicating extensive hydrophobic associations between protein and lipid (Steim et al., 1968).

Well-resolved, high-resolution NMR spectra indicate that the lipid of serum low-density lipoproteins must be in a highly mobile condition; the polar end of the phospholipid must be free and probably is in an aqueous environment (Leslie and Chapman, 1969). However, the nonpolar aromatic amino acids of the protein part are immobilized, perhaps by apolar interactions with lipids. Nevertheless, the lipid hydrocarbon chains are quite freely mobile, and the motion of the cholesterol nucleus in the cholesterol esters increases with temperature. An analogy is suggested between protein–lipid interactions in the low-density lipoprotein and in red blood cell membranes where the proton resonances of the aromatic amino acids of membrane protein are only observable after detergent treatment (see Section IV).

Mutual solubility of the lipid components of low-density lipoprotein from hen's egg yolk has been studied by Schneider and Tattrie (1968). Phospholipids are slightly soluble in anhydrous triglycerides but become insoluble when water is present. Cholesterol is preferentially soluble in the phospholipid phase in both presence and absence of water. The conclusion is reached that interactions with water are of importance in determining the spatial arrangement in the native lipoprotein. It is suggested that the lipoprotein particle consists of a nucleus of almost pure triglyceride surrounded by phospholipids and cholesterol.

Stewart (1967) has pointed out the lipoproteins and lipids of cells and tissues frequently exist as liquid–crystalline phases in the temperature range 4°–42°C. In this form, Stewart feels, they play important structural and functional roles where interfacial activity and limited flow are essential, such as in membranes, subcellular particles, and certain layered tissues.

D. BIMOLECULAR LIPID MEMBRANES

The presence of micelles in aqueous solutions of surface-active lipids is well established as discussed above. Less well established is the structure of micelles of which several kinds have been postulated. McBain, who first recognized the existence of micelles as early as 1913, postulated two types of micelles—spherical and lamellar micelles (McBain, 1950). In the latter case, the limiting structure of a lamellar micelle is depicted as comprising a layer of two fully extended hydrocarbon chains of lipid molecules placed end to end with an aqueous phase on each side of the bimolecular (bilayer) lipid leaflet. Further, the interior of the lamellar micelles is assumed to be in a liquid rather than in a crystalline state. Direct experimental evidence for the existence of lamellar micelles in aqueous solution was provided by the work of Mueller et al. (1962), who discovered a method for the formation of single isolated lipid membrane of bimolecular thickness (60–90 Å). Since the subject of bimolecular lipid membranes, also known as bilayer or black lipid membranes (BLM), has been reviewed several times recently (Bangham, 1968; Rothfield and Finkelstein, 1968; Henn and Thompson, 1969; Castleden, 1969; Tien and Diana, 1968), the following section is written for those who are unfamiliar with the BLM system as a new approach to the study of lipid–lipid and lipid–protein interactions. The lipid–protein interaction using BLM will be considered in Section III, A.

1. Techniques of Formation and Composition of Lipid Solutions

The basic method of BLM formation can be described by referring to Fig. 1. The cell assembly consists of two concentric chambers. The wall of the inner chamber made of Teflon or polyethylene contains a small hole (1–2 mm, diameter) in which the BLM is formed. Prior to BLM formation, both chambers are filled with an electrolyte solution (either 0.1 N NaCl or KCl) slightly above the hole. Lipid solution of appropriate composition (see below) is introduced in the hole. This can be done either with the aid of a sable hair brush or a Hamilton microsyringe. Since the size of the BLM formed is small (usually less than 50 mm²), the formation characteristics of the membrane are observed using a ×20–40 microscope with reflected light. The formation characteristics are similar to the generation of "black" soap films in air (Tien, 1968a).

In the manner described above, BLM have been formed using a variety of natural and synthetic lipids. The most widely used natural lipids have been either purified phosphatidyl choline (lecithin) or oxidized cholesterol, although complex mixtures of lipids derived from natural sources have been preferred by a number of investigators (Tien and Diana, 1968). Surface-active substances readily available commercially have also been employed in the BLM studies (Tien, 1967). It should be mentioned that the presence of at least one neutral hydrocarbon has been found necessary in the formation of a stable BLM. The composition of a number of typical

BLM-forming solutions is given in Table I. It should be noted that, although the composition of the BLM-forming solution can be given precisely, these compositions are not likely to be the same as the composition of the BLM itself. This point will be considered later in connection with the structure of BLM.

2. Intrinsic Properties of BLM

The intrinsic properties of an unmodified BLM generated from either lecithin or oxidized cholesterol in an alkane solvent are strikingly similar

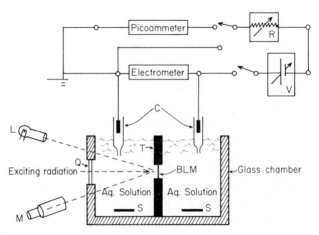

Fig. 1. Diagrammatic representation of an experimental arrangement for studying bimolecular or black lipid membranes (BLM), (R) resistor bank (0–10^{10} ohms); (V) polarizing voltage source; (L) light for observing membrane formation; (M) viewing tube or microscope (×20–40); (Q) quartz window for UV excitation; (C) calomel electrodes with salt bridge; (T) Teflon chamber or partition; (S) stirring bar.

to those expected of a layer of liquid hydrocarbon of equivalent thickness. These properties are reviewed in some detail in the following paragraphs.

a. *Thickness.* As a rough estimation, the thickness of BLM can be readily deduced from optical interference phenomena. Quantitative measurements have been made by electron microscopy (Mueller *et al.*, 1964; Henn *et al.*, 1967), by reflectance methods (Thompson and Huang, 1966; Tien *et al.*, 1966; Tien, 1968a), and by electrical capacitance (Mueller *et al.*, 1964; Hanai *et al.*, 1964). Some representative values obtained by these methods are given in Table II.

The reader interested in experimental details is referred to the original articles mentioned above and to a paper dealing with experimental techniques on BLM by Tien and Howard (1971).

TABLE I

COMPOSITION OF SOME TYPICAL BIMOLECULAR LIPID MEMBRANE-FORMING SOLUTIONS

Lipid	Hydrocarbon and additive	Solvent	Ref.
Natural sources			
Brain lipids (2%)	α-Tocopherol (1.5%); cholesterol (0.3%)	2:1 Chloroform–methanol	Mueller *et al.* (1964)
Egg lecithin (2%)	*n*-Tetradecane	3:2 Chloroform–methanol	Huang *et al.* (1964)
Saturated	*n*-Decane	—	Hanai *et al.* (1964)
Oxidized cholesterol (4%)	*n*-Octane	—	Tien *et al.* (1966)
Chloroplast extract (5%)	*n*-Octane (65%)	*n*-Butanol (30%)	Tien *et al.* (1968)
Synthetic surfactants			
Dioctadecyl phosphite (0.08%) + cholesterol (0.8%)	*n*-Dodecane	—	Tien (1967)
Dodecyl acid phosphate (0 35%) + cholesterol (0.93%)	*n*-Dodecane	—	Tien (1967)
Cholesterol (1%) + 0.008% HDTAB[a]	*n*-Dodecane	—	Tien (1967)

[a] HDTAB = hexadecyltrimethylammonium bromide.

TABLE II

THICKNESS OF BIMOLECULAR LIPID MEMBRANE OBTAINED BY VARIOUS METHODS

Membrane from	Thickness (Å)	Method	Ref.
Brain lipids	60–90	Electron microscopy	Mueller *et al.* (1964)
Egg lecithin	62–77	Optical	Thompson and Huang (1966); Tien (1967); Cherry and Chapman (1969)
Egg lecithin	38–116	Electron microscopy	Henn *et al.* (1967)
Egg lecithin	48[a]	Capacitance	Hanai *et al.* (1964)
Chloroplast lipids	105 ± 50	Optical	Tien *et al.* (1968)

[a] Hydrocarbon portion only.

b. Electrical Properties. The electrical properties of BLM can be easily measured by placing suitable electrodes in each chamber (see Fig. 1). A BLM can be represented by a parallel capacitance–resistance circuit. Both dc and ac methods have been used, although the former method is favored by most investigators (Tien and Howard, 1971).

The intrinsic resistance of an unmodified BLM is extremely high and is generally of the order of 10^8 ohm-cm^2 or more. The measurement of BLM resistance at times is difficult and nonreproducible. This variation is most likely a result of leakage at the BLM support and/or trapped emulsified droplets. The dc conductivity is linear up to a potential difference of about 50 to 100 mV although a potential difference as high as 300 mV has been reported (Bradley and Dighe, 1969). For a BLM 70 Å thickness, a 70-mV potential difference corresponds to a field strength of 10^5 volts/cm. This high field intensity can be withstood by most BLM for a long period of time without any apparent effect. At higher potential difference (>100 mV), the conductivity is no longer ohmic and increases with field strength. The increase of conductivity at higher field strengths may be a result of combinations of several factors. These include: (1) an increase in the number of charge carriers, (2) an increase in charge carrier mobility, (3) the formation of pathways by trapped emulsified droplets, and (4) the formation of bridges by induced dipoles in strong fields. Any one of these factors may contribute to the eventual dielectric breakdown of the BLM. Partial breakdown or electrical transients (excitability) in BLM observed by several investigators (Mueller and Rudin, 1968; Bean *et al.*, 1969) may be explained in terms of these factors. It is conceivable that a periodic buildup and release of space charges at the biface (i.e., the two coexisting solution–membrane interfaces) can give rise to complex electrokinetic phenomena.

The electrical capacitance is readily measured by the so-called charge leakage method (Tien and Diana, 1968). Briefly, the BLM is charged and then discharged through a resistor of known value. The decay of the BLM voltage is recorded as a function of time; the capacitance is calculated from the $\tau = RC$ formula, where τ is the time constant (37% discharge) and R and C are the parallel resistance and BLM capacitance. Most of the reported values lie between 0.3 and 0.6 μF/cm^2 (Tien and Diana, 1968). If a BLM is considered as an infinite parallel condenser and the aqueous solutions on both sides of the BLM act as electrical contacts, the thickness of the hydrocarbon layer (t_h) may be calculated according to the formula:

$$t_h = \frac{\epsilon_h}{4\pi C_m} \tag{1}$$

where ϵ_h is the dielectric constant of the hydrocarbon layer, and C_m is the

BLM capacitance per unit area. The use of Eq. (1) in estimating the thickness depends upon a judicious choice of the dielectric constant. Values in the range of 2 to 4 for ϵ_h have been used, which correspond to the dielectric constants of liquid hydrocarbons and oils. The values of t_h calculated with the aid of Eq. (1) are also given in Table II.

In contrast to resistance measurements, the measurement of the capacitance of BLM is more reproducible. The C_m values are found to be independent of frequency in the range 0–10^7 cycles/second and linear function of BLM area (Hanai et al., 1964; Haydon, 1968).

c. *Bifacial Tension.* Bimolecular lipid membranes made from a variety of natural and synthetic surface-active compounds possess appreciable interfacial tension. Since two solution–membrane interfaces (or a biface) are involved, the interfacial tension of BLM is better referred to as "bifacial" tension. The bifacial tension of BLM can be determined by measuring the pressure difference across the BLM using the maximum bubble pressure technique (Adamson, 1967). Values between 0.1 and 6.0 dynes/cm have been reported (Huang et al., 1964; Tien, 1968a). The energy of formation is independent of the BLM extension (Tien, 1967). This is consistent with the suggestion that the Plateu–Gibbs border, in addition to supporting the BLM, serves as a reservoir of lipid molecules. The measurements of bifacial tension as a function of temperature permit the evaluation of various thermodynamic properties (Tien, 1968a). The results show that the formation of BLM at the biface involves a spontaneous reduction of the free energy. As long as the integrity of the BLM is maintained, the limiting structure represents the lowest free-energy configuration.

d. *Permeability.* From the results of conductivity measurements, unmodified BLM exhibit a very low permeability to ions. For example, Pagano and Thompson (1968) have reported a sodium flux of 0.4×10^{-9} mole/cm²-second and a chloride flux of 90×10^{-9} mole/cm²-second for a BLM produced from a solution of egg lecithin and n-tetradecane in a chloroform–methanol solution. However, upon suitable modification, BLM can exhibit not only high permeability to certain ions but also marked selectivity and specificity. The effects of modifiers on ion permeability and other intrinsic properties will be discussed later.

The permeabilities of various nonelectrolytes through BLM have been investigated by several investigators, and, by far, water has been the most studied nonelectrolyte. Using both tagged water and osmotic gradient methods, Huang and Thompson (1966) reported permeability coefficients of 4.4 μ/second for tagged water and 17.3 to 104 μ/sec for the osmotically driven flux. Using the latter method, Mueller et al. (1964) earlier gave a value of 24 μ/second.

An extensive study of the permeability of lecithin BLM to organic compounds has been made by Vreeman (1966). This investigator reports the following permeability coefficients: 4.6×10^{-2} μ/second for glycerol; 7.5×10^{-3} μ/second for erythritol; and 4.2×10^{-2} μ/second. Similar results have been obtained by Lippe (1968) for thiourea on a hexadecyltrimethylammonium bromide (HDTAB)–cholesterol BLM.

The permeability of BLM to glucose has been studied by Wood *et al.* (1968) and by Jung and Snell (1968). The latter workers used a spherical BLM in their experiments and reported values in the range 1.8 to 3.6×10^{-4} μ/second which are comparable to the value 5×10^{-4} μ/second reported by Wood *et al.* In contrast to the above results, Bean *et al.* (1968) have obtained significantly higher permeability coefficients for indole and its related compounds (e.g., 2.3 to 3.9 μ/second for indole-3-ethanol). These large values for indoles may be explained in terms of their high solubility in organic solvents.

3. Structure of BLM

Apart from the material thickness and bifacial tension of BLM, at present our knowledge concerning other mechanical properties is practically nonexistent. The lack of precise chemical composition for even the simplest lecithin BLM has prevented a quantitative discussion of the structure of BLM. Nevertheless, a three-dimensional model of a lecithin BLM has been given (Tien, 1968a). The suggested model is akin to the concept first postulated by Gorter and Grendel (1925) and is similar to the so-called smectic or neat mesophase investigated by Luzzati (1968). The suggested model is essentially similar to that of a liquid crystalline structure. The hydrocarbon chains are, to a large extent, perpendicularly oriented to the plane of the biface and facing inward toward each other. On the basis of energetics, the polar groups are in contact with the aqueous phases. In the fashion depicted, the lipid molecules can thereby minimize their free energy by aggregating their hydrocarbon chain through van der Waals' interactions, and the polar portion of the molecules can maximize their coordination energy with each other and with the aqueous solutions.

Experimental evidence in support of the aforementioned model for the BLM can best be summarized as follows:

1. Optical and electron-microscopic findings indicate that all BLM thus far studied, irrespective of lipids used, possess a thickness of the order of two molecules.

2. The orientation of constituent molecules is deduced from energetic considerations; the value of measured bifacial tension indicates the absence of hydrocarbon moiety at the biface.

3. Low values of ion permeability and relatively high permeability coefficient for water are consistent with a liquid hydrocarbon interior of the BLM.

4. A BLM may be probed with a fine object without rupturing the structure thus indicating its liquidity.

5. The electrical capacitance measurements are in accord with an ultrathin structure of low dielectric constant.

TABLE III

COMPARISON OF PROPERTIES OF UNMODIFIED BIMOLECULAR LIPID MEMBRANE AND A LAYER OF LIQUID HYDROCARBON OF EQUIVALENT THICKNESS[a]

Property	Unmodified membrane (experimental)	Liquid hydrocarbon (extrapolated)
Thickness ($\mathring{A}$)	40–130	100
Resistance (ohm-cm^2)	10^8–10^9	10^8
Capacitance (μF-cm^{-2})	0.3–1.3	$\sim$1.0
Breakdown voltage (V/cm)	10^5–10^6	10^5–10^6
Dielectric constant	2–5	2–5
Refractive index	1.4–1.6	1.4–1.6
Water permeability (μ/second)	8–24	$\sim$35
Interfacial tension (dynes/cm)	0.2–6	$\sim$50
Potential difference per tenfold concentration of KCl (mV)	$\sim$0	$\sim$0
Electrical excitability	None	None
Photoelectric effects	None	None

[a] Data taken from Tien (1970).

4. Effects of Modifiers

The intrinsic properties of BLM discussed above are strikingly similar to those expected of a layer of liquid hydrocarbon of equivalent thickness. For example, the conductivity of most liquid hydrocarbon saturated with water is about 10^{-14} mhos/cm. Similarly, the dielectric breakdown voltage of purified long-chain liquid hydrocarbons is usually in the range of 10^5 to 10^6 volts/cm. In Table III a comparison is made between the intrinsic properties of BLM and the extrapolated properties of a layer of liquid alkane 100 $\mathring{A}$ thick.

The passive properties of BLM listed in Table III can be drastically altered by a large number of materials. Some of these materials are not chemically well defined, e.g., such as excitability-inducing material (EIM). Other compounds include surfactants, inorganic ions, pigments and dyes, macrocyclic antibiotics, polypeptides and proteins, and a number of organic compounds. To facilitate the discussion, the BLM modifying agents (modifiers) may be conveniently grouped into five categories: (a) those altering the passive electrical properties, (b) those changing the mechanical properties, (c) those conferring ion selectivity, (d) those inducing excitability, and (e) those generating photoelectric effects. The effects of these modifiers will be considered under separate headings in the following paragraphs.

a. Modifiers Altering Electrical Properties. The initial investigators (Mueller *et al.*, 1964) have found that addition of various materials to one or both sides of the BLM substantially reduces the electrical resistance. Examples are a large number of commercial detergents. Among simple electrolytes, KI has been found by Lauger *et al.* (1967) to lower lecithin BLM resistance by a factor of 10^3 or more. In HDTAB–cholesterol BLM, the order of effectiveness in reducing resistance follows a lyotropic anion squence: $I^- > Br^- > SO_4{}^{2-} > Cl^- > F^-$ (Tien and Diana, 1968). Cations such as Fe^{3+} at aqueous concentrations below 10^{-5} M can also decrease the resistance of lecithin BLM (Miyamoto and Thompson, 1967). On similar BLM, in contrast to Fe^{3+}, Cd^{2+}, Cu^{2+}, and Mn^{2+} increase the resistance of the membrane in concentrations less than 10^{-3} M. Bielawski and co-workers (1966) have found that 2,4-dinitrophenol when added in small amounts to the aqueous solution causes a marked reduction in BLM resistance. Other compounds related to nitrophenols which produce similar effects on BLM have also been reported (Liberman *et al.*, 1968; Sotnikov and Melnik, 1968). The effects of uncouplers of oxidative phosphorylation have been examined by several workers. Liberman and associates (1968) treated their BLM with tetrachloro-2-trifluoromethylbenzimidazole (TTFB), carbonylcyanide-*p*-trifluoromethoxyphenylhydrazone (FCCP), carbonylcyanide-*m*-chlorophenylhydrazone (CCCP), and other compounds such as tetraphenylboron (TϕB). They have found pronounced effects on the BLM conductivity which increases linearly or with the square of the concentration of additives. Current–voltage curves of BLM treated with some of these compounds exhibit a region with negative resistance. It is suggested that current–voltage characteristics with negative resistance can be related to diffusion processes not only in the BLM itself but at bifacial regions as well.

The effect of surfactants on the electrical properties of BLM has been investigated by Hashimoto (1968). Some of his findings are consistent with those reported earlier (Tien, 1967). In addition, Hashimoto finds that the resistance is nearly independent of the temperature (7–45°C) and pH. The effect of lipid composition on the electrical conductivity has been studied by Noguchi and Koga (1969). They have found that the BLM conductivity is dependent on lipid composition but little on the concentration and kind of the electrolyte used. However, the membrane resistance was found to be pH dependent.

b. *Modifiers Changing Mechanical Properties.* The mechanical properties of BLM are affected by a number of surfactants, polyene antibiotics, and proteins. Mueller *et al.* (1964) have shown that surfactants such as digitonin and sodium oleate tend to break BLM formed from brain lipids. At concentrations higher than $4.5 \times 10^{-4} M$, HDTAB effectively prevents the formation of stable BLM (Tien, 1967). Both the formation and the stability of certain BLM are dependent upon the bifacial tension and the critical micelles concentration of the surfactant used. Van Zutphen *et al.* (1966) have reported that polyene antibiotics (filipin) interact preferentially with cholesterol-containing BLM and reduce appreciably their mechanical stability. The effect of filipin on BLM stability has also been examined by Finkelstein and Cass (1968). These workers used BLM made from ox brain lipids plus *dl*-α-tocopherol and cholesterol. In addition, Finkelstein and Cass have also found that both nystatin and amphotericin B rupture their BLM at high concentrations. Papahadjopoulous and Ohki (1969) have examined certain phospholipid BLM. They have found that these BLM are more stable and possess higher electrical resistance when formed in the presence of both Ca^{2+} and Na^+ ions than the ones formed in Na^+ ions alone. The BLM instability becomes more evident when they are formed under conditions of asymmetric distribution of Ca^{2+}. Ohki and Papahadjopoulos (1969) suggest that the instability of their BLM is due to the difference in interfacial energy across the biface.

c. *Modifiers Conferring Ion Selectivity.* The intrinsic properties of lecithin or oxidized cholesterol BLM show little ion selectivity (apart from I^- noted above), which is consistent with the expected properties of the ultrathin hydrocarbon layer of the BLM. These poor ion-selective properties of BLM, however, can be modified by a number of neutral macrocyclic molecules. The cyclic peptide valinomycin, for instance, has been found to confer a selective order among the alkali metal cations as follows: $Rb^+ > K^+ > Cs^+ > Na^+ > Li^+$ (Mueller and Rudin, 1967; Lev and Buzhinsky, 1967).

As mentioned earlier, Lauger *et al.* (1967) have found that the resistance of the lecithin BLM can be reduced drastically when formed in KI solution. Lauger *et al.* have further shown that in addition to resistance reduction, the iodide–iodine-modified BLM are highly selective to I^-. That is, a 60-mV potential difference appears across the membrane per tenfold concentration gradient of iodide, the presence of iodine being required however. Recently (Verma *et al.*, 1971), iodine has been incorporated directly into the BLM-forming solutions (oxidized cholesterol or chloroplast extract). The BLM modified in this fashion have been found to respond ideally to iodide. The presence of other ions, such as Cl^-, SO_4^{2-}, or F^- did not interfere with the "electrode" response to iodide ions. The behavior of I_2-containing BLM is reminiscent of a metallic electrode reversible to its ion (e.g., Ag electrode and Ag^+). The development of the membrane potential in this case does not seem to require a concentration gradient across the BLM. It appears, therefore, that I^- or its complexes need not be transported across the membrane in order to give rise to a membrane potential.

d. Modifiers Inducing Electrical Excitability. The best-known but still incompletely understood modifier, EIM, can induce electrical excitability (Mueller *et al.*, 1962, 1964). That is, an EIM-modified black lipid membrane of which the resistance has been lowered by 3 to 4 orders of magnitude will respond to constant current pulses. The membrane resistance is voltage-dependent, and at a definite threshold potential increases regeneratively by a factor of 5 to 10. These unusual properties are reminiscent of the action potentials observed in frog nerve and alga *Valonia*. A further description of EIM-induced phenomena will be given in a later section on BLM–protein interaction.

Besides EIM, a cyclic peptide antibiotic, known as alamethicin, (Mueller and Rudin, 1968) can also produce similar electrical transient phenomena in BLM formed from oxidized cholesterol and dodecyl acid phosphate (Tien, 1967); the action potentials resemble those occurring in frog skin.

e. Modifiers Generating Photoelectric Effects. The photoelectric effects are observed when BLM (also thin lipid membranes 1000 Å to 0.1 mm thick) formed from photoactive pigments, such as chlorophylls and retinenes, are illuminated (Tien, 1968b, 1970). A typical curve showing the change in short-circuit current on irradiation of a thin membrane is shown in Fig. 2. The light response of this thin lipid membrane is completely reversible. When a chloroplast BLM is illuminated by light of various wavelengths, a photoaction spectrum can be obtained which is practically identical to the absorption spectrum (Van and Tien, 1970). The observation of the photoelectromotive force (open-circuit voltage induced by light) together with the photoelectric action spectrum of the chloroplast BLM provides strong evidence for the separation of electronic charges (electrons and holes) in the membrane.

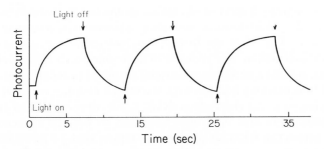

FIG. 2. Light-induced current change on irradiation of a reconstituted chloroplast bimolecular lipid membrane (BLM) with filtered light (3″ CuSO₄ solution). The BLM was formed in 0. 1 N KCl containing 10^{-3} M benzoquinone.

E. MICROVESICLES (LIPOSOMES)

In the presence of an aqueous solution, anhydrous phospholipids undergo a series of spontaneous molecular arrangements leading eventually to the production of varigated structures known as myelin figures (Dervichian, 1946). These cylindrical and spherical vesicles of minute size, ranging from several hundred Angstroms to fractions of a millimeter, have recently been studied by a number of investigators (Rendi, 1964; Bangham *et al.*, 1965a). These vesicles or microvesicles, also known as liposomes, are of special interest since their limiting structure is also of bimolecular thickness; they offer a much larger surface area than planar BLM described in the previous section.

Since the work on microvesicles has recently been reviewed by one of its proponents (Bangham, 1968), only a brief summary of the properties of these liposomes is given here.

The initial work on microvesicles was reported by Rendi (1964) who demonstrated water extrusion from swollen phospholipid suspensions in the presence of serum albumin by electron microscopy. Vatter (1963) has reported earlier that these suspensions exhibit irregular forms which can be described as vesicular and multilamellar. Bangham *et al.* (1965a) and Bangham and his colleagues (1968) have carried out extensive investigations on these microvesicles.

1. Method of Preparation

Detailed procedure for the preparation of microvesicles has been given by Bangham *et al.* (1965a), by Papahadjopoulos and Miller (1967), and by Huang (1969). In brief, dried phospholipids are allowed to swell in an electrolyte solution with gentle agitation at room temperature. Mechanical agitation facilitates fragmentation of the micelles, producing microvesicles of varying sizes in the range 5–50μ in diameter. Instead, mechanical agitation ultrasonication can be used, which produces microvesicles much smaller in size ($\sim$500 Å). The procedure is as follows: lyophilized phospholipids are suspended in a 0.1 M NaCl buffer with tris at pH 8.5. The suspension at a 3% concentration of phospholipids is ultrasonicated under nitrogen for 160 minutes at 2°C. The resulting suspension is then isolated by molecular sieve chromatography on large-pore agarose gels. By this method (Huang, 1969) a homogenous fraction of microvesicles can be obtained.

2. Structure of the Microvesicles

The structural characteristics of microvesicles formed from various phospholipids have been investigated by optical, electron microscopy, and X-ray diffraction techniques (Bangham, 1968). The phosholipid microvesicles in the light microscope exhibit birefringence. The size of birefringence depends upon the surface charge of microvesicles and ionic strength of the aqueous phase. X-ray diffraction of microvesicles shows the presence of multilamellar structures with repeating distances varying from 54 to 75 Å, depending upon the lipid used (Papahadjopoulous and Watkins, 1967). The electron micrographs of microvesicles stained negatively with ammonium molybdate show concentric lamellar layers, each approximately 50 Å thick (bilayer).

3. Osmotic Properties and Water Permeability

Using photometric and volumetric methods, Rendi (1967) has shown that the water present in the microvesicles is extruded by the addition of osmotically active compounds. The loss of water obeys the Boyle–Van't Hoff law. Bangham et al. (1967) have found that microvesicles made from charged phospholipids behave as almost perfect osmometers when alkali metal salts, glucose, sucrose, or mannitol are used as solutes. Other solutes, such as erythriotol, malonamide, urea, glycerol, propionamide, ammonium acetate, ethylene glycol, methyluria, and ethyluria, show graded permeability with ethylurea being most permeable (Bangham et al., 1967). The rapid volume changes combined with the measured total external surface areas are used to calculate water permeability coefficients. Values in the range 0.8–16μ/second have been obtained, which are seen to be in reasonable agreement with those obtained for BLM (see Table III).

4. Permeability to Ions

The minute size of microvesicles has thus far precluded any measurements of electrical properties. Nevertheless, Papahadjopoulous and Watkins (1967) have estimated a resistance value of 10^7 ohms-cm^2 using their Cl^- flux data. This estimated value, although 1 to 2 orders lower than that of BLM, is consistent with the postulated lipid interior and the osmotic properties discussed in the preceding sections. In general, these vesicles are more permeable to anions than cations. Bangham *et al.* (1965b) have found that the diffusion rate for anions follows the sequence: $I^- > Cl^- > F^- > NO_3^- > SO_4^{2-} > HOP_4^{2-}$. The diffusion rate for cations is very largely controlled by the size and magnitude of the surface charge. However, in contrast to anions, no significant differences are found for the cation series Li^+, Na^+, K^+, Rb^+, and choline. The effect of divalent cations has also been reported (Bangham, 1968).

5. Effects of Modifiers

The permeability of the microvesicles to ions and nonelectrolytes can be significantly increased by a variety of modifying agents, such as surfactants, steroids, antibiotics, and certain proteins. The effect of proteins on these microvesicles will be considered in a later section (Section III,*B*).

The effect of a very well-known surfactant, Triton X-100 [a mixture of *p-t*-octyl poly(phenoxyethoxy) ethanols; Union Carbide Chemical Co.], has been studied by Weissmann *et al.* (1966). These workers found that Triton X-100 accelerates anion release from microvesicles and causes substantial ultrastructural changes. The effects of a large number of steroids and other lytic agents, studied by Bangham *et al.* (1965b), can be categorized as those agents that increase the permeation and those that retard the permeation. The first group includes diethylstilbestrol, deoxycorticosterone, testosterone, progesterone, filipin, nystatin, and amphotericin B (Sessa and Weissmann, 1968). Compounds that reduce the permeability or stabilize the microvesicles are cortisone, cortisol, chloroquine, and 17β-estradiol.

Other modifiers that exhibit pronounced effects are bacterial toxins. Weissmann *et al.* (1965) have found that these hemolysins rupture microvesicles as well as certain natural membranes. It is suggested that bacterial toxins possess lipidlike structure (amphipathic). Insertion of such molecules into lamellar membranes could produce weak spots and eventually lead to structural failure.

III. Lipid–Protein Interactions in Model Membrane Systems

A. Bimolecular Lipid Membrane–Protein Interactions

To date, only a few studies have been made to examine the effect of proteins on the intrinsic properties of BLM. This lack of activity is perhaps owing partly to the fact that available experimental techniques are generally too crude to be of use in investigating the BLM–protein interactions. The following paragraphs give a summary of the results obtained in a few isolated experiments.

The first BLM–protein interaction experiments were reported by Mueller et al. (1962, 1964). A proteinaceous material, EIM, has been found to lower the resistance of BLM constituted from brain lipids from 10^7–10^8 to 10^3–10^4 ohms-cm^2. This EIM-modified system then responded to a suprathreshold dc pulse by abruptly shifting its resistance from one stable value to another after a latent period. The nature of EIM has further been investigated by Kushnir (1968), who isolated the EIM by adsorption on kieselgel and subsequent elution with 1% ammonia solution. Two moieties which constitute the total analytically detectable material of EIM have been identified. These are protein and ribonucleic acid (RNA). The basic properties of the protein moiety, not evident with total EIM, become apparent after its separation from RNA. In a more recent paper, Bean et al. (1969) have reported that discrete fluctuations in BLM conductivity may be observed during the initial stages of membrane interaction with EIM. It is suggested that potential-dependent conductivity changes may be due to changes in lipid–lipid and lipid–EIM interactions. Because of the complex and often contradictory reaction kinetics and behavior of the system, no unique explanation has thus far been proposed (Bean et al., 1969). The composition of BLM-forming solutions does not appear to be crucial in these experiments, since similar results have been obtained with BLM formed from brain lipids, lecithin, sphingomyelin, or oxidized cholesterol.

Albumin-modified BLM have been studied by two different groups of investigators with conflicting results. Hanai et al. (1965) have investigated lecithin BLM, in addition to egg albumin, insulin, and bovine serum albumin, and found no effect on the BLM capacitance. In contrast, Tsofina et al. (1966) have found that the electrical properties of the BLM modified with albumin are significantly altered. In their experiments, Tsofina and associates found that albumin-treated BLM have a resistance 2 orders of magnitude lower and a higher capacitance than those of unmodified membranes.

An interesting technique has been devised by del Castillo et al. (1966) to detect antigen–antibody reactions using BLM. These investigators prepared BLM from the bovine brain lipid extract in 0.1 N NaCl. The potential across the membrane was recorded with an oscilloscope the image of which was photographed fast enough for a 5-millisecond change in membrane impedance to be seen. In immunological experiments, albumins were used as the antigen and immune serums as the antibody. A large number of enzymes with appropriate substrates were also tested (e.g., trypsin + BAEE, urease + urea, chymotrypsin + ovalbumin). Addition of albumin or serum alone had no observable effect on the BLM. However, the addition of one material after the other had been introduced caused a sudden reduction in membrane resistance, then a slow return to control level. Adding more of the second compound had no effect. A similar response was obtained when small amounts of substrate were added to a solution in which the membrane had been previously exposed to the appropriate enzyme. The decrease in resistance lasted only about 0.7 second or less. The response of an enzyme-treated membrane to its substrate could be demonstrated repeatedly.

The interaction of BLM with the total protein derived from bovine erythrocytes has been carried out by Maddy et al. (1966) and by van den Berg (1968). On adding bovine protein extract to the bathing solution, the latter worker reported a sudden reduction in the BLM resistance. This decrease in resistance, however, was not observed by Maddy and co-workers. Yet, Maddy et al. (1966) reported that bovine protein caused a decrease in the bifacial tension of the BLM. An increase in the intensity of light reflectd from the lecithin BLM after the addition of protein was also observed. More recently, Smekal and Tien (1971) studied the interaction of a variety of proteins with oxidized cholesterol BLM using a reflectance technique (Tien and Howard, 1971). In their experiments an oxidized cholesterol BLM is first formed on the Teflon loop and its reflectivity measured. A solution of protein is then injected into the cell, and the reflectivity of the membrane is again measured. The following proteins were tested in this manner: alcohol dehydrogenase, insulin, γ-globulin, chymotrypsin, trypsin, and ribonuclease. In the case of alcohol dehydrogenase, the thickness of the membrane increased from 50 to about 100 Å. The other proteins increased the membrane thickness to about 80 Å. In most of these BLM–protein interaction experiments, the pH of the bathing solution had a marked effect on membrane reflectivity (Smekal and Tien, 1971).

Another interesting finding using BLM has been reported recently by Jain et al. (1969). These investigators formed BLM from a mixture of oxidized cholesterol and a synthetic surfactant. In these experiments the BLM is formed in the usual manner. To one side of the BLM, adenosine triphosphate (ATP) is added which is followed by the addition of a preparation of membrane-bound ATPase (Na–K-dependent fraction). After these additions, a decrease in the membrane resistance is observed. The current flowing across the BLM is monitored as described by Ussing and Zerahn (1951). Jain et al. observed a positive current flow across the BLM. This is interpreted as due to the transport of Na$^+$ which is analogous to that found in the natural membrane. The current flow in their reconstituted system is said to be dependent on the presence of Na$^+$ in the bathing solution and the magnitude of current on ionic strength. Further, the presence of ouabain (a cardiac glucoside), a compound known to block Na$^+$ transport in red cells, inhibits the current flow across the ATPase-modified BLM. It seems probable that, by introducing ATPase into the bathing solution, "molecular sodium pumps" could have been inserted into the BLM, which may account for the observation. If so, this implies that ATPase (or BLM-modified ATPase) is actually the sodium pump which has been described by numerous investigators in the literature (Skou, 1965).

Before concluding this section, mention should be made of the results obtained recently with the use of a fluorescent probe for studying BLM–protein interactions (Smekal and Tien, 1971). Recently Azzi et al. (1969) and Rubalcana et al. (1969) have used 8-anilino-1-naphthalene sulfonate (ANS) in studies of biological membranes. It has been known that the fluorescence properties of ANS and its derivatives are dependent upon the polarity and steric structure of the environment (Edelman and McClure, 1968). In aqueous solution, ANS is practically nonfluorescent, but it fluoresces with appreciable quantum yields when dissolved in organic solvents and when associated with proteins (Stryer, 1968). These facts make ANS an ideal probe for investigating BLM-protein interactions. In Smekal's experiments, BLM were formed from (1) oxidized cholesterol or (2) a mixture of oxidized cholesterol and egg lecithin on a Teflon loop. A low-pressure Hg lamp with a selected interference provided excitation at 365 mμ. The protein chosen for this investigation was bovine serum albumin (BSA). The results obtained may be described as follows. The intensity of the fluorescence response depends on which of the two compounds (BSA or ANS) is introduced into the aqueous phase first. When ANS is adsorbed onto the BLM first, followed by the addition of BSA, the fluorescence response is about 3 times larger than in the reverse case. Washing the BLM with 0.1 M KCl decreases the fluorescence to half of its original intensity in the first case (ANS followed by BSA), whereas in the reverse case the fluorescence returns almost to the base line. The maximum of pH-dependent fluorescence falls between pH 5.6 and 7 with a sharper rise of the curve on the acidic side. However, the change is not appreciable. When BLM has been previously saturated with ANS, addition of BSA does not show significant change in fluorescence. Also BLM formed from oxidized cholesterol alone does not give any fluorescent emission following addition of ANS and BSA (or vice versa). The following conclusions can be drawn from Smekal's results. First, the presence of phospholipids is necessary for ANS adsorption. Second, the lack of enhanced fluorescence when ANS is added after BSA suggests that this protein must interact with the phospholipid portion of the BLM in such a way that the hydrophobic regions are not available for ANS penetration.

B. LIPOSOME–PROTEIN INTERACTIONS

Interactions of microvesicles (liposomes) with proteins have been reported only in a few papers. Papahadjopoulous and Miller (1967) and Papahadjopoulous and Watkins (1967) have examined the effects of proteins on the properties of microvesicles. They reported that the presence of cytochrome c and to a less extent bovine plasma albumin tends to produce finer microvesicles (200–1000 Å). The presence of protein on the vesicles appeared to increase salt capture in the case of phosphatidylcholine. However, salt capture was decreased in the case of acidic phospholipids, such as phosphatidylserine, in the presence of cytochrome c. This phenomenon is explained in terms of complex formation diminishing the available ionic groups which presumably are involved in the salt-capture process.

Additional work on phospholipid microvesicles or liposomes has been reported (Reeves and Dowben, 1969; Sweet and Zull, 1969). It is reported that BSA, when adsorbed upon the exterior boundary of these structures, enhances glucose diffusion from the interior through the bilayer-like walls. A pH-dependent conformational change in the albumin is assumed, but the properties of the system including its detailed structure are not defined.

Before ending this paragraph on interactions of proteins with microvesicles, it may be useful to point out the relative merit of the liposome and BLM systems. First, the microvesicles (liposomes) provide large surface areas and, therefore, are well suited for the study of transport properties of ultrathin lipid membrane (<100 Å). Second, the uniformity of size of microvesicles, as demonstrated recently (Huang, 1969), offers an opportunity for investigating lipid–protein interaction using well-developed physical methods such as light scattering and ultracentrifugation. Other techniques such as electron microscopy, X-ray diffraction, nuclear magnetic resonance (NMR), and electron spin resonance (ESR) using spin labels (Griffith and Waggoner, 1969) may also be applied. The main drawback of microvesicles as an experimental model for the biological membrane (see Section IV) is that, owing to their minute size, the sophisticated electrical methods cannot be applied at the present time. Perhaps one of the best approaches is to carry out the studies concurrently using both BLM and microvesicles. These two model systems are not mutually exclusive. On the contrary, they are complementary to each other, since both are derived from a common substance—the amphipathic lipid.

IV. Lipid–Protein Interactions in Biological Membranes

Applications of physical techniques to studies of biological systems have revealed the ubiquitous presence of "membranes." Many vital cellular functions such as energy transduction, selective permeability, material transport, enzymic reactions, and conduction of nerve impulses are some of the outstanding examples (Sitte, 1969). Chemical analyses on a variety of biological membranes have clearly shown that lipids and proteins are the major structural constituents. At the present it is difficult to state to what extent that biological functions are the results of lipid–protein interactions. On the other hand, one can state confidently the obvious that no biological membranes can result in the absence of lipid–lipid and lipid–protein interactions. This section will discuss biological membranes in terms of lipid–protein interactions.

A. Types of Biological Membranes

Biological membranes may be classified in various ways. In terms of their functions, five basic types of biological membranes can be recognized. These are the plasma membrane of cells, the thylakoid membrane of chloroplasts, the cristae membrane of mitochondria, the nerve membrane of axon, and the visual receptor membrane. Some of the known facts of these five types of membranes are given in Table IV.

Alternately, biological membranes may be divided into two major groups—plant-cell membranes and animal-cell membranes. Obviously, the grouping of biological membranes into different classes is artificial. We shall see presently that all the biological membranes insofar as known possess a common design consisting of a lipid bilayer core with sorbed proteins. For the discussion, a brief summary of plant-cell and animal-cell membranes is presented below.

1. Plant-Cell Membranes

Among the important plant-cell membranes, chloroplast membranes are most unique. The internal membranes of the chloroplast are believed to be the site of photophysical and photochemical reactions. They are commonly known as grana, a stack of thylakoid membranes (Menke, 1967). Closely related to chloroplast membranes is the membrane of plastids which is simpler in structure and development. In addition to the membranes of chloroplasts and plastids, there are numerous other membranous organelles that have been characterized. These are the plasma membrane, Golgi bodies, endoplasmic reticulum, nuclear membrane, and tonoplast (Northcote, 1968). The familiar plasma membrane often called plasmalemma is the major diffusion barrier of cells. Golgi bodies (or ap-

TABLE IV

Some Basic Types of Biological Membranes[a]

Membrane type	Composition	Structure	Function[b]
Plasma	Lipids (40–50%)—phospholipids, cholesterol; proteins (60–50%)	Two opaque layers separated by a lighter layer (overall thickness ~75 Å), i.e., bilayer type	Diffusion barrier; active transport; antigenic properties
Thylakoid (chloroplasts)	Lipids (40%); pigments (8%), protein (35–55%)	Bilayer type (50–100 Å)	Quantum conversion; ATP synthesis
Mitochondrial (inner or cristae)	Lipids (20–40%)—phospholipids; protein (80–60%)	Triple-layered arrangement (thickness about 75 Å); bilayer type	Energy transduction; ATP synthesis
Rod outer segment (retina)	Lipids (40%) of which 80% are phospholipids; retinene (4–10%); opsin	Layered discs or sacs; bilayer type	Light detection and energy transfer
Nerve (axon)	Lipids (80%); proteins (20%)	Bilayer type	Conduction of nervous impulse

[a] Data taken from the following sources: van Deenen (1965), Eichberg and Hess (1967), Green and Fleischer (1964), Krinsky (1958), Menke (1967), Moretz et al. (1969); Mühlethaler (1966), O'Brien (1967), Park and Pon (1963), Salton (1967), Sjöstrand (1959), Wolken (1968), and Zahler (1969).
[b] ATP = adenosine triphosphate.

paratus) are usually made up of a stack of lamellar membranes. Nuclear membrane surrounding the nucleus is seen under the electron microscope to be perforated with channels. Further details on plant-cell membranes may be found in the recent issues of *Annual Review of Plant Physiology* (see also Salton, 1967).

2. Animal-Cell Membranes

These membranes may be classified as external (plasma), internal (cytoplasmic), and organelle membranes (Whittaker, 1968). Plasma membranes of animal cells are frequently covered with a surface layer of mucopolysaccharide in addition to proteins (Weinstein et al., 1969). Endoplasmic reticulum and Golgi apparatus are types of cytoplasmic membranes, which may have ribosomes attached. Membranes of mitochondria are complex, and they are further divided into the inner (cristae) membrane and the outer (plasma) membrane. There are a number of excellent reviews on the subject of animal-cell membranes (Rouser *et al.*, 1968; Zahler, 1969).

B. COMPOSITION OF BIOLOGICAL MEMBRANES

As indicated in Table IV, the essential components of biological membranes are lipid and protein. The ubiquitous presence of water in membranes is often assumed but not explicitly stated. In most cells, membranous organelles constitute anywhere from 40 to 80% of the dry weight. O'Brien (1967) has summarized the chemical composition of the first four types of membranes mentioned above. The composition of the visual receptor sac membranes is not precisely known at present (Abrahamson and Ostroy, 1967; Eichberg and Hess, 1967).

1. Plasma Membrane of Red Blood Cells

This is by far the most extensively studied and best characterized of all biological membranes (Zahler, 1969). Depending upon the species, the gross composition of the red blood cell plasma membrane is about 60 to 80% protein and 20 to 40% lipid (van Deenen, 1965; Maddy, 1966). The major lipids are phospholipids and cholesterol. Phosphatidylcholine (PC), phosphatidylethanolamine (PE), phosphatidylserine (PS) together with sphingomyelin (SM) account for about 50% of the phospholipids. Cholesterol content for different species varies from 22 to 32% of the total weight of the lipids (van Deenen, 1965). Other lipids are glycolipids and gangliosides (Nelson, 1967). Remarkable differences exist in the fatty acid compositions of the plasma membranes. In general, there are about 40% polyunsaturated C_{18}–C_{22} fatty acids (O'Brien, 1967).

The protein of red blood cell plasma membranes has been very difficult to isolate and study. Maddy (1966) has extracted proteins from red cell ghosts using aqueous butanol. The proteins showed one major and three minor peaks by electrophoresis. By gel filtration the molecular weight was estimated to be between 2 and 7 × 10⁵. Morgan and Hanahan (1966) showed that the protein can be separated from the lipid by sonication of ghosts in 15% butanol.

2. Thylakoid Membrane of Chloroplasts

The gross composition of thylakoid membranes consists of 45% protein and 55% lipids (Park and Pon, 1963; Park and Biggins, 1964). Phospholipids and phosphatidyl glycerol make up 75% of the total lipids. The remaining 25% are chlorophylls, carotenoids, and other pigments. Sterols and glycerides are very minor constituents (Lichtenthaler and Park, 1963). The major lipids are digalactosyldiglyceride and monogalacto-syldiglyceride (Carter et al., 1956; Weier and Benson, 1966). The structural proteins have been obtained by Criddle (1969). Only one major peak was seen in the analytical ultracentrifuge. The molecular weight was 23,000. Other studies on the protein component have been carried out by Lockshin and Burris (1966) and by Thornber et al. (1967). The latter workers used sodium dodecyl sulfate in their preparations.

3. Mitochondrial Membranes

Two types of mitochondrial membranes are generally recognized—the outer membrane and the inner (cristae) membrane. The outer membrane contains about 55% protein and 45% lipids (Parson et al., 1966). Of the total lipids, a small portion is cholesterol (O'Brien, 1967). The inner membrane contains all the cytochromes and other enzymes and less lipid than the outer membrane. Green and associates (1967) have obtained four protein fractions. Pullman et al. (1960) and Racker (1967) have isolated the so-called F fraction, an oligomycin-sensitive ATPase, and shown it to be a sphere of diameter 85 Å. In recent years, a number of investigations have been carried out to reconstitute the isolated membrane fractions (Green and Perdue, 1966; Fleischer et al., 1967). Finally, a novel technique recently developed by DePury and Collins (1966) for studying the binding of lipid micelles by mitochondrial protein has been reported.

4. Nerve Membranes

Unlike other membranes, the nerve membrane as typified by myelin, contains the smallest percentage of protein. For example, human myelin consists of 80% lipid and only 20% protein (O'Brien, 1967). The fatty acids of the lipids are the monounsaturated oleic and saturated stearic acid together with some polyunsaturated C_{28}–C_{24} acids. The proteins of myelin are mainly nonenzymic (Finean, 1969).

5. Visual Receptor Membranes

The inner layers of the retina or double-membrane sacs of the outer segments of the retinal receptor cells are believed to have their origin in the plasma membrane (Sjöstrand, 1959). Chemically, the visual pigments, the integral part of the visual membrane structure, are well characterized (Wald, 1968). The gross composition of the rod outer segments is about

40 to 60% protein, 20 to 40% lipids, and 4 to 10% retinenes (Wolken, 1968; Eichberg and Hess, 1967). The visual pigment, rhodopsin, accounts for about 35% of the dry weight, the rest being mainly lipids (Krinsky, 1958). The protein moiety of the visual pigment, known as opsin, is less stable when isolated from rhodopsin (Radding and Wald, 1956), which has a molecular weight between 40,000 and 60,000. Of the retinenes, 11-*cis* plays the most crucial role in visual excitation (Wald, 1968).

C. STRUCTURE OF BIOLOGICAL MEMBRANES

At the present level of knowledge on the chemical composition of membranes, the data thus far gathered are fragmentary and at best available for only a few types of membranes (see Section III, B). Therefore, it is fair to state at the outset that any discussions on the structure of membranes will be highly speculative.

The current theory of membrane structure has its origin in surface and colloid chemistry. A succinct account of the role played by surface chemistry in biology has been given by Lawrence (1958). The study of monomolecular layers (monolayers) at air–water interfaces introduced by Langmuir (1917) and others (see Adamson, 1967) had provided a physical picture of the structure and orientation of lipid molecules at interfaces. The monolayer technique was applied by Gorter and Grendel (1925) to estimate the thickness of erythrocyte plasma membrane. Notwithstanding their experimental artifacts, which are mainly of historical interest, Gorter and Grendel's contribution lies in their recognition that the postulated lipid barrier (Overton, 1899) consisted of a bimolecular lipid leaflet. With modern surface-film balance the classic experiment has been repeated recently and fully confirmed Gorter and Grendel's conclusion (Bar *et al.*, 1966). The bimolecular lipid layer with adsorbed protein layers was put into pictorial form by Davson and Danielli (1952) who deduced their model from the concept of Gorter and Grendel (1925) and from the results of interfacial tension measurements on cells (Harvey and Shapiro, 1934; Cole, 1932). The bimolecular leaflet model (Fig. 3) for the structure of biological membranes may be simply stated as follows. The basic construct of all biological membranes consists of a bimolecular lipid leaflet (bilayer) with adsorbed nonlipid layers (mainly proteins). The lipid molecules are oriented with their polar groups facing outward and their hydrocarbon chains away from interfaces forming the interior of the membrane. In the interior of the bilayer, hydrocarbon chains are held together by London–van der Waals forces and are in a liquidlike state. More recently, Lucy and Glauert (1968), based on their experimental observations, have developed an alternative model for lipoprotein membranes. It is assumed that some of the lipid molecules are arranged in globular micelles in a dynamic equilibrium with the lamellae. In either of these models the bonding between lipids and proteins and other nonlipid materials is not known but is mostly likely either ionic or hydrophobic or a combina-

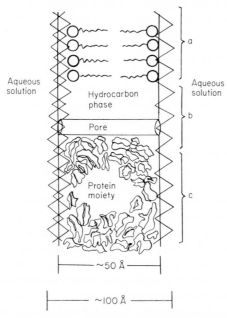

FIG. 3. A composite model for the structure of biological membranes based on the bimolecular leaflet concept. (a) Illustrating a simple lipid bilayer with adsorbed protein coats. (b) A region of lipid bilayer perforated by a pore. (c) A complex lipoprotein region resulting from lipid–protein interaction. Protein coats are shown asymmetrically in accordance with the "unit membrane" concept. The proteins may be in α-helical configuration. The hydrocarbon phase is in a liquid state and may contain globular micelles and solubilized water. The overall thickness may vary from 50 to 100 Å.

tion of both (Willmer, 1961). The absence of covalent bonding is supported by the fact that membrane lipids and proteins can be separated with ease using organic solvents and surfactants. The role played by ubiquitous water is obscure but must be crucial in maintaining the integrity of the membrane structure. For further details concerning the growth of the concept of the bimolecular leaflet model, the article by Davson (1962) is of interest.

In recent years the bilayer leaflet model has been popularized by

Robertson (1966) who, based upon his extensive electron-microscopic observations, has come to the conclusion that all biological membranes, irrespective of their functions or composition, are lipid bilayers or a modification thereof. The modified bilayer lipid models include the models of Benson (1966), Green and Perdue (1966), Lenard and Singer (1966), Lucy (1968), Mühlethaler (1966), and others (see Kavanau, 1965; Castleden, 1969). A composite model for the bimolecular lipid leaflet and its modifications is shown in Fig. 3.

It should be mentioned that the bilayer lipid model for the structure of biological membranes is not universally accepted by all interested investigators (Green and Perdue, 1966; Korn, 1966). A brief summary of some of the major arguments against the bilayer lipid model is given below.

Korn (1966), a leading critic of the bimolecular leaflet model, has argued that neither the intracellular nor plasma membranes are passive structural elements but, instead, are sites of important metabolic reactions. He has further stated that it is more reasonable to consider protein as the fundamental structural unit—it is the substance providing control of selectivity and specificity. Korn has suggested that protein might be laid down first and lipids then added in an arrangement governed by the structure of the protein; the resultant lipoproteins interact to form the membrane. The unit membrane model seems to require that the lipid bilayer is formed first, and this specifies the protein arrangement. The shift in optical rotatory dispersion (ORD) and circular dichroism (CD) can be interpreted in terms of hydrophobic interactions between lipid and protein or in terms of interactions between α-helical regions of adjacent protein molecules. Whether it is a requisite of these models that the protein helices be buried in a hydrophobic region is at the present time uncertain.

The second argument against the bilayer lipid model has been based upon the uncertainty of electron microscopy (Korn, 1966). The interpretation of electron micrographs requires a great deal of skill and imagination. There is no agreement among authorities in the field on what constitutes an artifact under the electron microscope (Finean, 1969; Sjöstrand, 1959; Green and Perdue, 1966; Korn, 1966; Elbers, 1964; Stoeckenius, 1966; Glauert, 1968). Thus, until better experimental techniques are available, the validity based upon the results of electron microscopy remain uncertain.

The third argument against the bilayer model is a chemical one. It has apparently been found that the lipid-to-protein ratio in isolated membranes varies widely. For example, the lipid contents vary all the way from 30% of the dry weight (in mitochondria) to about 80% in nerve myelin. The remaining weight is made up mostly by proteins (see Table

IV and Section IV, A). The argument based on the ratio of lipid to protein constituents is a weak one, since at present the techniques used in the isolation of membranes from the rest of cellular components are far from ideal. In many cases, meaningful chemical data simply do not exist (van Deenen, 1965). The problem of isolating membrane fraction of high purity from cells must be first solved. The fragmentary data now available, therefore, should be used with discretion.

Perhaps one of the main shortcomings (or its strength) is the all-inclusive nature of the bilayer lipid model which boldly claims that all biological membranes are alike. A brief reflection on the diversity and functions of biological membranes, as they are understood today, would seem to indicate otherwise. This argument although plausible is not central to the point. The bilayer lipid model sets the limit as to what is the most likely structure of the biological membrane. A proper interpretation, perhaps, is that the bilayer lipid model should be viewed as a basic matrix upon which the diversity and functions of biological membranes are built. The perturbation and modification of the basic bilayer structure by a host of lipid and nonlipid materials would appear to be more than adequate to accommodate all the functional diversities that are likely to arise in nature. On energetic considerations the bilayer model provides not only an overall symmetry but configurations that are natural for lipids and proteins. At the present time it is difficult to conceive a more basic structure in two dimensions than the bilayer leaflet model.

It must be concluded, therefore, that the bilayer lipid model, together with its modifications, at the present time constitutes a valid approach to understanding the structure and function of biological membranes. As a general framework, the bilayer lipid model is useful not only theoretically but suggests positive and constructive experments as now being carried out with BLM and with lipid microvesicles (Bangham, 1968; Castleden, 1969; Henn and Thompson, 1969).

D. Methods of Investigating Lipid–Protein Interactions

In addition to approaches using model systems (Bangham, 1968; Castleden, 1969; Tien, 1970), there are a number of sophisticated physical techniques that have been used in the study of lipid–lipid and lipid–protein interactions. Three spectroscopic techniques have been recently applied to the study of lipid–protein interactions in membranes. The first is NMR spectroscopy (Chapman, 1968). The fundamentals of NMR spectroscopy are that certain nuclei (e.g., ^1H and ^{15}N) behave like minute magnetic bars and can be lined up in the presence of a magnetic field. The absorption of radiation energy can cause the magnets to flip from one energy level to another. The protons in methylene, amine, and double bonds, for example, give separate lines associated with the characteristic resonance energy. Interactions between lipid and protein can be investigated by comparing the spectrum with those of the lipid and protein alone. Chapman (1968) and associates (Ladbrooke *et al.*, 1968; Leslie and Chapman, 1969) have reported NMR spectra from phospholipids, bile salts, serum lipoproteins, and red blood cell ghosts. Serum lipoprotein has also been studied by Steim *et al.* (1968). The conclusions drawn from these studies are that interaction between protein and lipid resulted in the restriction of the mobility of the hydrocarbon chains which presumably were in a liquid state. The nature of interaction between lipid and protein is believed to be hydrophobic.

The second spectroscopic technique is ORD and its variation, CD. Both ORD and CD spectra of a variety of membranes have been reported (Lenard and Singer, 1966; Wallach and Zahler, 1966; Urry *et al.*, 1967; Steim and Fleischer, 1967). Both types of spectra may indicate that a part of the protein is in an α-helical configuration. In addition a red shift is also observed as compared with the standard spectrum of poly-L-lysine (Steim and Fleischer, 1967). It is interesting to note that infrared studies of erythrocyte and ascites tumor cells give no evidence of a β-configuration in these membranes (Maddy and Malcolm, 1966; Wallach and Zahler, 1966).

More recently, the so-called "spin-labeling" technique has been extended to the study of biological membranes (McConnell and Hubbell, 1968; Griffith and Waggoner, 1969). The basis of spin-labeling technique is that a tag molecule (spin label) on a macromolecule such as a protein can be measured with an ESR spectrometer. The spin labels thus far used are nitroxide-free radicals such as di-t-butyl nitroxide and 2,2,6,6-tetramethylpiperidine-1-oxyl (TEMPO). Except in detecting instrumentation, the use of a spin label is very much similar to that of introducing a dye into a biological material. The advantages of spin labels are sensitivity to the local environment and ability to measure very rapid molecular motion (Griffith and Waggoner, 1969). To date, the spin-label technique has been used on rabbit nerves, excitable membranes, bovine red blood cell ghosts, and model systems. The conclusion drawn from these experiments is that biological membranes possess hydrophobic regions consistent with the current bilayer lipid model (see Fig. 3). Whether the technique can provide other information concerning the state of aggregation and the molecular structure of biological membranes remains to be seen. However, the technique is unique in many ways and undoubtedly will be further exploited in the study of biological membranes.

The other physical techniques currently being used are gel filtration, nonaqueous dialysis, differential scanning calorimetry (DSC), and microcalorimetry (heat-burst calorimetry). A brief review of the literature is given below.

Gel filtration technique has been used by Fleischer *et al.* (1962, 1967) and by Scanu *et al.* (1969). The latter investigator used Sephadex G-200 to separate the products of the high-density lipoprotein (HDL). Fleischer *et al.* used the method to investigate the interaction of γ-lipoprotein with a phospholipid. This method is of value in determining the stoichiometry of binding between lipids and protein. Similarly, nonaqueous dialysis will provide information about the nature of lipid–protein interaction. The technique involves a partition of lipid in a nonaqueous solvent separated by a dialyzing membrane immersed in an aqueous protein solution. Products of chloroplast lamellae were studied by this technique (Ji *et al.*, 1968). Interactions between fatty acids and serum albumin have been investigated by Spector *et al.* (1969) and by Arvidson and Belfrage (1969).

Differential scanning calorimetry is a classic method but application to biological systems is a recent event (Steim, 1968; Ladbrooke *et al.*, 1968). The basis of DSC is to heat a sample in a reference compound of known melting point and measure the difference in power input.

If a transition takes place, more heat must be supplied to the sample than to the known compound. The heat of transition is proportonal to the area of recording. The phase transitions observed in the DSC can be correlated with changes in biological properties. For example, Steim *et al.* (1968) have observed that the transition temperature of intact membranes of *M. laidlawii* occurs at the same temperature as in the extracted lipids dispersed in water. Using DSC technique, Ladbrooke *et al.* (1968) have studied the effect of cholesterol on phase transitions of phospholipids.

The last technique to be outlined is the so-called heat-burst calorimetry. The technique has been used in biochemistry since the work of Kitzinger and Benzinger (1960). The technique has been used by Lovrien and Anderson (1969) to study the interaction between sodium dodecyl sulfate and γ-lactoglobulin, and by Klopfenstein (1969) to determine the enthalpy change on binding of lysolecithin by BSA. The technique is of potential value in providing quantitative thermodynamic data useful in the understanding of the nature of binding forces between the lipid and protein.

E. INTERACTIONS IN BIOLOGICAL MEMBRANES

1. Myelin

Turning to biological membranes in quest of information about protein–lipid interactions at interfaces, we find initially some confirmation of the classic unit membrane concept (Robertson, 1966).

With myelin sheaths of nerves and retinal rods and cones, the experimental evidence coming from polarized light studies, X-ray diffraction, chemical analysis, and electron microscopy confirms that the structures present are multilamellar. Finean (1969) has reviewed the subject extensively. From lipid analyses useful generalizations can be made. Myelin and probably the plasma or cell-surface membranes are characterized by high cholesterol and sphingolipid content together with a predominance of saturated and monounsaturated fatty acids. In contrast, mitochondria, endoplasmic reticulum, and perhaps cytoplasmic membranes as a group possess generally low levels of cholesterol and sphingolipids and contain a relatively high proportion of polyunsaturated fatty acids.

Myelin seems to be the exceptional structure which provides the best experimental support for the bilayer lipid model. Studies of birefringence of the myelin sheath of peripheral nerve led Schmidt (1936) to postulate a structure of alternating protein and lipid layers encircling the nerve axon. The lipid chains are considered to be oriented radially, and successive lipid layers are separated by thin layers of protein. The process of myelin formation has been seen to arise from a kind of spiral wrapping up of the axon by an extension of the surface membrane of a Schwann cell (Finean, 1969). The repeat period of myelin of different origins varies from 100 to 120 Å in the central nervous system to about 160 Å in optic nerve and white matter of brain and spinal cord to nearly 180 Å in mammalian peripheral nerve. Each myelin layer, of course, encompasses two cell membranes of opposite orientation. Thus the thickness of myelin supports the concept that it includes two lipid bilayers alternating with two nonlipid regions (perhaps largely protein in the α-configuration) of thickness perhaps 30 to 45 Å each. Although myelin is a rather atypical membrane system, low in protein and metabolically rather static, studies of spinal cord in vitro do show higher metabolic activity for the protein portion. This is taken as further evidence for the lipid bilayer or unit membrane structure for myelin. It must be mentioned that the first study of lipoprotein interaction in nerve myelin was reported by Schmitt et al. (1941) and has recently been reviewed (Schmitt, 1970).

2. Plasma Membranes

Formally similar to myelin in lipid content and in appearance on electron micrographs are plasma or cell surface membranes. The most studied of these is the red blood cell membrane (Zahler, 1969). It was, in fact, from this membrane that Gorter and Grendel (1925) first extracted lipid, spread it as a monolayer at the air–water interface, and calculated that the area occupied was about twice that of the original red blood cells leading to the now familiar bilayer concept as mentioned earlier.

In recent years, despite the generally obtained unit membrane pattern on electron micrographs of the red cell membrane or in red cell ghosts, a dispute has arisen about the detailed membrane structure.

Low-frequency impedance studies on beef red cell suspensions have failed to show any dispersion below 1 kc which casts doubt on the possibility of lipids being present in a continuous bimolecular layer in these red cell membranes. Human red cell membranes upon treatment with the enzyme phospholipase C lose about 70% of their total phosphorus by hydrolysis (Lenard and Singer, 1968). However, the membrane remains intact under phase microscopy and the protein content, as measured by CD, is virtually unchanged. The model proposed is stabilized by hydrophobic interactions and locates polar and ionic heads of lipids on the outer surface of the membrane accessible to enzyme. Nonpolar interactions between hydrocarbon chains of lipid and protein are also suggested on the basis of NMR studies of red blood cell membrane fragments after sonic disruption (Chapman, 1968). Absorption peaks of methylene protons of hydrocarbon residues of proteins are unresolved until after treatment with urea or trifluoroacetic acid.

Wallach and Zahler (1966) conclude that red cell ghosts do not contain large amounts of protein in the extended β-configuration based upon infrared and fluorescence spectroscopy. Optical rotatory dispersion measurements suggest that portions of the membrane protein exist in regions of high refractive index, supporting the suggestions of hydrophobic interactions. A model is proposed in which hydrophilic peptide portions lie at the membrane surface connected by hydrophobic helical rods penetrating the apolar core of the membrane normal to its surface. The interior of these rods could be polar providing for hydrophilic "pores" through the membrane (Wallach and Gordon, 1968). Other authors agree that protein extracted from the plasma membrane is not in the β-configuration and consider, instead, that it is a typical globular protein with low α-helical content (Maddy and Malcolm, 1966). Spin-labeling experiments indicate that red blood cells, their ghosts, and mitochondrial membranes among others do not contain liquidlike, hydrophobic regions of low viscosity capable of binding the TEMPO spin label (Hubbell and McConnell, 1968). These regions are presumably phospholipid bilayers similar to those found in artificially prepared phospholipid vesicles.

Additional difficulties arise from interpretation of electron micrographs of plasma membranes. Recent work is summarized by Chapman (1968). Among the findings which must either be made to fit the membrane model or be explained away as artifacts are more than twofold variations in thickness and granular or "pebbly" substructure revealed principally by shadowing techniques. Korn (1966), as mentioned earlier, has been highly critical of attempts to consider unit membranes ubiquitous in nature and, indeed, has raised serious questions about the basis for interpreting electron micrographs in terms of the triple-layered structure required for the unit membrane. The assumption that OsO_4 fixation stains only double bonds is in question currently and so is the assumption that the membrane structure is unchanged by such fixation. Recent small-angle X-ray diffraction studies indicate that standard electron microscopy fixatives, OsO_4 and $KMnO_4$, did not prevent rearrangement of the structure of the myelin membrane of frog sciatic nerve during subsequent acetone and alcohol dehydration (Moretz et al., 1969). The observed changes in periodicity and intensity are said to be associated with a considerable extraction of cholesterol and a lesser extraction of polar lipids. It should be clear that the bimolecular lipid leaflet with accompanying protein layers has been based to a significant extent on studies of myelin, which may be greatly different from other membranes, and upon studies of the lamellar liquid–crystalline phase of lipids formed in water. The important matter is the configuration adopted when lipids and proteins are present together in the membrane.

More recently, the effect of tranquilizers and anesthetics on red cell plasma membrane ghosts has been reported (Seeman et al., 1969).

3. Mitochondrial Membranes

Electron micrographs of sectioned preparations of mitochondria show two dark layers, one representing the smooth, continuous, outer mitochondrial membrane, and the other is invaginated to form cristae. When fixed with potassium permanganate, the dense lines appear granular and in some preparations the membrane has a segmented appearance. Fernández-Morán (1962) and Sjöstrand (1959) reported finding globular subunits in the inner opaque layer of the mitochondrial membrane.

Extraction of 90% or more of the mitochondrial lipid with acetone–water mixtures does not materially alter the gross appearance of the OsO_4-

fixed specimen in electron micrographs. This suggests that protein may be primary in determining mitochondria membrane structure (Green and Fleischer, 1964). Respiratory activity is lost following such delipidation but regained when phospholipid is restored even though the added phospholipid may be greatly different in composition from the original (Fleischer et al., 1962). It is suggested that ionic, hydrogen, and hydrophobic bonding are involved in anchoring the lipid within the protein particulate.

Optical rotatory dispersion spectra of mitochondria and submitochondrial particles show features similar to those of the red blood cell and Ehrlich ascites carcinoma membranes—a characteristic α-helix curve shifted to longer wavelengths (Urry et al., 1967). With mitochondrial membranes the Cotton effect is displaced in a positive background and the amplitude of the effect is very low; the latter implying that much of the optical activity of the membrane protein is masked. The red shift of ORD absorption and short-wavelength crossover is attributed to transfer to a less polar environment. The long-wavelength shift in the ORD trough and the low amplitude of the Cotton effect are explained in terms of the optical activity of the membrane phospholipid.

However, ORD spectra of membranes are said to be characteristic of partially helical polypeptides (about one-third helical) except that the Cotton effect is red-shifted by several millimicrons. Based upon studies of lipid-free membrane proteins, it is suggested (Steim, 1968) that this red shift is due to protein–protein association. Differential scanning calorimetry applied to bacterial membranes (M. laidlawii) and to dispersions of membrane lipids shows that both undergo thermal phase transitions characteristic of lipids in the bilayer configuration. The red shift found in ORD on lipid-free mitochondrial structural protein and red cell ghost protein is attributed to protein–protein interactions leading to aggregated protein.

Wallach and Gordon (1968) have summarized the evidence for hydrophobic lipid–protein interactions in mitochondrial and other biological membranes. They propose a membrane model compatible with such interactions along with a high α-helical content in the membrane protein. Essentially their suggestion is that peptides are located on the membrane surfaces and also within the apolar membrane core. The peptide located in the surface is irregularly coiled but that portion which penetrates the membrane is helically coiled in rods and presents apolar amino acid residues to the other surfaces of the rodlike assemblies. These interact through specific binding sites with hydrocarbon residues of tightly bound membrane lipids. This proposal is similar to that of Lenard and Singer

(1966) mentioned earlier, except that they placed all ionic groups, protein and lipid, on the exterior surfaces of the membrane.

In spite of the variety of original and occasionally provocative new models, a number of papers continue to appear tending to support the bimolecular leaflet model. For example, the β-conformation in extracted mitochondrial structural protein is reported to predominate (Wallach *et al.*, 1969). Infrared spectroscopy on rat liver mitochondria, their "inner" and "outer" membranes and extracts, reveals a distinct shoulder near 1630 cm^{-1}, said to be at the position of the amide-I band of the β-structure. Also seen is another shoulder near 1690 cm^{-1} suggesting that the β-structure is an antiparallel pleated sheet.

Crane and Hall (1969) have reinterpreted electron micrographs of stained mitochondrial sections in terms of a binary membrane consisting of a double layer of lipoprotein globules. The mitochondrial membrane appears as a five-layered structure—three dark lines separated by two unstained regions. The proposal had its origin in fragmentation studies of cristae. It is said that fracture through the central polar regions of the membrane is consistent with the lipoprotein bilayer membrane but not with the unit membrane consisting of a single layer of lipoprotein particles as described by Green and Perdue (1966). Crane and Hall assert that the Schwann cell membrane appears to be binary, whereas the myelin derived from it has the unit structure; thus, there must be some sort of transition in membrane structure in which one of the binary layers seems to be lost during myelin formation. These authors list additional electron micrographs which they consider to show the binary membrane structure.

It has also been found that mitochondrial cristae membranes can be split by detergent treatment into two fractions each of which shows a globular structure (Prezbindowski *et al.*, 1968). The thicker membrane is said to be re-formed when the two fractions are recombined. From a consideration of amino acid composition of soluble and membranous lipoproteins, Hatch and Bruce (1968) conclude that the marked differences in the types of lipoproteins are related to the different environments to which they are exposed.

4. Light-Sensitive Biological Membranes

The best known light-sensitive membranes are the thylakoid membrane of chloroplasts and membranes in retinal rods. Lipid–protein interactions in these membranes are not at all well characterized as compared with the other types described in the preceding paragraphs. Studies of chloroplast membranes by polarized light have shown birefringence (Frey-Wyssling, 1957; Menke, 1967; see also Branton, 1969). The birefringence was explained as a result of alternating lipid and protein lamellae. Park and Pon (1963) proposed a "quantasome" model based on the granular appearance of some areas of chloroplast membranes under the electron microscope. However, this model for the thylakoid membrane gives no information concerning the interaction between lipids and proteins. It is assumed nonetheless that the quantasomes are embedded in a matrix of lipid (bilayer?). Menke (1967), who suggested the term "thylakoid" membrane, proposed a different model in that a continuous lipid layer, only one molecule thick, is envisioned having a protein coat on one side only. The Menke model has many conceptual difficulties. More consistent with the classic bimolecular layer concept is the model of Mühlethaler (1966) as already mentioned in Section IV, C.

The photosensitive visual pigments are located in the outer segment of retinal rods. They are complexes of proteins (opsins) and retinals forming so-called rhodopsins. Phospholipids, especially phosphatidyl ethanolamine, are implicated in the visual receptor membranes (Abrahamson and Ostroy, 1967). Whether 11-*cis*-retinal, the principal chromophore, is linked directly to opsin or bound by a Schiff base linkage to the phospholipid is uncertain at the present time (Poincelot *et al.*, 1969). The suggestion that the photolytic transfer of the retinal from lipid to protein may be involved in visual excitation is intriguing and needs to be tested experimentally. Perhaps the use of BLM for such studies may prove helpful (Tien, 1970).

V. Conclusions

Since the main components of biological membranes are lipids and proteins, the importance of their study cannot be overemphasized. In this review an attempt has been made to stress the usefulness of model systems in the investigation of lipid–lipid and lipid–protein interactions that are relevant to an understanding of complicated biological membranes.

In the past years, colloid and surface chemistry have contributed enormously toward our understanding of interfacial phenomena. Molecular orientation of amphipathic molecules at interfaces and the shape of molecules are some of the outstanding examples. Recently, the development of new model systems such as BLM and liposomes has provided fresh impetus and new approaches for studying lipid–protein interactions. As an experimental model for the biological membrane, BLM have demonstrated that a stable structure in the order of bimolecular thickness is capable of separating two aqueous solutions. Certain properties of modified BLM have been found resembling those of biological membranes.

Although at the present time monolayer techniques are much less used, they still constitute a powerful tool for obtaining information concerning the manner in which lipid and protein molecules orient and interact at interfaces. In the bulk phase, liposomes appear to be a better system for studying lipid–protein interaction than micelles. However, that micelles, globular micelles in particular, may play an important role in biological membranes should not be overlooked.

In addition to model membrane systems, sophisticated physical techniques such as ESR, NMR, and DSC, are being applied to a variety of biological membranes. These membranes are typified by the plasma membrane, the thylakoid membrane, the cristae membrane, the nerve membrane, and the visual receptor membrane. It is currently accepted that all biological membranes have a basic construct comprising a bimolecular lipid core with sorbed protein layers. It is hoped that the lipid–lipid and lipid–protein interaction studies reviewed in this chapter provide a useful basis for further investigations.

Acknowledgments

This chapter was to have been written in collaboration with Dr. Leroy G. Augenstein. This was not possible because of his untimely death. This work was supported by National Institutes of Health Grants CA-006-30, and GM-14971 and by Atomic Energy Commission Grant AT-11-1-1155 (to L. G. Augenstein).

REFERENCES

Abrahamson, E. W., and Ostroy, S. E. (1967). *Progr. Biophys. Mol. Biol.* **17**, 179.

Adamson, A. W. (1967). "Physical Chemistry of Surfaces." Wiley (Interscience), New York.

Arvidson, E. O., and Belfrage, P. (1969). *Acta Chem. Scand.* **23**, 232.

Azzi, A., Chance, B., Radda, G. K., and Lee, C. P. (1969). *Proc. Nat. Acad. Sci. U.S.* **62**, 612.

Bangham, A. D. (1968). *Progr. Biophys. Mol. Biol.* **18**, 29–95.

Bangham, A. D., Standish, M. M., and Watkins, J. C. (1965a). *J. Mol. Biol.* **13**, 238.

Bangham, A. D., Standish, M. M., and Weissmann, G. (1965b). *J. Mol. Biol.* **13**, 253.

Bangham, A. D., de Gier, J., and Greville, G. D. (1967). *Chem. Phys. Lipids* **1**, 225.

Bar, R. S., Deamer, D. W., and Cornwell, D. G. (1966). *Science* **153**, 1010.

Bean, R. C., Shepherd, W. C., and Chan, H. (1968). *J. Gen. Physiol.* **52**, 495.

Bean, R. C., Shepherd, W. C., Chan, H., and Eichner, J. (1969). *J. Gen. Physiol.* **53**, 741.

Benson, A. A. (1966). *J. Am. Oil Chemists' Soc.* **43**, 265.

Bielawski, J., Thompson, T. E., and Lehninger, A. L. (1966). *Biochem. Biophys. Res. Commun.* **24**, 948.

Blank, M. (1968). "Biological Interfaces: Flows and Exchanges." Little, Brown, Boston, Massachusetts.

Bradley, J., and Dighe, A. M. (1969). *J. Colloid Interface Sci.* **29,** 157.

Branton, D. (1969). *Annu. Rev. Plant. Physiol.* **20,** 209.

Bruckdorfer, K. R., and Green, C. (1967). *Biochem. J.* **104,** 270.

Cadenhead, D. A., and Phillips, M. C. (1968). *Advan. Chem. Ser.* **84,** 131.

Carter, H. E., McClure, R. H., and Slifer, E. D. (1956). *J. Amer. Chem. Soc.* **78,** 3735.

Castleden, J. A. (1969). *J. Pharm. Sci.* **58,** 149.

Chapman, D. (1968). *Lipids* **4,** 251.

Cherry, R., and Chapman, D. (1969). *J. Theor. Biol.* **24,** 137.

Colacicco, G. (1969). *J. Colloid Interface Sci.* **29,** 345.

Colacicco, G., and Rapport, M. M. (1968). *Advan. Chem. Ser.* **84,** 157–168.

Cole, K. S. (1932). *J. Cell. Comp. Physiol.* **1,** 1.

Cook, W. H., and Martin, W. G. (1962). *Can. J. Biochem. Physiol.* **40,** 1273.

Crane, F. L., and Hall, J. D. (1969). *Biochem. Biophys. Res. Commun.* **36,** 174.

Criddle, R. S. (1969). *Annu. Rev. Plant. Physiol.* **20,** 239.

Davies, J. T., and Rideal, E. (1963). "Interfacial Phenomena." Academic Press, New York.

Davson, H. (1962). *Circulation* **26,** 1022.

Davson, H., and Danielli, J. F. (1952). "Permeability of Natural Membranes." Cambridge Univ. Press, London and New York.

Dawson, R. M. C. (1968). *In* "Biological Membranes" (D. Chapman, ed.), p. 203. Academic Press, New York.

del Castillo, J., Rodriguez, A., Remero, C. A., and Sanchez, V. (1966). *Science* **153,** 185 (1966).

DePury, G. G., and Collins, F. D. (1966). *Chem. Phys. Lipids* **1,** 20.

Dervichian, D. G. (1946). *Trans. Faraday Soc.* **42,** 180.

Edelman, G. M., and McClure, W. O. (1968). *Accounts Chem. Res.* **1,** 65.

Eichberg, J., and Hess, H. H. (1967). *Experientia* **23,** 993.

Elbers, P. F. (1964). *Recent Progr. Surface Sci.* **2,** 443.

Eley, D. D., and Hedge, D. G. (1956). *Discuss. Faraday Soc.* **21,** 221.

Faraday society. (1949). *Discuss. Faraday Soc.* **6,** 000.

Fernández-Morán, H. (1962). *Circulation* **26,** 1039.

Finean, J. B. (1969). *Quart. Rev. Biophys.* **2,** 1.

Finkelstein, A., and Cass, A. (1968). *J. Gen. Physiol.* **52,** 145s.

Fisher, W. R., and Gurin, S. (1964). *Science* **143,** 362.

Fleischer, S., Brierley, G., Klouwen, H., and Slautterback, D. B. (1962). *J. Biol. Chem.* **237,** 3264.

Fleischer, S., Fleischer, B., and Stockenius, W. (1967). *J. Cell Biol.* **32,** 193.

Fraser, M. J., Kaplan, G., and Schulman, J. H. (1955). *Discuss. Faraday Soc.* **20,** 44.

Frey-Wyssling, A. (1957). "Macromolecules in Cell Structure." Harvard Univ. Press, Cambridge, Massachusetts.

Ghosh, S., and Bull, H. B. (1962). *Biochim. Biophys. Acta* **66,** 150.

Glauert, A. M. (1968). *J. Roy. Microsc. Soc.* [3], **88,** 49.

Gorter, E., and Grendel, F. (1925). *J. Exp. Med.* **41,** 439.

Gotto, A. M., Levy, R. I., and Fredrickson, D. S. (1968). *Proc. Nat. Acad. Sci. U.S.* **60,** 1436.

Green, D. E., and Fleischer, S. (1964). *In* "Metabolism and Physiological Significance of Lipids" (R. M. C. Dawson and D. N. Rhodes, eds.), p. 238. Wiley, New York.

Green, D. E., and Perdue, J. (1966). *Proc. Nat. Acad. Sci. U.S.* **55**, 1295.

Green, D. E., Allman, D. W., Bachmann, E., Baum, H., Kopaczyk, K., Korman, E. F., Lipton, S., MacLennan, D. H., McConnell, D. G., Perdue, J. F., Rieske, J. S., and Tzagoloff, A. (1967). *Arch. Biochem. Biophys.* **119**, 312.

Griffith, O. H., and Waggoner, A. S. (1969). *Accounts Chem. Res.* **2**, 17.

Gurd, F. R. N. (1960). *In* "Lipide Chemistry" (D. J. Hanahan, ed.), p. 135. Wiley, New York.

Hanai, T., Haydon, D. A., and Taylor, J. (1964). *Proc. Roy. Soc., Ser. A* **281**, 377.

Hanai, T., Haydon, D. A., and Taylor, J. (1965). *J. Theor. Biol.* **9**, 422.

Harkins, W. D., and Boyd, G. E. (1941). *J. Phys. Chem.* **45**, 20 (1941).

Harvey, E. N., and Shapiro, H. (1934). *J. Cell. Comp. Physiol.* **5**, 255.

Hashimoto, M. (1968). *Bull. Chem. Soc. Jap.* **41**, 2823.

Hatch, F. T., and Bruce, A. L. (1968). *Nature (London)* **218**, 1166.

Haydon, D. A. (1968). *In* "Membrane Models and the Formation of Biological Membranes" (L. Bolis and B. A. Pethica, eds.), pp. 91–97. North-Holland Publ., Amsterdam.

Henn, F. A., and Thompson, T. E. (1969). *Annu. Rev. Biochem.* **38**, 241.

Henn, F. A., Decker, G. L., Greenawalt, J. W., and Thompson, T. E. (1967). *J. Mol. Biol.* **24**, 51.

Huang, C. (1969). *Biochemistry* **8**, 344.

Huang, C., Wheeldon, L., and Thompson, T. E., (1964). *J. Mol. Biol.* **8**, 148.

Hubbell, W. L., and McConnell, H. M. (1968). *Proc. Nat. Acad. Sci. U. S.* **61**, 12.

Jain, M. K., Strickholm, A., and Cordes, E. H. (1969). *Nature (London)* **222**, 871.

James, L. K. (1958). Ph.D. Thesis. University of Illinois, Urbana, Illinois.

James, L. K. (1968). *J. Colloid Interface Sci.* **28**, 260.

James, L. K., and Augenstein, L. G. (1966). *Advan. Enzymol.* **28**, 1–40.

Ji, T. H., Hess, J. L., and Benson, A. A. (1968). *Biochim. Biophys. Acta* **150**, 676.

Jung, C. Y., and Snell, F. M. (1968). *Fed. Proc., Fed. Amer. Soc. Exp. Biol.* **27**, 286.

Kavanau, J. L. (1965). "Structure and Function in Biological Membranes," Vol. I. Holden-Day, San Francisco, California.

Kendrew, J. C. (1962). *Brookhaven Symp. Biol.* **15**, 216.

Khaiat, A., and Miller, I. R. (1969). *Biochim. Biophys. Acta* **183**, 309.

Kitzinger, C., and Benzinger, T. H. (1960). *In* "Methods of Biochemical Analysis" (D. Glick, ed.), Vol. 8. (Interscience) Wiley, New York.

Klopfenstein, W. E. (1969). *Biochim. Biophys. Acta* **181**, 323.

Korn, E. D. (1966). *Science,* **153**, 1491.

Krinsky, N. I. (1958). *AMA Arch. Ophthalmol.* **60**, 688.

Kushnir, L. D. (1968). *Biochim. Biophys. Acta* **150**, 285.

Ladbrooke, B. D., Williams, R. M., and Chapman, D. (1968). *Biochim. Biophys. Acta* **150**, 333.

Langmuir, I. (1917). *J. Amer. Chem. Soc.* **39**, 1848.

Lauger, P., Lesslauer, W., Marti, E., and Richter, J. (1967). *Biochim. Biophys. Acta* **135**, 20.

Lawrence, A. S. C. (1958). *In* "Surface Phenomena in Chemistry and Biology" (J. F. Danielli, K. G. A. Pankhurst, and A. C. Riddiford, eds.), pp. 9–17. Pergamon Press, Oxford.

Lenard, J., and Singer, S. J. (1966). *Proc. Nat. Acad. Sci. U. S.* **56**, 1828.

Leslie, R. B., and Chapman, D. (1969). *Chem. Phys. Lipids* **3**, 152.

Lev, A. A., and Buzhinsky, E. P. (1967). *Zh. Evol. Biokhim. Fiziol.* **9**, 102.

Liberman, E. A., Mokhova, E. N., Skulachev, V. P., and Topaly, V. P. (1968). *Biofizika* **13**, 188.

Lichtenthaler, H. K., and Park, R. B. (1963). *Nature (London)* **198**, 1070.

Lippe, C. (1968). *J. Mol. Biol.* **35**, 635.

Lockshin, A., and Burris, R. H. (1966). *Proc. Nat. Acad. Sci. U. S.* **56**, 1564.

Lovrien, R., and Anderson, W. (1969). *Arch. Biochem. Biophys.* **131**, 139.

Lucy, J. A. (1968). *In* "Biological Membranes" (D. Chapman, ed.), p. 233. Academic Press, New York.

Lucy, J. A., and Glauert, A. M. (1968). *Symp. Int. Soc. Cell Biol.* **6**, 19–37.

Luzzati, V. (1968). *In* "Biological Membranes" (D. Chapman, ed.), p. 71. Academic Press, New York.

McBain, J. W. (1950). "Colloid Science." Heath, Boston, Massachusetts.

Maddy, A. H. (1966). *Int. Rev. Cytol.* **20**, 1.

Maddy, A. H., and Malcolm, B. R. (1966). *Science* **153**, 213.

Maddy, A. H., Huang, C., and Thompson, T. E. (1966). *Fed. Proc., Fed. Amer. Soc. Exp. Biol.* **25**, 933.

Malcolm, B. R. (1968). *Proc. Roy. Soc., Ser. A* **305**, 363.

Matalon, R., and Schulman, J. H. (1949). *Discuss. Faraday Soc.* **6**, 27.

Matsubara, A. (1965). *Bull. Chem. Soc. Jap.* **38**, 1254.

Menke, W. (1967). *Brookhaven Symp. Biol.* **19**, 328.

Miller, I. R. (1968). *J. Gen. Physiol.* **52**, 209s.

Miyamoto, V. K., and Thompson, T. E. (1967). *J. Colloid Interface Sci.* **25**, 16.

Moretz, R. C., Akers, C. K., and Parsons, D. F. (1969). *Biochim. Biophys. Acta* **193**, 1.

Morgan, T. E., and Hanahan, D. J. (1966). *Biochemistry* **5**, 1050.

Mueller, P., and Rudin, D. O. (1967). *Biochem. Biophys. Res. Commun.* **26**, 398.

Mueller, P., and Rudin, D. O. (1968). *J. Theor. Biol.* **18**, 222.

Mueller, P., Rudin, D. O., Tien, H. T., and Wescott, W. C. (1962). *Nature (London)* **194**, 979.

Mueller, P., Rudin, D. O., Tien, H. T., and Wescott, W. C. (1964). *Recent Progr. Surface Sci.* **1**, 379–393.

Mühlethaler, K. (1966). *In* "The Biochemistry of Chloroplasts" (T. W. Goodwin, ed.), pp. 49–64. Academic Press, New York.

Noguchi, S., and Koga, S. (1969). *J. Gen. Appl. Microbiol.* **15**, 41.

Northcote, D. H. (1968). *In* "Plant Cell Organelles" (J. B. Pridham, ed.), p. 213. Academic Press, New York.

O'Brien, J. S. (1967). *J. Theor. Biol.* **15**, 307.

Oncley, J. L. (1963). *In* "Brain Lipids and Lipoproteins and the Leucodystrophies" (J. Folch-Pi and H. Bauer, eds.), p. 161. Elsevier, Amsterdam.

Overton, E. (1899). *Vierteljahresschr. Naturforsch. Ges. Zuerich* **44**, 88.

Pagano, R., and Thompson, T. E. (1968). *J. Mol. Biol.* **38**, 41.

Papahadjopoulos, D., and Miller, N. (1967). *Biochim. Biophys. Acta* **135**, 624.

Papahadjopoulos, D., and Ohki, S. (1969). *Science* **164**, 1075.

Papahadjopoulos, D., and Watkins, J. C. (1967). *Biochim. Biophys. Acta* **135**, 639.

Park, R. B., and Biggins, J. (1964). *Science* **144**, 1009.

Park, R. B., and Pon, N. G. (1963). *J. Mol. Biol.* **6**, 105.

Parson, D. F., William, G. R., and Chance, B. (1966). *Ann. N. Y. Acad. Sci.* **137**, 643.

Poincelot, R. P., Millar, P. G., Kimbel, R. L., and Abrahamson, E. W. (1969). *Nature (London)* **221**, 256.

Prezbindowski, K. S., Ruzicka, F. J., Sun, F. F., and Crane, F. L. (1968). *Biochem. Biophys. Res. Commun.* **31,** 164.

Pullman, M. E., Penefsky, H. S., Datta, A., and Racker, E. (1960). *J. Biol. Chem.* **235,** 3322.

Racker, E. (1967). *Fed. Proc., Fed. Amer. Soc. Exp. Biol.* **26,** 1335.

Radding, C. M., and Wald, G. (1956). *J. Gen. Physiol.* **39,** 923

Reeves, J. F., and Dowben, R. M. (1969). *J. Cell. Physiol.* **73,** 49.

Rendi, R. (1964). *Biochim. Biophys. Acta* **84,** 694.

Rendi, R. (1967). *Biochim. Biophys. Acta* **135,** 333.

Richards, F. M. (1963). *Annu. Rev. Biochem.* **32,** 269.

Robertson, J. D. (1966). *Principles Biomol. Organ., Ciba Found. Symp., 1965,* p. 357.

Rothfield, L., and Finelstein, A. (1968). *Annu. Rev. Biochem.* **37,** 463.

Rouser, G., Nelson, G. J., Fleischer, S., and Simon, G. (1968). *In* "Biological Membranes" (D. Chapman, ed.), pp. 5–64. Academic Press, New York.

Rubalcana, B., DeMunoz, D. M., and Gitler, C. (1969). *Biochemistry* **8,** 2742.

Salton, M. R. J. (1967). *Annu. Rev. Microbiol.* **21,** 417.

Scanu, A., Pollard, H., Hirz, R., and Kothary, K. (1969). *Proc. Nat. Acad. Sci. U. S.* **62,** 171.

Schmidt, W. J. (1936). *Z. Zellforsch. Mikrosk. Anat.* **23,** 657.

Schmitt, F. O. (1970). *In* "Symposium: Neurosciences Research" (F. O. Schmitt ed.), Vol. 4, pp. 399–514. MIT Press, Boston.

Schmitt, F. O., Bear, R. S., and Palmer, K. J. (1941). *J. Cell. Comp. Physiol.* **18,** 31.

Schneider, H., and Tattrie, N. H. (1968). *Can. J. Biochem.* **46,** 979.

Schulman, J. H., and Rideal, E. K. (1937). *Proc. Roy. Soc., Ser. B* **122,** 46.

Searcy, R. L., and Bergquist, L. M. (1962). "Lipoprotein Chemistry in Health and Disease." Thomas, Springfield, Illinois.

Seeman, P., Kwant, W. O., and Sauks, T. (1969). *Biochim. Biophys. Acta* **183,** 499.

Sessa, G., and Weissmann, G. (1968). *J. Lipid Res.* **9,** 310.

Shah, D. O. (1969). *Biochim. Biophys. Acta* **193,** 217.

Shinoda, K., Nakagawa, T., Tamamushi, B., and Isemura, T. (1963). "Colloidal Surfactants." Academic Press, New York.

Sitte, P. (1969). *Ber. Deut. Bot. Ges.* **82,** 329.

Sjöstrand, F. S. (1959). *Rev. Mod. Phys.* **31,** 301.

Skou, J. C. (1965). *Physiol. Rev.* **45,** 596.

Small, D. M. (1968). *J. Amer. Oil Chem. Soc.* **45,** 108.

Smekal, E., and Tien, H. T. (1971). To be published.

Sotnikov. P. S., and Melnik, E. I. (1968). *Biofizika* **13,** 185.

Spector, A. A., John, K., and Fletcher, J. E. (1969). *J. Lipid Res.* **10,** 56.

Steim, J. M. (1968). *Advan. Chem. Ser.* **84,** 259.

Steim, J. M., and Fleischer. S. (1967). *Proc. Nat. Acad. Sci. U. S.* **58,** 1292.

Steim, J. M., Edner, O. J., and Bargoot, F. G. (1968). *Science* **162,** 909.

Stewart, G. T. (1967). *Advan. Chem. Ser.* **63,** 141.

Stoeckenius, W. (1966). "Principles of Bimolecular Organization." Little, Brown, Boston, Massachusetts.

Stryer, L. (1968). *Science* **162,** 526.

Sweet, C., and Zull, J. E. (1969). *Biochim. Biophys. Acta* **173,** 94.

Thompson, T. E., and Huang, C. (1966). *J. Mol. Biol.* **16,** 576.

Thornber, J. P., Gregory, R. P. F., Smith, C. A., and Bailey, J. L. (1967). *Biochemistry* **6,** 391.

Tien, H. T. (1967). *J. Phys. Chem.* **71,** 3395.

Tien, H. T. (1968a). *J. Gen. Physiol.* **52**, 125s.

Tien, H. T. (1968b). *J. Phys. Chem.* **72**, 4512.

Tien, H. T. (1970). *Advan. Exp. Med. Biol.* **7**, 135–154.

Tien, H. T., and Diana, A. L. (1968). *Chem. Phys. Lipids* **2**, 55.

Tien, H. T., and Howard, R. E. (1971). *In* "Techniques of Surface Chemistry and Physics" (R. J. Good, R. R. Stromberg, and R. L. Patrick, eds.), Vol. 1. Marcel Dekker, New York (to be published).

Tien, H. T., Carbone, S., and Dawidowicz, E. A. (1966). *Nature (London)* **212**, 718.

Tien, H. T., Huemoeller, W. A., and Ting, H. P. (1968). *Biochem. Biophys. Res. Commun.* **33**, 207 (1968).

Tsofina, L. M., Liberman, E. A., and Babakov, A. V. (1966). *Nature (London)* **212**, 681.

Urry, D. W., Mednieks, M., and Bejnarowicz, E. (1967). *Proc. Nat. Acad. Sci. U.S.* **57**, 1043.

Ussing, H. H., and Zerahn, K. (1951). *Acta Physiol. Scand.* **23**, 110.

Van, N. T., and Tien, H. T. (1970). *J. Phys. Chem.,* **74**, 3559.

van Deenen, L. L. M. (1965). *Progr. Chem. Fats Other Lipids* **8**, 1.

van den Berg, H. J. (1968). *Advan. Chem. Ser.* **84**, 99.

Van Zutphen, H., van Deenen, L. L. M., and Kinsky, S. C. (1966). *Biochem. Biophys. Res. Commun.* **22**, 393.

Vatter, A. E. (1963). *J. Cell Biol.* **19**, 72A

Verma, S. P., Kobamoto, N., and Tien, H. T. (1971). To be published.

Vilallonga, F., Altschul, R., and Fernandez, M. S. (1967). *Biochim. Biophys. Acta* **135**, 406.

Vreeman, H. J. (1966). *Proc. Kon. Ned. Akad. Wetensch., Ser. C* **69**, 542.

Wald, G. (1968). *Science* **162**, 238.

Wallach, D. F. H., and Gordon, A. (1968). *Fed. Proc., Fed. Amer. Soc. Exp. Biol.* **27**, 1263.

Wallach, D. F. H., and Zahler, P. (1966). *Proc. Nat. Acad. Sci. U. S.* **56**, 1552.

Wallach, D. F. H., Donald, F. H., Graham, J. M., and Fernbach, B. R. (1969). *Arch. Biochem. Biophys.* **131**, 332.

Weier, T. E., and Benson, A. A. (1966). *Amer. J. Biol.* **54**, 389.

Weinstein, D. B., Marsh, J. B., and Glick, M. C. (1969). *J. Biol. Chem.* **244**, 4103.

Weissmann, G., Keiser, H., and Bernheimer, A. W. (1965). *J. Exp. Med.* **118**, 205.

Weissmann, G., Sessa, G., and Weissmann, S. (1966). *Biochem. Pharmacol.* **15**, 1537.

Whittaker, V. P. (1968). *Brit. Med. Bull.* **24**, 101.

Willmer, E. N. (1961). *Biol. Rev.* **36**, 368.

Wolken, J. J. (1968). *J. Amer. Oil Chem. Soc.* **45**, 251.

Wood, R. E., Wirth, F. P., and Morgan, H. E. (1968). *Biochim. Biophys. Acta* **163**, 171.

Zahler, P. (1969). *Experientia* **25**, 449.

CHAPTER V

The Kinetics of Supported Enzyme Systems

K. J. LAIDLER AND P. V. SUNDARAM

I. Introduction

The vast majority of the purely kinetic work on enzyme systems has been carried out with both the enzyme and the substrate present in free solution. Investigations of this kind have been of great value in leading to a detailed knowledge of the reaction mechanisms for a number of enzyme-catalyzed reactions. For example, for a considerable number of enzymes we now know the various reaction steps that occur and we know which functional groups on the enzyme molecule are involved in the various stages of the reaction. Investigations of this kind provide a sound basis for understanding how enzymes perform their functions in the living system.

In biological systems the enzymes are usually not in free solution but are attached to structural material, and, as a result, they do not have the mobility that they have in free solution. Sometimes they are attached to the membranes of cells, and in this case the substrate molecules must diffuse through the membrane material in order to be acted upon by these enzymes. The kinetic behavior is then different from the behavior in free solution.

In view of this, it is of importance for kinetic investigations to be made on systems in which the enzyme molecules have become attached to a support. Up to the present time only a few kinetic studies of this kind have been made, but enough has been done for some of the principles to be fairly well understood. The object of this chapter is to discuss some of this work.

The review consists of four main topics: (1) a general account of the various ways in which enzymes can become attached to membranes and other supports (Section II); (2) general kinetic principles which arise in systems of this kind (Section III); (3) these principles as applied to some of the experimental results that have so far been obtained (Section IV); and (4) the rather special aspect of flow systems involving supported enzymes (Section V).

An understanding of the kinetics of supported enzyme systems will lead to a deeper knowledge of the behavior of enzymes in living systems. In addition, investigations of this kind have some technological significance, in view of certain applications of supported enzyme systems; some of these are referred to in Section V.

II. General Aspects of Supported Enzyme Systems

A. Classification of Supported Enzymes

The terminology used in this field has hitherto been somewhat varied; the expressions "solid-supported enzymes," "insoluble enzyme derivatives," "insolubilized enzymes," and "matrix-bound enzymes" have been used. We have decided, for the purpose of this review, to use the term "supported enzyme" (or "solid-supported enzyme" if the support is a solid) when we are referring to an enzyme which is either distributed more or less uniformly through a support or is attached to its surface. We think that the term "insolubilized enzyme" is best reserved for an enzyme which has been modified and rendered less soluble by a reagent which acts upon certain groups on the enzyme—for example, enzymes treated with glutaraldehyde to yield an insoluble product.

Figure 1 shows a possible classification of supported enzyme systems.

B. ATTACHMENT OF ENZYMES TO SUPPORTS

Only in the last 10 years or so have efforts been directed toward developing methods for making supported enzymes and studying their properties. The main systems that have been studied so far are listed in

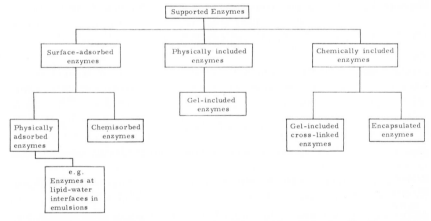

FIG. 1. A classification of supported enzyme systems. This classification applies to both solid and liquid (e.g., emulsion) supports.

Table I. Most of the enzymes studied are hydrolytic enzymes which in general do not require coenzymes; their molecular weights range from 50,000 to about 500,000.

There are a few cases in which enzymes have been studied when present at an interface, such as a lipid–water interface in an emulsion. This has been done, for example, with catalase by Fraser et al. (1955). Enzymes have been confined in tiny collodion and nylon capsules and studied, particularly by Chang (1967; Chang et al., 1966).

The methods employed for attaching enzymes to solids by covalent bonds will not be discussed here, since they have recently been reviewed fully by Silman and Katchalski (1966) and by Kay (1968). For the enzyme to remain active, it must obviously be attached by means of groups that are not part of the active center or are not even indirectly involved in the catalysis. In most cases it is not known which groups are involved in the

TABLE I

SUPPORTED ENZYME SYSTEMS ON WHICH ACTIVITY STUDIES HAVE BEEN MADE[a]

Enzyme	Support	Reference
A. Covalent attachment of enzyme to support		
ATPase	Cellulose	Brown et al. (1966)
ATP deaminase	DEAE cellulose	Chung et al. (1968)
Bromelain	CM cellulose	Wharton et al. (1968a; b)
Catalase	Cellulose ester (neutral)	Surinov and Manoilov (1966)
Chymotrypsin	Cellulose ester (neutral)	Surinov and Manoilov (1966)
Chymotrypsin	p-Amino-DL-phenylalanine-L-leucine copolymer	Katchalski (1962)
Chymotrypsin	Maleic anhydride–ethylene copolymer (polyanionic)	Goldstein et al. (1964)
Chymotrypsin	Polycationic	Pecht (1966)
Chymotrypsin	CM cellulose	Lilly and Sharp (1968)
Ficin	CM cellulose	Hornby et al. (1966, 1968), Lilly et al. (1966)
Lactic dehydrogenase	Anion exchange cellulose particles and sheet	Wilson et al. (1968a)
Papain	p-Amino-DL-phenylalanine-L-leucine copolymer(neutral)	Katchalski (1962); Silman (1964)
Papain	Collodion membrane	Goldman et al. (1968a, b)
Peroxidase (horseradish)	CM cellulose	Weliky, et al. (1969)
Pyruvate kinase	Filter-paper discs	Wilson, et al. (1968b)
Ribonuclease	Cellulose ester	Surinov and Manoilov (1966)
Trypsin	p-Amino-DL-phenylalanine-L-leucine copolymer (neutral)	Bar-Eli and Katchalski (1963)
Trypsin	Maleic anhydride–ethylene copolymer (polyanionic)	Goldstein et al. (1964); Levin et al. (1964)
Trypsin	Cellulose ester	Surinov and Manoilov (1966)
B. Enzyme adsorbed on support		
Amino acylase	DEAE cellulose	Tosa et al. (1960)
Asparaginase	Ion-exchange cellulose	Nikolaev and Mardashev (1961); Nikolaev (1962)
Catalase	Lipid–water interface in emulsions	Fraser et al. (1955)
Chymotrypsin	Kaolinite (anionic)	McLaren and Estermann (1957)
Ribonuclease	Ion-exchange resin	Barnett and Bull (1959)
Urease	Kaolinite	Sundaram (1967)

TABLE I (*continued*)

Enzyme	Support	Reference
	C. Other types of attachment	
β-Fructofurano- side	Glass beads (crude enzyme particles held between glass beads in column)	Arnold (1966)
Urease	Enclosed in nylon microcap- sules	Sundaram (1969)

a Abbreviations: ATPase—adenosinetriphosphatase; DEAE—diethylaminoethyl;
CM—*O*-carboxymethyl.

attachment to the solid. If metals and cofactors are required for the enzymic activity, these should be attached to the support at their optimum concentrations.

The whole problem of the attachment of enzymes to solids is a very complex one, and the factors are by no means well understood. The remainder of this subsection is concerned with some aspects that have been found to be important.

The ease of attachment of an enzyme to a solid may be very strongly affected by various factors such as pH, solvent composition, and salt concentration. The presence or absence of acidic or basic groups on the support has a profound effect on absorptive capacity. Synthetic ion-exchange resins and cellulose ion-exchange absorbents are often tailored to suit the purpose. Under proper conditions, certain celluloses can absorb up to 70 to 80 % of their own weight of protein. In general, the larger the available surface area of the solid support, the better the absorption.

External factors that can be varied and which influence the case of absorption of an enzyme are strongly limited by their action on the enzyme itself; thus one cannot vary pH too widely because of the possible denaturation of the enzyme. There is always a possibility that protein denaturation will be brought about by the environment of the enzyme when it is supported. For example, proteins are sometimes denatured in a nonpolar environment, and enzyme inactivation might, therefore, be encouraged by attachment to a solid having nonpolar groups.

Another factor is that the presence of too many binding sites on the support may compete with the secondary and tertiary bonds responsible for maintaining the structural integrity of the native protein. This may well be a major factor in causing an irreversible change in the conformation of the enzyme molecule. Some evidence exists that a native protein is distorted into a new conformation by an increase in the availability on adsorption of groups which are hidden in the native protein; this was found for SH groups in catalase (Fraser et al., 1955). For carboxymethyl (CM) cellulose–ficin, Hornby et al. (1968) found a similar effect which they attributed to electrostatic interference. Some enzymes, however, show increased stability on storage in the bound form, presumably as a result of a conformational change; an example is urease which is very unstable when pure. This high molecular weight globulin has a strong tendency to polymerize in solution, with inactivation; this is probably due to S—S bond formation from the numerous SH groups available in the native protein. Anchoring of the protein to the support helps to prevent this polymerization by keeping the SH groups apart (Sundaram, 1967).

Inorganic supports show an additional type of complication. Examples are the clays which can be classified as (1) those with a lattice that expands when adsorption occurs (e.g., bentonite) and (2) those with a lattice that does not expand (e.g., kaolinite). In the latter, only the surface ionization sites are believed to take part in adsorption, whereas the expanding lattice clays are capable of forming intercalation compounds (Weiss, 1961). Clays undergo ionization slowly, and the conventional rules that govern spontaneous ionization of soluble compounds do not apply to them. Their ionizations involve hysteresis, and this makes it difficult to control the pH of the system as an enzyme becomes attached; it is necessary to equilibrate the enzyme and the clay at the same pH prior to bringing them together.

It is often difficult to achieve reversible adsorption of proteins in these clay systems. Changes in pH have a marked effect on the charge distribution on the protein and the adsorbent, and small changes in salt concentration can influence the strengths of the electrostatic salt linkages between the protein and the clay. The process is reversed with particular difficulty when the adsorbent has a high charge density, as in the case of a clay, and when the protein is of high molecular weight. This was found by Sundaram (1967) for urease adsorbed on kaolinite, and by McLaren (1954) for lysozyme adsorbed on kaolinite. However, easily reversible adsorption was obtained by Hirs et al. (1953) using relatively stable and low molecular weight proteins with high isoelectric points.

Only a few studies of the kinetics and mechanism of the attachment of proteins to supports have been carried out. McLaren (1954) studied the adsorption characteristics of proteins covering a wide range of isoelectric points (between 3 and 11), the adsorbent being kaolinite; this is hydrated aluminum silicate with ionizable OH groups attached to Si and Al atoms, the broken bonds or unsatisfied valences usually being found around the edges of the hexagonal crystals. These are the groups that can take part in bond formation with the active spots on the protein molecule. A study with kaolinite and urease was made by Sundaram (1967). These investigations of McLaren and Sundaram showed that the attachment of the enzyme to the kaolinite is very rapid. The rates were not measured with sufficient precision to permit any reliable conclusions to be drawn about the dependence of rate on pH and other factors.

The total amounts of enzyme adsorbed were, however, measured over a pH range and exhibited some interesting features. In the region of the isoelectric point of the enzyme, the amount adsorbed increases as the pH is lowered. This is readily understood in view of the fact that kaolinite is negatively charged. Decreasing the pH increases the net positive charge on the protein and, therefore, tends to increase the total amount adsorbed; at pH values higher than the isoelectric point, when the enzyme molecules are negatively charged, the amount of adsorption is relatively small. However, the results reveal that the matter is more complex than this simple account would suggest. One fact leading to complexity is that adsorption will involve specific sites on the protein molecules, so that the net charge on the molecule and the isoelectric point cannot be expected to determine the total amount of adsorption. Similar complexities in rates of adsorption are also to be expected.

A few investigations have been made on the reversibility of the adsorption of proteins on supports. Some degree of reversibility has usually been observed. Mitz and Schleuter (1959) studied the adsorption and desorption of several enzymes (pepsin, trypsin, and chymotrypsin) on diethylamino-ethyl (DEAE) cellulose, and found rapid adsorption and desorption in some cases. Usually most of the bound enzyme could be desorbed by the use of specific buffers; water alone could not bring about desorption. Sundaram and Crook (1968) studied adsorption and desorption in urease–kaolinite systems and found that some (about 10 %) of the adsorbed urease could be desorbed; the remainder could not be removed even at high pH values and high salt concentrations. Evidently two kinds of adsorption are involved; possibly the easily desorbed protein is attached to protein molecules that are already strongly bound, or, perhaps, it is attached to less active sites on the kaolinite.

C. ENZYMES AT INTERFACES

Enzyme molecules can be present at various types of interface, such as solid–liquid, air–liquid, and liquid–liquid (emulsion) interfaces. In nature, many enzyme reactions occur at solid–liquid or liquid–liquid interfaces, the liquid phase being aqueous and the solid phase frequently being of hydrophobic character. The attachment of an enzyme at a hydrophilic–hydrophobic interface is a matter of considerable complexity. Protein molecules contain polar and nonpolar regions, and at an interface will tend to orient themselves so that the polar groups are in the hydrophilic phase and the nonpolar groups in the hydrophobic phase. One result may be some unfolding of the protein molecule, and this may result in deactivation of the enzyme.

Relatively little work has been done in this area; only a brief reference will be made to some of the relevant publications. Jonxis (1939) made some early observations of the spreading of oxymyoglobin at interfaces. Cheeseman and Davies (1954) and Cheeseman and Schuller (1954) studied the adsorption of proteins and polypeptides at air–water and oil–water interfaces and drew some conclusions about the orientations of the molecules at these interfaces. They suggested that there is always some unfolding at interfaces. Davies and Rideal (1963) have discussed the rigidity of the structures of proteins and other polymers as affected by different interfacial conditions.

The adsorption of catalase at oil–water interfaces was studied by Fraser *et al.* (1955). By using a variety of stabilizers, they succeeded in forming protein layers of various thicknesses adsorbed on emulsion droplets. They studied how the enzymic activity was affected and correlated it with the degree of unfolding.

D. Supported Substrates

A small number of investigations have been done on systems in which the substrate, rather than the enzyme, has been attached to a support. As with supported enzymes, the substrate may be attached, physically or covalently, to a surface, may be physically included in the support, or may be covalently attached throughout the support.

Work of this kind has been done by Trurnit (1953) using bovine serum albumin fixed on glass slides and by McLaren and Estermann (1957) using denatured lysozyme adsorbed on kaolinite; in both cases the effect of chymotrypsin on these supported substrates was studied.

E. Methods of Studying Supported Enzymes

Brief reference will be made to a few of the ways in which the kinetic aspects of supported enzymes have been investigated.

1. Slurry

In a few investigations (Hornby *et al.*, 1966; Money and Crook, referred to by Crook, 1968; Sundaram, 1967) the supported enzyme has been present as a slurry, i.e., as a stirred solution of substrate containing tiny particles of the support and enzyme. The kinetic methods employed are much the same as when the enzyme is in homogeneous solution.

A modification of this method involves causing the substrate solution to flow into a reaction vessel containing the slurry, the course of reaction being followed continuously, e.g., using a pH-stat.

2. Column Catalysis or Reaction in a Packed Bed

In this technique the solid-supported enzyme particles are packed in a column and the substrate solution is perfused through at the desired rate. The effluent is analyzed for the amount of product formed. This method has the advantage over the slurry method in that samples for product estimation do not contain small amounts of the enzyme particles and that the complications of stirring are not encountered. The kinetic analysis of reactions in flow systems is considered in Section V.

3. Enzyme Sheets or Membrane-Bound Enzymes

Enzymes attached to sheets or membranes (e.g., 2.5-cm diameter cellulose sheets; cf. Wilson et al., 1968a,b) are packed between filter paper of similar size and put in a holder. Substrate is perfused through at a definite rate and the effluent analyzed, e.g., spectrophotometrically.

4. Enzymes Included in or Covalently Attached to Gels

Work has been done (Wieland et al., 1966; Updike and Hicks, 1967a,b) with polyacrylamide gels in which enzymes (glucose oxidase, lactic and alcohol dehydrogenases, and trypsin) have either been mechanically trapped or have become attached by covalent bonds (cross-linking). These supported enzymes can be prepared from the monomer by inducing polymerization in the presence of the enzyme. In the case of physically included enzymes the pores in the gels are sometimes so large that the enzyme can escape. This is avoided if the enzyme is cross-linked with the polymer as the polymerization is taking place, but this process sometimes leads to the inactivation of the enzyme.

5. Encapsulated Enzymes

A rather special case of supported enzymes involves trapping the enzymes in microcapsules. Chang (1967; Chang et al., 1966) in particular has made nylon and collodion microcapsules, of membrane thicknesses varying from 200 to 700 Å, containing various enzymes. The free amino groups of the protein may be cross-linked with groups in the capsule membrane. The membrane thickness and size can be controlled, as can the concentration of enzyme within the microcapsules. These capsules have been used in a slurry and in columns. The latter have been used in extracorporeal shunts for the treatment of patients with inborn errors of metabolism; enzymic reactions can be brought about without actually introducing the enzyme into the body, and in this way immunological complications are avoided. Kinetic studies on encapsulated urease have been initiated by Sundaram (1969).

III. Basic Kinetic Laws for Solid-Supported Enzymes

The present section is concerned with the general kinetic principles relating to the action of solid-supported enzymes on substrates that are in free solution. In particular, equations will be developed showing the variation of the reaction rate with the substrate concentration. The experimental results are discussed in Section IV in the light of these basic kinetic equations. The final section, Section V, deals with the special problem of the kinetic equations applicable to flow systems when, for example, the substrate is allowed to flow through a solid bed that contains enzymes.

A. KINETIC LAWS IN HOMOGENEOUS SYSTEMS

It will be useful first to write down the equations applicable to homogeneous enzyme systems, since it will be convenient to compare the behavior of solid-supported enzymes with that of enzymes present in free solution. Also, the basic equation applicable to many homogeneous enzyme systems, the Michaelis–Menten equation, applies to a good approximation to systems involving solid-supported enzymes.

The Michaelis–Menten equation (Michaelis and Menten, 1913) can be written as

$$v = \frac{k_c[\text{E}][\text{S}]}{K_m + [\text{S}]} \tag{1}$$

where [E] and [S] are the total enzyme and substrate concentrations, and k_c and K_m are constants; the former, sometimes written as k_{cat}, is generally known as the catalytic constant, the latter as the Michaelis–Menten or Michaelis constant. This equation was first derived by Michaelis and Menten with reference to the following mechanism:

$$\text{E} + \text{S} \underset{k_{-1}}{\overset{k_1}{\rightleftarrows}} \text{ES} \overset{k_2}{\longrightarrow} \text{E} + \text{P}$$

According to this scheme, the enzyme and substrate molecules first form a binary addition complex ES, which in a subsequent stage breaks down (perhaps with the participation of a solvent molecule) into enzyme and product. Briggs and Haldane (1925) showed that application of the steady-state principle to the above scheme leads to Eq. (1) with the kinetic parameters k_c and K_m having the following significance:

$$k_c = k_2 \qquad (2)$$

$$K_m = \frac{k_{-1} + k_2}{k_1} \qquad (3)$$

A kinetic equation of exactly the same form as Eq. (1) is also obtained for any number of intermediates, provided that they are all formed from the preceding one by first-order reactions. A relatively simple situation is when there are two intermediates,

$$E + S \underset{k_{-1}}{\overset{k_1}{\rightleftharpoons}} ES \overset{k_2}{\underset{P_1}{\longrightarrow}} ES' \overset{k_3}{\longrightarrow} E + P_2$$

This type of mechanism is commonly found in the hydrolysis of an ester, RCOOR', in which case the ES' complex is a species in which the acyl residue of the ester RCO has become attached to the enzyme with the formation of an acyl enzyme, RCO-E; the product P_1 is now the alcohol R'OH, formed by the transfer of a proton from the enzyme molecule to the ester molecule when it is present in the Michaelis complex ES. The final stage in this scheme represents the hydrolysis of the acyl enzyme with the formation of the acid RCOOH and the regeneration of the free enzyme molecule. A similar mechanism applies to the hydrolysis, by many enzymes, of the amide and peptide linkages and appears to be of quite wide applicability.

The steady-state treatment gives in this case

$$v = \frac{\dfrac{k_2 k_3}{k_2 + k_3}\,[E]\,[S]}{\dfrac{k_{-1} + k_2}{k_1}\cdot\dfrac{k_3}{k_2 + k_3} + [S]} \qquad (4)$$

This equation is seen to have the same form as Eq. (1) for the dependence of v on [S]. Equation (4) may be written as

$$v = \frac{k_c[E]\,[S]}{K_m + [S]} \qquad (5)$$

where

$$k_c = \frac{k_2 k_3}{k_2 + k_3} \qquad (6)$$

and

$$K_m = \frac{k_{-1} + k_2}{k_1} \cdot \frac{k_3}{k_2 + k_3} \tag{7}$$

It will be seen later in this section that Eq. (1) also applies to a close approximation to solid-supported enzyme systems.

B. FACTORS INFLUENCING ACTIVITY OF SUPPORTED ENZYMES

There are four main reasons why an enzyme may behave differently when supported than when present in free solution.

1. The enzyme may be conformationally different in the supported state as compared with free solution. It is well known that kinetic behavior depends very critically on enzyme conformation, certain changes leading to a complete loss of activity.

2. In the support the interaction between the enzyme and the substrate takes place in a different environment from that existing in free solution. Studies of enzyme reactions in different solvents have shown that environmental effects can be very profound (Barnard and Laidler, 1952; Laidler and Ethier, 1953; Findlay et al., 1963). Some of these effects have been explained satisfactorily (Laidler, 1958) in terms of the influence of the dielectric constant of the medium on the electrostatic interactions; such explanations are, however, undoubtedly too simple, although they may be very helpful.

3. There will be partitioning of the substrate between the support and the free solution so that the substrate concentration in the neighborhood of the enzyme may be different from what it is in free solution. For example, a relatively nonpolar substrate will be more soluble in a support which contains a number of nonpolar groups than in aqueous solution; a polar substrate will be less soluble. In addition to this, profound effects may arise if both the substrate and the support are electrically charged; this particular situation has been considered in some detail by Goldstein et al. (1964).

4. The reaction in the solid support may be to some extent diffusion-controlled. In homogeneous aqueous solution, even the fastest of enzyme-catalyzed reactions appear not to be diffusion-controlled, but this is no longer the case when the substrate has to diffuse toward the enzyme through the solid support. The indications are that only in the case of an exceedingly slow enzyme reaction will the reaction be other than diffusion-controlled.

Little information is at present available about factors 1 and 2 for solid-supported enzymes. The kinetic effects of partitioning of substrate between support and solution have been considered by Goldman et al. (1968b), but only for the case of electrostatic interactions. These workers have also given a treatment of the diffusion problem in a number of special cases; they did not, however, consider the limiting behavior at high substrate concentrations. Hornby *et al.* (1968) have also developed a theoretical treatment of the diffusional effects.

Sundaram *et al.* (1970) have developed a general treatment of the kinetic laws applicable to the situation in which an enzyme-containing membrane is present in a solution of a substrate, such that the solution is in contact with the opposite faces of the membrane. To a good approximation this

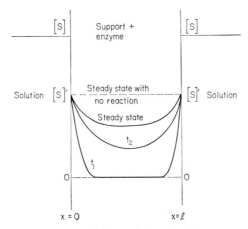

FIG. 2. Schematic representation of the variations in substrate concentrations when an enzyme-containing support, of thickness l, is immersed in a substrate solution of concentration [S].

treatment will apply to discs suspended in the substrate solution, provided that the diameter of the discs is large compared with the thickness. Sundaram *et al.* were especially concerned with the limiting behavior at low and high substrate concentrations and with how the kinetic parameters are affected by the characteristics (e.g., the thickness) of the membrane. They also gave some computer solutions in order to confirm the general conclusions. The following subsection is based on this treatment.

C. GENERAL THEORY OF KINETICS OF SUPPORTED ENZYME SYSTEMS

The variation of substrate concentration through an enzyme-containing support is shown schematically in Fig. 2. The substrate concentration in the free solution is [S] and that just inside the interface is [S]′; the ratio

[S]'/[S] is the partition coefficient P for the substrate between the two phases. After a short period of time t_1, the concentration will fall sharply to zero close to each interface; after a longer period of time the variation may be as shown by the curve marked t_2. Eventually a steady state will be reached. If no reaction were occurring, the steady state would correspond to a horizontal line, and this situation is approached if the enzyme is extremely inactive or the membrane very thin.

1. Attainment of the Steady State

It is important to consider first the time it takes for steady state to be essentially established. An exact solution of the presteady-state diffusion equations, for the case in which reaction is occurring as well as diffusion, does not seem to be possible. A reasonably reliable estimate of the time for the establishment of equilibrium will, however, be given by a treatment in which only diffusion is occurring. The nonsteady-state solution (Barrer, 1941) for the system represented in Fig. 2 is that the concentrations of substrate within the membrane varies with the distance x and the time t according to

$$s = [S]' \left[1 - \frac{2}{\pi} \sum_{n=1}^{\infty} \frac{1 - \cos n\pi}{n} \sin \frac{n\pi x}{l} \exp\left(-\frac{Dn^2\pi^2 t}{l^2} \right) \right] \tag{8}$$

where D is the diffusion constant. At the larger t values at which the steady state is more or less established, the leading term, with $n = 1$, will be most important, so that approximately

$$s = [S]' \left[1 - \frac{4}{\pi} \sin \frac{\pi x}{l} \exp\left(-\frac{D^2\pi^2 t}{l^2} \right) \right] \tag{9}$$

In order to have an arbitrary criterion for the establishment of the steady state, we may ask at what time, at $x = l/2$, does the concentration s differ by 10 % from the steady-state value [S]'. This time τ is defined by

$$\frac{4}{\pi} \exp\left(-\frac{D\pi^2 t}{l^2} \right) = \frac{9}{10} \tag{10}$$

which reduces to

$$\tau = 0.257 \frac{l^2}{D} \tag{11}$$

A typical value of D for a low molecular weight substrate is 3×10^{-6} cm^2 second^{-1} and hence

$$\tau = 8.5 \times 10^4 \, l^2 \text{ seconds} \tag{12}$$

If the membrane is 1 cm thick, the time to come within 10 % of steady-state conditions is thus of the order of 10^5 seconds or 28 hours. For a membrane 1 mm thick the time required is 10^3 seconds or 17 minutes.

When chemical reaction is occurring in addition, the times will be not as great since the steady-state concentrations that are attained in the membrane are not so high. It seems safe to conclude that under ordinary circumstances, with membranes less than 1 mm thick, steady state will be essentially established within a few minutes. With membranes much thicker than this, however, it is unsafe to assume that steady-state conditions apply.

2. Slow Enzymic Reaction

If the enzymic reaction is slow (i.e., the enzyme is very inactive or is present in low concentrations), there will be little depletion of substrate through the support and the reaction will not be diffusion-controlled; the treatment is then very simple. The rate of reaction per unit volume of support is

$$v = \frac{k_c' \, [E]_m [S]'}{K_m' + [S]'} \text{ moles liter}^{-1} \text{ second}^{-1} \tag{13}$$

where $[E]_m$ is the number of moles of enzyme present per liter of support, and k_c' and K_m' are the true catalytic constant and Michaelis constant, respectively, for the reaction within the membrane (they may differ from the values k_c and K_m in the free solution because of factors 1 and 2). Suppose that E is the total number of moles of enzyme present in a support of thickness l (centimeters) and cross-sectional area A (square centimeters); $[E]_m$ is then $10^3 E/Al$ moles liter^{-1}, and the total rate of reaction in the support of volume Al (milliliters) is given by

$$v' = \frac{k_c'[E]_m[S]' Al/10^3}{K_m' + [S]'} = \frac{k_c' E[S]'}{K_m' + [S]'} \text{ moles second}^{-1} \tag{14}$$

The concentration $[S]'$ is equal to $P[S]$, where P is the partition coefficient; the rate is thus

$$v = \frac{k_c' E[S]}{(K_m'/P) + [S]} \tag{15}$$

The apparent Michaelis constant is K_m/P. The rate at low substrate concentrations is

$$v'_{\text{low}} = \frac{P k_c'}{K_m'} E[S] \tag{16}$$

If the same amount of enzyme were distributed in 1 liter of free solution, the rate at low substrate concentrations would be $k_c E/K_m$; the rate on the support thus differs by the factor P, the partition coefficient, apart from differences between k_c' and k_c and between K_m' and K_m, brought about by factors 1 and 2.

At high substrate concentrations, however, Eq. (14) leads to the result that

$$v'_{\text{high}} = k_c'E \tag{17}$$

so that the rate will be the same on the supported enzyme as in free solution, provided that k_c' is the same as k_c.

It is evident that systems of this kind, where the enzyme is of low activity (by keeping the enzyme at sufficiently low concentrations), will be valuable for studying the influence of factors 1 and 2 on the activity of the enzyme.

3. More Active Enzyme Systems

If the rate of the enzymic reaction is greater (enzyme more active or present at higher concentrations), there will be a significant variation in substrate concentration through the support, and the effects of diffusion will then be more important. According to Fick's second law of diffusion, the rate of accumulation of substrate at any cross section, due to diffusion, is given by

$$D \left(\frac{\partial^2 s}{\partial x^2} \right)_t \text{ moles liter}^{-1} \text{ second}^{-1}$$

where D (square centimeters per second) is the diffusion constant, and the concentration s is in moles per liter. In the steady state this must exactly balance the rate of removal of substrate by reaction, which is equal to

$$\frac{k_c'[\text{E}]_m s}{K_m' + s} \text{ moles liter}^{-1} \text{ second}^{-1}$$

The equation to be solved is, therefore,

$$D \left(\frac{\partial^2 s}{\partial x^2} \right)_t = \frac{k_c'[\text{E}]_m s}{K_m' + s} \tag{18}$$

An explicit solution of this equation cannot be obtained except in the limiting cases $s \ll K_m'$ and $s \gg K_m'$. These limiting solutions will be obtained first, after which some numerical solutions will be given.

Case I. Low Substrate Concentrations. When $s \ll K_m'$, Eq. (18) may be written as

$$\frac{\partial^2 s}{\partial x^2} = \alpha^2 s \tag{19}$$

where α (per centimeter) is equal to $(k_c'[E]_m/DK_m')^{1/2}$. The solution of this equation, subject to the boundary conditions $x = 0$, $s = [S]'$ and $x = l$, $s = [S]'$, is

$$s = [S]' \frac{\sinh \alpha x + \sinh \alpha(l - x)}{\sinh \alpha l} \tag{20}$$

The gradients ds/dx at $x = 0$ and $x = l$ are

$$\frac{ds}{dx} = \mp \alpha[S]' \frac{\cosh \alpha l - 1}{\sinh \alpha l} \tag{21}$$

The rate of entry of substrate at each of the surfaces is, therefore,

$$v = -AD \left(\frac{ds}{dx}\right)_{x=0} = AD \left(\frac{ds}{dx}\right)_{x=l} \tag{22}$$

$$= \alpha AD[S]' \frac{\cosh \alpha l - 1}{\sinh \alpha l} \tag{23}$$

This is equal to the rate of exit of product, since in the steady state there is no accumulation of substrate or product within the membrane; the net rate of formation of product in a membrane of cross-sectional area A is thus

$$v' = 2\alpha AD[S]' \frac{\cosh \alpha l - 1}{\sinh \alpha l} \tag{24}$$

The rate per unit volume of support is

$$\frac{v'}{Al} = \frac{2\alpha D[S]'}{l} \cdot \frac{\cosh \alpha l - 1}{\sinh \alpha l} \tag{25}$$

$$= \frac{k_c'[E]_m}{K_m'} [S]' \cdot \frac{2}{\alpha l} \cdot \frac{\cosh \alpha l - 1}{\sinh \alpha l} \tag{26}$$

This result had previously been obtained by Goldman et al. (1968b). The function $F = 2(\cosh \alpha l - 1)/\alpha l \sinh \alpha l$ is plotted in Fig. 3; it has its maximum value of unity when αl approaches zero. The maximum rate attainable when a given amount of enzyme $[E]_m$ is present in 1 liter of support is thus obtained when the thickness l approaches zero and is equal to

$$v = \frac{k_c'[E] [S]'}{K_m'} = \frac{k_c'[E]P[S]}{K_m'} \tag{27}$$

If instead this amount of enzyme were present in 1 liter of free solution and if the substrate concentration were sufficiently low that the kinetics were first order, the rate would be

$$v = \frac{k_c[E]_m[S]}{K_m} \text{ moles second}^{-1} \tag{28}$$

The limiting rate for the supported enzyme, when the membrane is made thinner, therefore, differs from that in homogeneous solution by the factors

$$\frac{k_c{}'}{k_c}, \qquad \frac{K_m}{K_m{}'}, \qquad P = \frac{[S]'}{[S]}$$

The form of the function F is of some interest; the value is very close to unity (within 1 %) for αl values less than about 0.3, after which there is a substantial fall. The value of 0.3 for αl, therefore, represents something of a critical value, below which the rates on the supported enzyme will correspond closely to those in free solution (apart from factors 1 and 2), and

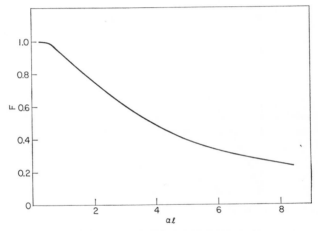

Fig. 3. Function $F = 2(\cosh \alpha l - 1)/\alpha l \sinh \alpha l$ plotted against αl.

above which they will have lower values. The quantitative significance of this is discussed below.

Case II. High Substrate Concentrations. When $s \gg K_m{}'$, Eq. (18) becomes

$$\frac{\partial^2 s}{\partial x^2} = \beta \qquad (29)$$

where β (units of moles cm^{-5}) is equal to $10^{-3} K_c{}' [\text{E}]_m/D$. The solution of this, subject to the boundary conditions $x = 0$, $s = [S]'$ and $x = l$, $s = [S]'$, is

$$s = [S]' - \frac{1}{2} \beta x (l - x) \qquad (30)$$

The concentration gradients at $x = 0$ and $x = l$ are

$$\frac{ds}{dx} = \pm \frac{1}{2} \beta l \qquad (31)$$

The rate of entry of substrate through area A at each side of the support is, therefore,

$$v' = \tfrac{1}{2}\beta AD \tag{32}$$

The total rate of product formation in a membrane of cross-sectional area A (square centimeters) is thus

$$v' = \beta A Dl \text{ moles second}^{-1} \tag{33}$$

$$= 10^{-3}k_c'[E]_m Al \text{ moles second}^{-1} \tag{34}$$

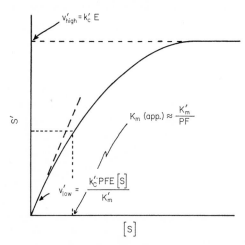

Fig. 4. Plot of v' against substrate concentration in the free solutions. The rate v' is the rate when E moles of enzyme are present in unit volume of support. Function $F = 2(\cosh \alpha l - 1)/\alpha l$; it approaches unity as the thickness of the membrane is decreased (cf. Fig. 3).

The rate per liter of support is thus

$$v' = k_c'[E]_m = k_c'E \text{ moles liter}^{-1} \text{ second}^{-1} \tag{35}$$

If the same amount of enzyme is present in 1 liter of solution, the rate is

$$v = k_c E \text{ moles liter}^{-1} \text{ second}^{-1} \tag{36}$$

The limiting, high substrate concentration rates now differ only by the factor k_c'/k_c. Experimental studies at high substrate concentrations will thus reveal this factor, uncomplicated by the partitioning effect.

Case III. The General Case. Solutions of the general equation (11) can only be obtained by numerical means. One can, however, infer the approximate form of the rate vs. substrate concentration dependence on the basis of the solutions for Cases I and II. This is shown schematically in Fig. 4. The limiting rate at high substrate concentrations is k_c' E [Eq. (36)] and at low concentrations, $k_c'PFE$ [S]$/K_m'$, where P is the partition coefficient and F the function plotted in Fig. 3. If Michaelis–Menten kinetics are obtained for the supported enzyme, then the rate expression will be

$$v' = \frac{K_c(\text{app.})E[\text{S}]}{K_m(\text{app.}) + [\text{S}]} \tag{37}$$

where

$$k_c(\text{app.}) = k_c' \tag{38}$$

and

$$K_m(\text{app.}) = \frac{K_m'}{PF} \tag{39}$$

The value of K_m(app.) (apparent Michaelis constant) extrapolated to low membrane thickness is thus K_m'/P, and if P is measured directly, K_m' can be evaluated and compared with the value in free solution.

Solution of Eq. (18) can be carried out numerically by reducing the equation to two first-order equations. The equation can be written as

$$\frac{d^2s}{dx^2} = \frac{as}{b + s} \tag{40}$$

where a is $k_c'[\text{E}]_m/D$ and b is K_m'. The substitution $z = dy/dx$ leads to

$$\frac{1}{2} z^2 = \int \frac{as}{b + s} \, ds + B \tag{41}$$

where B is a constant of integration. This equation can be integrated directly and gives

$$z = \frac{ds}{dx} = \pm 2^{1/2} \left[s - b \ln \left(1 + \frac{s}{b} \right) + B \right]^{1/2} \tag{42}$$

This can be integrated numerically over the required interval of x, subject to the boundary conditions

$$x = 0, s = [\text{S}]' \quad \text{and} \quad x = l, s = [\text{S}]'$$

Since these do not permit the evaluation of A in Eq. (42), a rough estimate of ds/dx at $x = 0$ is made, and Eq. (42) is integrated until $x = l$. The initial value ds/dx can then be adjusted, by trial and error, until $[\text{S}]'_{x=l} = [\text{S}]'_{x=0}$ to the desired degree of accuracy.

For computational purposes the negative sign in Eq. (42) is taken until $ds/dx = 0$, after which the positive sign is used; the changeover point is taken as that value of x at which the argument of the square root term in Eq. (42) becomes negative.

A number of computer calculations have been carried out, and they confirm the general conclusions drawn above. Examples are shown in Figs. 5 and 6; the left-hand curves show concentration–distance profiles, and the right-hand curve is a plot of rate against the logarithm of $[S]'$, the substrate concentration just inside the surface. Figure 5 relates to the following condi-

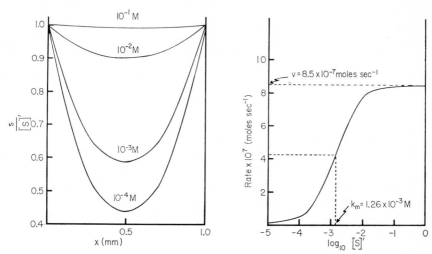

FIG. 5. Computer solutions for a membrane 1.0 mm thick, using the constants given in the text. The left-hand curves show substrate concentration variations through the membranes for the four surface concentrations indicated. The rates in the right-hand figure correspond to 1 liter of support containing 10^{-7} mole of enzyme.

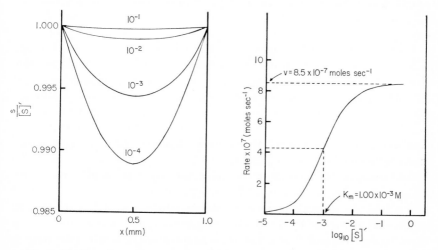

FIG. 6. Similar plots to those shown in Fig. 5 for a membrane 0.1 mm thick.

tions:

$$l = 1 \text{ mm}; \qquad K_m' = 10^{-3} \, M; \qquad a = 0.85 \text{ mole cc}^{-5}\text{cm}^{-5}$$

The latter condition corresponds, for example, to the following values:

$$k_c' = 8.5 \text{ second}^{-1}; \qquad [E]_m = 10^{-7} \, M; \qquad D = 10^{-6} \text{ cm}^2 \text{ second}^{-1}$$

It is seen from Fig. 5 that if the substrate is equally soluble in the support and in the solution, the K_m(app.) value would be $1.26 \times 10^{-3} \, M$, i.e. slightly greater than K_m'. The k_c(app.) value, however, is 8.5 second^{-1}, exactly the same as k_c'. The rate constant at low substrate concentrations, k_c(app.)/ K_m (app.), is $6.75 \times 10^{-3} \, M^{-1}$ second^{-1}, slightly less than the value of 8.5×10^3 for the enzyme in free solution. If the partition coefficient is not unity but is P, the K_m(app.) values will be reduced by the factor P, and the k_c(app.)/ K_m(app.) values raised by the factor P.

Figure 6 shows the results of similar calculations for a membrane of thickness 0.1 mm, other conditions being the same. In line with the earlier discussion, the K_m(app.) value (for $P = 1$) is now much closer to the K_m' value; it is indeed, almost exactly $10^{-3} \, M$. Similarly, the k_c(app.) and k_c(app.)/ K_m(app.) values are indistinguishable from the values for the enzyme in free solution. Again, if P is other than unity, the k_c(app.) value remains unchanged, but K_m(app.) and k_c(app.)/K_m(app.) are altered.

4. General Conclusions

The main points that have been revealed by the treatment are summarized below.

1. If k_c and K_m are unchanged by the support and $P = 1$, the rates at high substrate concentrations will be the same for a given amount of supported enzyme as for the same amount of enzyme per liter of free solution. At lower substrate concentrations the rates will tend to be less on the supported enzyme, and the Michaelis constants higher because of diffusional effects.

2. However, as the membrane is made thinner (at constant membrane volume), the lower-concentration rates for the support will tend to approach those for the enzyme in free solution. The computer calculations have shown that, under typical conditions, identical rates have been obtained for a membrane 0.1 mm thick.

3. The higher the enzymic activity (i.e., the higher k_c' or $[E]_m$) the lower will be the membrane thickness at which the rates become the same.

4. If $k_c' \neq k_c$, the high substrate concentration rates will differ by the ratio k_c'/k_c but will not be affected by the partition coefficient P.

5. If $K_m' \neq K_m$ and $P \neq 1$, the low-concentration rates will differ by the

factor $Pk_c'K_m/K_m'k_c$ and the apparent Michaelis constants by the factor K_m'/PK_m.

Points 2 and 3 can be put into a more quantitative form with reference to Fig. 3. The value of F is very close to unity for values of αl less than about 0.3 and falls fairly strongly at higher values. The critical thickness below which the rates on the supported enzyme correspond to those in free solution, therefore, corresponds to an αl value of about 0.3. The value of α for the parameters used in the computer calculations is about 30 cm^{-1}, so that the critical l value should be about $0.3/30 = 0.01$ cm. The computer calculations did in fact show that at $b = 0.01$ cm the limit had been reached.

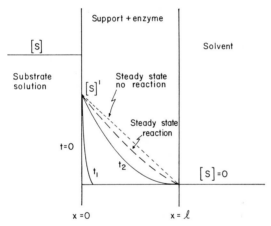

FIG. 7. Schematic representation of the substrate concentration changes when a substrate solution of concentration [S] is separated from pure solvent by a membrane containing enzyme.

5. The Unsymmetrical Case

The situation represented in Fig. 7 is also an important one to consider; this is the case in which substrate solution is on one side of the membrane and pure solvent on the other. The only difference between this case and the preceding one is in the boundary conditions, which are now

$$x = 0, s = [S]', \quad \text{and} \quad x = l, s = 0$$

Details will not be given here. A solution of the equations for the limiting, low substrate concentration case, where the kinetics are first order in substrate, was given by Goldman et al. (1968b). The equation for the concentration gradient is

$$\frac{ds}{dx} = -\frac{\alpha[S]' \cos \alpha(l - x)}{\sinh \alpha l} \tag{43}$$

and the rates at which the product passes from the membrane into the solution are

$$v_{x=0} = D[S]' \left[\frac{\alpha \cosh \alpha l}{\sinh \alpha l} - \frac{1}{l} \right] \tag{44}$$

$$v_{x=l} = D[S]' \left[\frac{1}{l} = \frac{\alpha l}{\sinh \alpha l} \right] \tag{45}$$

The ratio $v_{x=0}/v_{x=l}$ varies with αl in the manner shown in Fig. 8, the value

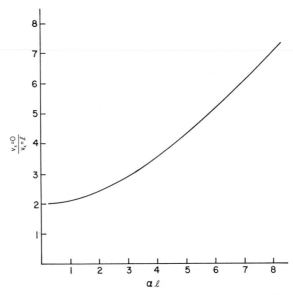

FIG. 8. The ratio of the rates of flow of product out of the membrane at two surfaces, as a function of αl, where l is the thickness of the membrane and $\alpha = (k/D)^{1/2}$; k is the first-order rate constant, and D the diffusion constant for the substrate in the membrane.

approaching 2 as αl approaches zero. It follows that if a solution of substrate is enclosed in a membrane bag containing enzyme and the kinetics are first order, at least twice as much product will flow into the bag as will flow out; the greater the thickness l of the membrane and the greater the enzymic activity (the greater is α) the larger will be the proportion of product that remains in the bag.

The corresponding solutions of the equation for zero-order kinetics (i.e., for high substrate concentrations) are

$$v_{x=0} = v_{x=l} = \frac{1}{2} AD\beta l \tag{46}$$

where $\beta = 10^{-3} k_c'[E]_0/D$ as previously. As in the symmetrical case, the result is again that the limiting, high substrate concentration rate is independent of the characteristics of the support and is the same as for an equivalent amount of enzyme in free solution.

It is easy to show that the solution for the symmetrical case is a superposition of two unsymmetrical cases—one, for $s = [S]'$ at $x = 0$ and $s = 0$ at $x = l$ and, the other, for $s = 0$ at $x = 0$ and $s = [S]'$ at $x = l$. This is illustrated schematically in Fig. 9, in which the expressions for the concentration dependence relate to the low substrate concentration case.

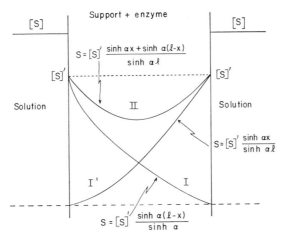

FIG. 9. Diagram showing that substrate concentration changes in Fig. 7 are a superposition of the two versions shown in Fig. 2. The expressions showing the concentration variations relate to the case of low substrate concentrations.

D. SIGNIFICANCE OF THE PARTITION COEFFICIENT P

In the foregoing treatments an important role has been played by the partition coefficient P, which is the ratio of the concentration of substrate just inside the support to that in the bulk solution. It is necessary to consider this coefficient in further detail. Two situations are to be distinguished —one in which the support can be regarded as homogeneous and the other in which it is heterogeneous, the enzymic action taking place in only part of the support (e.g., in pores).

In the *homogeneous* case no particular problem arises. The enzyme normally will be distributed uniformly throughout the support. The substrate will partition itself between the bulk solution and the surface of the support, and the distribution within the support will be as discussed. The magnitude of the partition coefficient P could be determined directly by separate experiments with the support to which no enzyme has been attached, since the enzyme will normally have little effect on the coefficient. The magnitude of P may be very different from unity in certain cases. For example, a substrate having a large number of nonpolar groups will be much more soluble in a support containing many nonpolar groups than in pure water; P will then be much greater than unity. A highly polar substrate, however, may be much more soluble in water than in the support, and P will be less than unity. Another important factor, to which Katchalski has called attention in particular (see, especially, Goldstein et al., 1964), relates to the electrostatic charges on the substrate and support. Thus, if the substrate and support bear opposite charges, P will tend to be large; if they bear like charges, P will tend to be small. Some examples of this are to be found in Section IV, B.

When the support is *heterogeneous* the above principles still apply, but certain complications arise. Suppose that a fraction Θ of the support consists of pores and the remainder of inert material. If E moles of enzyme are attached to Al milliliters of support the *effective* concentration will not be 10^3 E/Al since the enzyme will be present only in the pores; instead it will be $10^3 E/\Theta Al$. This factor alone will tend to produce an enhancement in rate, as compared with the homogeneous case, by a factor of $1/\Theta$. However, the fact that the substrate can only dissolve in the pores leads to a reduction in rate, the rate being Θ times what it would have been if the pore phase constituted the entire support. These two effects exactly cancel. The conclusion is, therefore, that the rate in a porous support will be the same as in a homogeneous support of the same dimensions, if the composition of the material in the pores is the same as that in the homogeneous support.

However, it is evident that the partition coefficient to be used in the theoretical treatment must relate to the distribution of substrate between the bulk solution and the volume within the porous phase, and not between the bulk solution and the support as a whole. The effective partition coefficient P_{eff}, equal to the ratio of the concentrations of substrate in the support and in the bulk solution, will be smaller than the true coefficient P by the factor Θ (i.e., $P_{eff} = \Theta P$). If Θ can be determined by separate experiments, then P can be calculated and used in the theoretical treatment.

E. Electrostatic Effects

One factor that has an important effect on the magnitude of the partition coefficient P is the electrostatic one. If both the support and the substrate are electrically charged and have opposite signs, P will be abnormally large because of the electrostatic forces between them; if the signs are the same, the value of P will tend to be small.

A quantitative treatment of this effect has been developed by Goldstein et al. (1964) in terms of the electrical potential ψ of the support relative to the bulk solution. This potential is positive if the support bears an excess of positive charges, and negative if it bears an excess of negative charges. Suppose that the substrate molecules bear a charge $z\epsilon$, where ϵ is the charge on the H$^+$ ion and z is a positive or negative integer. The energy of a substrate molecule on the support is $z\epsilon\psi$ with respect to the bulk solution, and, according to the Boltzmann principle, the distribution of the substrate between the support and the bulk solution is given by

$$P_{el} = \frac{[S]'}{[S]} = e^{-z\epsilon\psi/kT} \tag{47}$$

Here $[S]'$ is the concentration of substrate in the support and $[S]$ that in the bulk solution; e is the base of the natural logarithms.

The P_{el} is the magnitude of the partition coefficient arising from electrostatic effects alone. If this were the only factor, Eqs. (37) to (39) will apply to a good approximation, with expression (47) used for P. At sufficiently low membrane thicknesses $F = 1$, so that the rate equation becomes

$$v = \frac{k_c' E[S]}{K_m' e^{z\epsilon\psi/kT} + [S]} \tag{48}$$

It is to be noted that, as discussed, the electrostatic interactions have no effect in the limit of high substrate concentrations.

The apparent Michaelis constant for the supported system is

$$K_m(\text{app.}) = K_m' e^{z\epsilon\psi/kT} \tag{49}$$

Thus when z and ψ are of the same sign (e.g., a positively charged substrate and a positively charged support) the apparent K_m is greater than the true value; when they are of opposite signs, $K_m(\text{app.})$ is less than the true K_m'.

Unfortunately, it is difficult to make a reliable estimate of the magnitude of ψ for a particular support. The procedure used by Goldstein et al. (1964) was to calculate ψ values from the experimental results; some examples are considered in Section IV, B.

F. Theory of pH Effects

The influence of pH on the rates of enzymic reactions will, in general, be different for a supported enzyme compared with an enzyme in free solution.

Goldstein et al. (1964) have treated the problem in terms of the distribution of hydrogen ions between the bulk solution and the support. They have developed a quantitative treatment for the situation in which the support contains ionizing groups and bears an electrostatic charge. This treatment follows the same lines as that for the distribution of substrate between support and free solution. It is well known that the relative concentration of hydrogen ions in the near neighborhood of a charged polyelectrolyte gel is different from that of the solution with which it is in equilibrium; an enzyme embedded in a gel will thus be exposed to a pH which is different from that in the bulk of the solution. If the polyelectrolyte is positively charged, the hydrogen-ion concentration will be lower than in the bulk of the solution, whereas, if the polyelectrolyte is negative, the hydrogen-ion concentration will be higher. The pH activity curve of an enzyme attached to a support should, therefore, be displaced toward lower pH values when the support is positively charged and toward higher pH values when the support is negatively charged.

A quantitative treatment of these effects proceeds as follows. Suppose that the support bears a positive charge which creates a positive potential ψ. The ratio of the concentration of hydrogen ions in the support $[H^+]'$ is then related to that in the bulk solution $[H^+]$ by the Boltzmann factor:

$$\frac{[H^+]'}{[H^+]} = e^{-\epsilon\psi/kT} \tag{50}$$

where ϵ is the charge on the H^+ ion, and e the base of the natural logarithms. Since for the free solution

$$pH = -\log_{10}[H^+] \tag{51}$$

and for the support,

$$pH' = -\log_{10}[H^+]' \tag{52}$$

it follows that

$$\Delta pH \equiv pH' - pH = \frac{\epsilon\psi}{2.303kT} \tag{53}$$

The factor 2.303 is for conversion from natural to common logarithms. These pH shifts cause corresponding shifts in the pK values, as shown schematically in Fig. 10. Some applications of this treatment are considered in Section IV, B.

The action of the electrostatic effects on the distribution of hydrogen ions and charged substrate molecules can be looked upon in a slightly different and more realistic way than that envisioned by Goldstein et al. (1964). These workers regard the potential ψ as extending uniformly throughout the bulk of the support, whereas actually on a microscopic scale there will be profound variations in potential through the support. The shifts in pK values are better interpreted in terms of local electrostatic effects than in terms of an average potential. Consider, for example, the ionization

$$-B^+\!\!-H \; \overset{K}{\underset{}{\rightleftharpoons}} \; -B + H^+$$

If a COO⁻ group is brought into proximity to this ionizing group, the

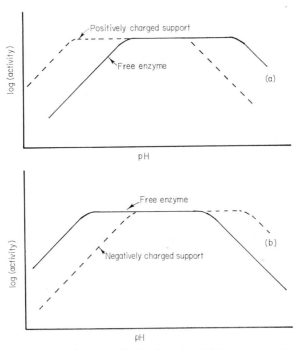

Fig. 10. Profiles of activity vs. pH showing the shifts brought about by (a) a positively charged support and (b) a negatively charged support.

ionization

$$\underset{COO^-}{\overset{B^+\!\!-H}{\underset{}{\gtrless}}} \; \overset{K'}{\underset{}{\rightleftharpoons}} \; \underset{COO^-}{\overset{B}{\underset{}{\gtrless}}} + H^+$$

will occur with greater difficulty because of the attraction of the —COO⁻ group for the proton. The dissociation constant K' will thus be less than K; i.e., $pK' > pK$. This way of looking at the problem leads to the same qualitative conclusion that an excess of negatively charged groups on the support will increase the pK values, whereas positive groups will reduce the pK values. These shifts will not, however, be related to bulk hydrogen-ion concentrations in the manner envisioned by Goldstein *et al.*

A similar point of view is to be preferred with reference to the K_m shifts.

G. Effects of Temperature and of Inhibitors

The theory of Section III, C could readily be extended to cover the effects of temperature and of inhibitors on the kinetics of supported enzyme systems. This will not be considered here, however, since hardly any experimental work has been done in these areas.

IV. Application to Experimental Results

A. Systems for Which Electrostatic Effects Are Unimportant

Several kinetic aspects of solid-supported enzyme systems have been investigated experimentally, but few systematic studies have been made, and the whole subject is still in its infancy. For example, no study has yet been made in which the thickness l of the support has been varied to try to establish the significance of the parameter αl which plays an important role in the theory. Many reports of the experimental studies do not indicate the thickness employed.

Table II gives a very brief summary of some of the kinetic investigations. It would appear that in most of the studies αl was below the critical value of 0.3 so that diffusional effects were unimportant. Partition coefficients between the support and the free solution have apparently never been measured, but one would expect them to be not too far different from unity in a number of cases where there are no electrostatic interactions between the enzyme and the support.

For example, Katchalski (1962), Bar-Eli and Katchalski (1963), Riesel and Katchalski (1964), and Silman et al. (1966) have studied a number of systems in which an enzyme has been attached to an uncharged support, and they have found the k_c(app.) and K_m(app.) values to be little changed. Similar evidence has been obtained by Hornby et al. (1968), who found that the neutral support p-aminobenzyl cellulose caused little change in the K_m-(app.) value for the action of adenosinetriphosphate (ATP)–creatine phosphotransferase on ATP.

Similarly, when the substrate is neutral there is generally little effect on K_m values, even when the support is charged. For example, Money and Crook (quoted by Hornby et al., 1968) found that with the uncharged substrate, N-acetyltyrosine ethyl ester, the attachment of chymotrypsin to CM cellulose-70, a negatively charged support, brought about only a small change in the K_m value. Table III gives some K_m values for systems such as these where there are no electrostatic interactions between the substrate and the support.

A study over a range of pH values has been made by Sundaram (1967) for urease adsorbed on kaolinite. For this system, in which the substrate has no net charge, the K_m values are always somewhat larger for the bound enzyme than for the attached enzyme. Figure 11 shows a plot of $V_{\max}$ $(=k_c(\text{app.})\ [\text{E}])$ values against pH. The pH profiles are similar in

TABLE II
ACTIVITY STUDIES ON BOUND ENZYMES[a]

Enzyme	Support	Remarks	Reference
Trypsin	p-Amino-DL-phenylala-nine-L-leucine copoly-mer	Neutral support; pH optimum unchanged	Bar-Eli and Katchalski (1963)
Chymotrypsin	p-Amino-DL-phenylala-nine-L-leucine copoly-mer	Neutral support; pH optimum unchanged	Katchalski (1962)
Papain	p-Amino-DL-phenylala-nine-L-leucine copoly-mer	Neutral support; pH optimum unchanged	Katchalski (1962)
	p-Amino-DL-phenylala-nine-L-leucine copoly-mer	K_m for BAEE unchanged	Silman (1964)
Trypsin	Maleic anhydride-ethylene copolymer	Negatively charged support; pH optimum raised by 2.5 units	Goldstein, et al. (1964)
Chymotrypsin	Maleic anhydride-ethylene copolymer	pH Optimum raised	Goldstein et al. (1964)
	DEAE cellulose	Positively charged support; pH optimum lowered	Pecht (1966)
Aminoacylase	DEAE cellulose	pH Optimum lowered	Tosa et al. (1967)
Chymotrypsin	Kaolinite	Negatively charged support; pH optimum raised by 2 units	McLaren and Estermann (1957)
Urease	Kaolinite	Marked activation, k_c raised; pH optimum unchanged	Sundaram (1967)
Catalase	Lipid-water interface	Reduced activity	Fraser et al. (1955)

[a] Abbreviations: DEAE—diethylaminoethyl; BAEE—N-benzoylarginine ethyl ester.

TABLE III
APPARENT MICHAELIS CONSTANTS FOR ENZYMES ATTACHED TO UNCHARGED SOLID SUPPORTS[a]

Enzyme	Support	Substrate	Apparent $K_m(M)$	Reference
ATP–creatine phospho-transferase	None	ATP	6.5×10^{-4}	Hornby et al. (1968)
	p-Aminoben-zylcellulose	ATP	8.0×10^{-4}	
Chymotrypsin	None	ATEE	2.7×10^{-4}	Money and Crook, quoted by Hornby et al. (1968)
	CM cellulose-70	ATEE	5.6×10^{-4}	

[a] Abbreviations: ATP—adenosine triphosphate; ATEE—N-acetyltyrosine ethyl ester; CM—O-carboxymethyl.

shape, with the adsorbed enzyme having a somewhat higher activity throughout the pH range investigated.

These results indicate that when either the support or the substrate is uncharged, so that electrostatic interactions are unimportant, the support does not bring about any important change in the kinetic parameters. However, there is often a small but significant increase in the K_m(app.) value when the enzyme is attached to a support in such systems. This effect is understandable in terms of the treatment of diffusion given in Section III, C. This point is essential: because of the necessity for diffusion, the concentra-

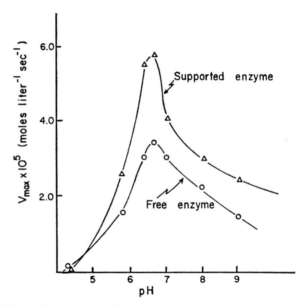

FIG. 11. Plots of V_{max} against pH for free urease and urease attached to kaolinite.

tion of the substrate close to the enzyme is somewhat less than that in the bulk phase, and this will lead to less binding and, therefore, to a higher K_m(app.) value.

The finding of Sundaram (1967) that the V_{max} values are all higher for the bound enzyme than for the unbound requires some effect in addition to diffusion. The V_{max} values relate to enzyme saturation, so that the depletion of substrate near the enzyme has no effect on these values. Several suggestions can be offered for the increased V_{max} values.

1. The attachment of the enzyme to the support may result in the removal of an inhibiting substance from the enzyme. In the urease–kaolinite system, however, tests of this hypothesis were made, but no evidence was obtained that inhibitors were being removed in this way.

2. Soluble substances washed out from the kaolinite might act as activators. In order to test this, washes from the clay were added to the enzyme but were not found to produce activation.

3. The only remaining possibility appears to be that there is a true environmental effect, the k_c(app.) value being greater when the enzyme is supported. This might be due to a change in the effective dielectric constant of the surroundings, which will have an effect on the electrostatic interactions between the enzyme and the substrate.

TABLE IV

APPARENT MICHAELIS CONSTANTS FOR ENZYMES ATTACHED TO SOLID SUPPORTS[a]

Enzyme	Support	Charge on support	Sub-strate	Charge on sub-strate	Apparent $K_m(M)$	Reference
ATP-creatine	None		ATP	—	6.5×10^{-4}	
transphos-	p-Aminobenzylcellulose	0	ATP	—	8.0×10^{-4}	Hornby et al.
phorase	CM cellulose-90	—	ATP	—	7.0×10^{-3}	(1968)
Ficin	None		BAEE	+	2.0×10^{-2}	Hornby et al.
	CM cellulose-70	—	BAEE	+	2.0×10^{-3}	(1966)
Chymotrypsin	None		ATEE	0	2.7×10^{-4}	Hornby et al.
	CM cellulose-70	—	ATEE	0	5.6×10^{-4}	(1968)
Trypsin	None		BAA	+	6.8×10^{-3}	Goldstein et al.
	Maleic acid–ethylene co-polymer	—	BAA	+	2.0×10^{-4}	(1964)
Papain	None		BAEE	+	1.9×10^{-2}	Katchalski
	p-Aminophenylalanine–L-leucine copolymer	0	BAEE	+	1.9×10^{-2}	(1964)
Bromelain	None		BAEE	+	0.11^b	Wharton et al.
	CM cellulose	—	BAEE	+	0.049^b	(1968b)

[a] Abbreviations: ATP—adenosine triphosphate; BAEE—N-benzoylarginine ethyl ester; ATEE—N-acetyltyrosine ethyl ester; BAA—N-benzoylarginineamide; CM—O-carboxymethyl.

[b] At an ionic strength of 0.2 M.

B. CHARGED SUPPORTS

When there are charges on both the substrate and the support, the Michaelis constant for the bound enzyme is significantly different from its value for the free enzyme. This is seen in Table IV. For the ATP–creatine phosphotransferase system, for example, there is a tenfold increase in K_m-(app.) when the enzyme is attached to the polyanionic support, CM cellulose-90, and little change when the enzyme is attached to the neutral p-aminobenzyl cellulose. This increase in K_m(app.) when the negatively charged support is used can be correlated with the electrostatic repulsion between the support and the negatively charged substrate, which will have the effect of decreasing the concentration of substrate in the neighborhood of the enzyme, and of decreasing the binding between enzyme and substrate. In the ficin system, however, where the substrate is positively charged, there is a tenfold decrease in K_m(app.) when the enzyme is attached to the negatively charged support. These observations can be readily explained in terms of electrostatic interactions between substrate and support, as discussed in Section III, E. Unlike charges on substrate and support will lead to enhancement of the substrate concentration close to the surface (P is increased), and this will lead to additional binding, i.e., to a lower $K_m(\text{app.})$. Conversely, like charges will decrease the local substrate concentration in comparison with that in the bulk solution (P is lowered) and will lead to a decrease in binding and to a higher $K_m(\text{app.})$.

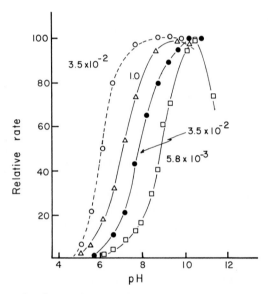

Fig. 12. Curves showing activity against pH for free trypsin and for trypsin attached to a negatively charged support (copolymer of maleic acid and ethylene). The curve for the free enzyme is dashed; the other curves are for the supported enzyme. Ionic strengths (in moles per liter) are shown beside each curve.

Support for the conclusion that these changes in $K_m(\text{app.})$ are due to electrostatic effects is provided by the observation of A. Katchalsky (1964) that the effects are almost entirely eliminated by increasing the ionic strength of the solution. Increasing the ionic strength, like increasing the dielectric constant, has the effect of decreasing the electrostatic forces between charges and will thus reduce the attractive or repulsive forces between the support and the substrate.

Another effect that can be clearly related to electrostatic interactions is the shift in the pH optimum when, for a charged substrate, the enzyme is attached to a charged support. An example of this is seen in Fig. 12; the enzyme is trypsin and the substrate is N-benzoylargininamide, which is

positively charged. When the enzyme is attached to a negatively charged
support, the pH optimum is displaced by approximately 2.5 pH units toward more alkaline values. As the ionic strength is increased, the pH-activ-
ity curve shifts back toward more acid pH values, approaching the curve
for pure trypsin.

As discussed in Section III, F, Goldstein *et al.* developed a theory of these
pH shifts in terms of an electrostatic potential ψ. They did not develop an
explicit treatment of ψ in terms of the properties of the polyelectrolyte sup-

TABLE V

CALCULATED ELECTROSTATIC POTENTIALS FOR TRYPSIN ATTACHED TO A SUPPORT[a]

Ionic strength (moles per liter)	ΔpH	$\psi \times 10^3$ (volts)
0.6×10^{-2}	-2.4 ± 0.2	-145 ± 15
1.0×10^{-2}	-2.0	-124 ± 13
3.5×10^{-2}	-1.6	-96 ± 12
0.2	-1.3	-81 ± 10
1.0	-0.4	-27 ± 8

[a] Support—copolymer of maleic acid and ethylene.

TABLE VI

CALCULATED ELECTROSTATIC POTENTIALS FOR TRYPSIN ATTACHED TO A SUPPORT[a]

Ionic strength (moles per liter)	K_m (app.) $\times 10^3$ (moles per liter)	$K'_m \times 10^3$ (moles per liter)	$\psi \times 10^3$ (volts)
0.04	0.2 ± 0.05	6.9 ± 1.0	-92 ± 12
0.50	5.2 ± 0.5	6.9 ± 1.0	-7 ± 3

[a] Support—copolymer of maleic acid and ethylene; substrate—benzoyl-L-arginine amide ($z = 1$).

port, but calculated ψ values from the ΔpH values observed experimentally
for supported and unsupported enzymes. Some values for trypsin supported
on a copolymer of maleic acid and ethylene are shown in Table V for various
ionic strengths. The pH values [cf. Eq. (53)] are those corresponding to 50%
of maximal activity for the free enzyme, whereas the pH' values are equal
to the pH at which the native enzyme shows a catalytic activity equal to
that of the bound enzyme. The ψ values obtained in this way are consistent
with values obtained in other ways for this polyelectrolyte. The ψ values
are seen to decrease with ionic strength.

Another estimate of the value of ψ can be made by seeing how the apparent Michaelis constant is changed when the enzyme is attached to the
charged support. The equation that applies is Eq. (49). Table VI shows some

results, at two ionic strengths, for the trypsin–maleic acid–ethylene system. The ψ values calculated from Eq. (49) are consistent with those given in Table V calculated from the pK shifts.

V. Kinetics of Flow Systems

A few investigations have been made of the kinetics of enzyme reactions in which the enzyme is attached to a water-insoluble polymer, and the substrate solution is allowed to flow through it. Besides providing useful fundamental information which may be valuable in connection with a study of enzymes *in situ*, these studies are also relevant to processes in which enzymes are allowed to catalyze reactions of industrial interest; the latter aspects have been discussed by Lilly and Sharp (1968).

Two situations are of particular interest. In the first, the substrate solution moves steadily through a packed bed of catalyst in such a way that a cross section of the fluid moves as does an imaginary piston; in this case there is said to be "pellet" or "plug" flow. In the second type of system the catalyst system is continuously stirred, and the reactor is then known as a "continuous feed stirred-tank" (CFST) reactor.

A. PELLET FLOW THROUGH A PACKED BED

The kinetic equations for pellet flow of substrate solution through a packed bed of catalysts attached to a water-insoluble polymer were first worked out by Lilly *et al.* (1966), whose treatment is as follows. The general rate equation for an enzyme-catalyzed reaction, e.g., Eq. (1), may be written as

$$-\frac{d[S]}{dt} = \frac{k_c[E]\,[S]}{K_m + [S]} \tag{54}$$

and integration of this equation subject to the boundary condition $[S] = [S]_0$ when $t = 0$ gives rise to

$$K_m \ln \frac{[S]_0}{[S]} + [S]_0 - [S] = k_c[E]t \tag{55}$$

This is an equation for the time course of the reaction, i.e., the variation of substrate concentration [S] with time; an equation of this form was first given by Victor Henri (1901). In a flow system where there is a pellet flow the time t that the system spends in contact with the catalyst is

$$t = \frac{V_l}{Q} \tag{56}$$

where V_l is the "void" volume or "free" volume and Q the rate of flow. The total enzyme concentration $[E]$ is equal to the total number of moles of enzyme, E, in the reaction system divided by the total volume V_t of the system:

$$[E]_0 = \frac{E}{V_t} \tag{57}$$

The fraction f of substrate converted is

$$f = \frac{[S]_0 - [S]}{[S]_0} \tag{58}$$

Insertion of Eqs. (56), (57), and (58) into Eq. (55) gives

$$f[S]_0 - K_m \ln (1 - f) = \frac{k_c E V_e}{Q V_t} \tag{59}$$

This may be written as

$$f[S]_0 - K_m \ln (1 - f) = \frac{k_c E \beta}{Q} \tag{60}$$

and

$$f[S]_0 - K_m \ln (1 - f) = \frac{C}{Q} \tag{61}$$

where β, equal to V_e/V_t is the voidage of the column, and C, equal to $k_c E\beta$ is referred to as the reaction capacity of the packed bed reactor.

These equations relate the fraction f of substrate converted during the passage through the column to the flow rate Q, in terms of the initial substrate concentration $[S]_0$, the Michaelis constant K_m, and the reaction capacity of the column. The way in which f varies with Q/C is shown in Fig. 13 for various values of K_m and $[S]_0$. For zero flow rate (infinite residence time), there is complete conversion; for infinite flow rate, there is no conversion.

In order to determine K_m and C/Q from experimental data, $f[S]_0$ may be plotted against $-\ln(1 - f)$; a schematic plot is shown in Fig. 14. The intercept on the $f[S]_0$ axis is C/Q and the slope, $-K_m$. From C/Q, if Q, B, and E are known, the kinetic constant k_c can be calculated. Figure 15 shows a plot given by Lilly et al. (1966) for the hydrolysis of N-benzoylarginine ethyl ester by a column of CM cellulose–ficin. If the theory were strictly applicable, the slopes of these plots, equal to $-K_m$, would be independent of flow rate, but some variation is seen. This may be due to the fact that the flow is more complex than envisaged; each cellulose–ficin particle is surrounded by a diffusion layer, and the reaction rate may be controlled to some extent by the rate of diffusion of the substrate through this layer.

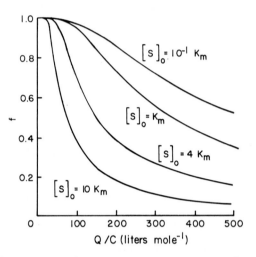

FIG. 13. Variation of fraction f of substrate converted with Q/C, for various values of $[S]_0/K_m$ [cf. Eq. (61)].

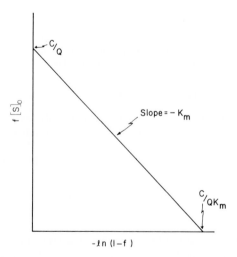

FIG. 14. Schematic plot of $f[S]_0$ against $-\ln(1-f)$ [cf. Eq. (61)]. The slopes and intercepts are shown.

B. Continuous Feed Stirred-Tank Reactors

Kinetic equations for flow of substrate solution through a CFST reactor have been given by Lilly and Sharp (1968). It is assumed that there is per-

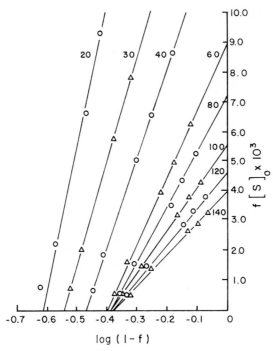

Fig. 15. Plot of $f[S]_0$ against ln $(1 - f)$, for the hydrolysis of N-benzoylarginine ethyl ester by a column of cellulose ficin. The number on the curves indicate the flow rates Q (in milliliters per hour). (From Lilly et al., 1966.)

fect micromixing throughout the vessel. The substrate balance equation is

Rate of entrance of substrate = rate of exit of substrate
+ rate of removal of substrate by the reaction

If Q is the rate of flow of the substrate (in liters per second), the rates of entrance and exit are

$$\text{Rate of entrance} = Q[S]_0 \tag{62}$$

$$\text{Rate of exit} = Q[S]_t \tag{63}$$

The rate of removal of the substrate in the stirred tank of liquid volume v_l is

$$v = \frac{k_c[E][S]_i V_l}{K_m + [S]_i} \tag{64}$$

the substrate concentration within the tank being $[S]_i$. Then

$$Q([S]_0 - [S]_f) = \frac{k_c[E]_0[S]_i V_l}{K_m + [S]_i} \tag{65}$$

If the total amount of enzyme present is E and the volume of the liquid and the stirred tank is U_T ,

$$[E]_0 = \frac{E}{V_1} \tag{66}$$

With the definitions

$$f = \frac{[S]_0 - [S]_i}{[S]_0} \tag{67}$$

and

$$\epsilon = \frac{V_l}{V_t} \tag{68}$$

Eq. (65) becomes

$$f[S]_0 Q = \frac{k_c E \epsilon}{K_m/[S]_0(1 - f) + 1} \tag{69}$$

or

$$f[S]_0 + \frac{f}{1 - f} K_m = \frac{k_c E \epsilon}{Q} = \frac{C_T}{Q} \tag{70}$$

where C_T , equal to $k_c E \epsilon$, is known as the reaction capacity for the stirred tank reactor. Equation (70) somewhat resembles Eq. (61) and gives a similar variation of P with Q/C_T . Lilly and Sharp (1968) have made comparisons of the behavior of the two types of reactors. It is evident that if the stirring is not complete in the CFST reactor, there will be deviations from Eq. (70). Lilly and Sharp have investigated the effect of stirrer speed on the kinetic behavior, using carboxymethylcellulose–chymotrypsin with acetyltyrosine ethyl ester as substrate, and found that the reaction rate increases with the rate of stirring.

REFERENCES

Arnold, W. N. (1966). *Arch. Biochem. Biophys.* **113,** 451.
Bar-Eli, A., and Katchalski, E. (1963). *J. Biol. Chem.* **238,** 1690.
Barnard, M. L., and Laidler, K. J. (1952). *J. Amer. Chem. Soc.* **74,** 6099.
Barnett, L. B., and Bull, H. B. (1959). *Biochim. Biophys. Acta* **36,** 244.
Barrer, R. M. (1941). "Diffusion in and through Surfaces," p. 15. Cambridge Univ. Press, London and New York.
Briggs, G. E., and Haldane, J. B. S. (1925). *Biochem. J.* **19,** 338.

Brown, H. D., Chattopadhyay, S. K., and Patel, S. K. (1966). *Biochem. Biophys. Res. Commun.* **25**, 304.

Chang, T. M. S. (1967). *Sci. J.* July, p. 62.

Chang, T. M. S., MacIntosh, F. C., and Mason, S. G. (1966). *Can. J. Physiol. Pharmacol.* **44**, 115.

Cheeseman, D. F., and Davies, J. T. (1954). *Advan. Protein Chem.* **9**, 440.

Cheeseman, D. F., and Schuller, H. (1954). *J. Colloid Sci.* **9**, 113.

Chung, S.-T., Hamano, M., Aida, K., and Uemura, T. (1968). *Agr. Biol. Chem.* **32**, 1287.

Crook, E. M. (1968). *Biochem. J.* **107**, 1P.

Davies, J. T., and Rideal, E. K. (1963). "Interfacial Phenomena," p. 246. Academic Press, New York.

Findlay, E., Mathias, A. P., and Rabin, B. R. (1963). *Biochem. J.* **85**, 139.

Fraser, M. J., Kaplan, J. G., and Schulman, J. H. (1955). *Discuss. Faraday Soc.* **20**, 44.

Goldman, R., Kedem, O., Silman, I., Chaplan, S. R., and Katchalski, E. (1968a). *Biochemistry* **7**, 486.

Goldman, R., Kedem, O., and Katchalski, E. (1968b). *Biochemistry* **7**, 4518.

Goldstein, L., Levin, Y., and Katchalski, E. (1964). *Biochemistry* **3**, 1913.

Henri, V. (1901). *Compt. Rend. Acad. Sci.* **133**, 891.

Hirs, C. W. H., Moore, S., and Stein, W. D. (1953). *J. Biol. Chem.* **200**, 493.

Hornby, W. E., Lilly, M. D., and Crook, E. M. (1966). *Biochem. J.* **98**, 420.

Hornby, W. E., Lilly, M. D., and Crook, E. M. (1968). *Biochem. J.* **107**, 669.

Jonxis, J. H. P. (1939). *Biochem. J.* **33**, 1743.

Katchalski, E. (1962). *Polyamino Acids, Polypeptides, Proteins, Proc. Int. Symp., 1st, 1961* p. 283.

Katchalsky, A. (1964). *Biophys. J.* **4**, 9.

Kay, G. (1968). *Process Biochem.*, August.

Laidler, K. J. (1958). "The Chemical Kinetics of Enzyme Action," Chapter 7. Oxford Univ. Press (Clarendon), London and New York.

Laidler, K. J., and Ethier, M. C. (1953). *Arch. Biochem. Biophys.* **44**, 338.

Levin, Y., Pecht, M., Goldstein, L., and Katchalski, E. (1964). *Biochemistry* **3**, 1905.

Lilly, M. D., and Sharp, A. K. (1968). *Chem. Eng. (London)* **215**, CE12.

Lilly, M. D., Hornby, W. E., and Crook, E. M. (1966). *Biochem. J.* **100**, 718.

McLaren, A. D. (1954). *J. Phys. Chem.* **58**, 129.

McLaren, A. D., and Estermann, E. F. (1957). *Arch. Biochem. Biophys.* **68**, 157.

Michaelis, M., and Menten, M. L. (1913). *Biochem. Z.* **49**, 333.

Mitz, M. A., and Schlueter, R. J. (1959). *J. Amer. Chem. Soc.* **81**, 4024.

Nikolaev, A. Ya. (1962). *Biokhimiya* **27**, 842.

Nikolaev, A. Ya., and Mardashev, S. R. (1961). *Biokhimiya* **26**, 641.

Pecht, M. (1966). Master's Thesis, Hebrew University, Jerusalem, Israel.

Riesel, E., and Katchalski, E. (1964). *J. Biol. Chem.* **239**, 1521.

Silman, I. H. (1964). Doctoral Thesis, Hebrew University, Jerusalem, Israel.

Silman, I. H., Albu-Weissenberg, M., and Katchalski, E. (1966). Biopolymers **4**, 441.

Silman, I. H., and Katchalski, E. (1966). *Annu. Rev. Biochem.* **35**, 873.

Sundaram, P. V. (1967). Doctoral Thesis, University of London, London.

Sundaram, P. V. (1969). Unpublished data.

Sundaram, P. V., and Crook, E. M. (1968). *Proc. 7th Int. Congr. Biochem., 1967* p. 801.

Sundaram, P. V., Tweedale, A., and Laidler, K. J. (1970). *Can. J. Chem.* **48**, 1498.

Surinov, B. P., and Manoilov, S. E. (1966). *Biokhimiya* **31**, 387.

Tosa, T., Mori, T., Fuse, N., and Chibata, I. (1967). *Enzymologia* **32,** 153.

Trurnit, H. J. (1953). *Arch. Biochem. Biophys.* **47,** 251.

Updike, S. J., and Hicks, G. P. (1967a). *Nature (London)* **214,** 896.

Updike, S. J., and Hicks, G. P. (1967b). *Science* **158,** 270.

Weiss, A. (1961). *Angew. Chem.* **73,** 736.

Weliky, N., Brown, F. S., and Dale, E. C. (1969). *Arch. Biochem. Biophys.* **131,** 1.

Wharton, C. W., Crook, E. M., and Brocklehurst, K. (1968a). *Eur. J. Biochem.* **6,** 565.

Wharton, C. W., Crook, E. M., and Brocklehurst, K. (1968b). *Eur. J. Biochem.* **6,** 572.

Wieland, T., Teterman, H., and Buening, K. (1966). *Z. Naturforsch. B* **21,** 1003.

Wilson, R. J. H., Kay, G., and Lilly, M. D. (1968a). *Biochem. J.* **108,** 845.

Wilson, R. J. H., Kay, G., and Lilly, M. D. (1968b). *Biochem. J.* **109,** 137.

Author Index

Numbers in italics refer to the pages on which the complete references are listed.

A

Aas, M., 137, 139, 154, *169*
Abe, K., 100, *169*
Abood, L. G., 87, 88, 89, 90, 92, 130, 131, 133, 135, 136, *169, 199*
Abraham, A. D., 133, *169*
Abrahams, H. E., 164, *194*
Abrahamson, E. W., 234, 247, *248, 251*
Adachi, K., 109, 110, 112, *190*
Adams, A., 120, *187*
Adamson, A. W., 208, 219, 236, *248*
Adelstein, S. J., 159, *169*
Adloff, E., 157, *178, 197*
Agar, H. D., 56, *68*
Agosin, M., 82, 84, *169*
Agostini, A., 82, *169*
Agranoff, B. W., 86, *171*
Ahmed, Z., 87, 88, *169*
Aida, K., 258, *295*
Aithal, H. N., 132, *169*
Akers, C. K., 233, 244, *251*
Akhtar, M., 80, 115, *169*
Aki, K., 161, *189, 199*
Akopyan, Zh. I., 132, *180*
Alberici, M., 85, 99, *176*
Albers, R. W., 80, 90, 107, 109, *169, 197*
Albu-Weissenberg, M., 284, *295*
Aldridge, W. N., 86, 87, 90, 132, 159, *169, 198, 200*
Alessio, L., 130, *169*
Alfin-Slater, R. B., 87, *196*
Alifano, A., 142, *192*
Allard, C., 87, 88, 89, 132, 136, 163, *169, 176*
Allen, A. K., 91, *169*
Allen, J. M., 162, *169*
Allende, J., 120, *169*
Allfrey, V. G., 38, 39, 44, *68*, 155, 156, *169, 189*
Allman, D. W., 128, 129, 132, 138, 139, 148, 152, 154, *169, 170, 180*, 235, *250*
Alonso, D., 100, *181*
Alpers, D. H., 164, *170*

Altschul, A. M., 90, 106, 107, 109, 110, *173*
Altschul, R., 210, *253*
Alvares, A., 84, 98, *187*
Amar-Costesec, A., 83, 88, *200*
Amberson, W. R., 82, 91, *170*
Amons, R., 137, 146, *170*
Anders, M. W., 84, *170*
Andersen, O. F., 100, *175*
Anderson, A. J., 164, *170*
Anderson, N. G., 38, 39, 40, 41, *68*
Anderson, R. C., 84, *182*
Anderson, W., 88, *177, 191*, 241, *251*
Andersson-Cedergren, E., 51, *71*
Andre, J., 128, 133, *187*
Andreoli, T. E., 92, 146, *170, 195*
Anfinsen, C. B., 163, *170*
Annison, E. F., 80, 85, 87, 115, *171*
Annlight, P., 139, *179*
Aparicio, S. G. R., 89, *202*
Apgar, J., 120, *182*
Appelmans, F., 83, 88, 89, 92, 131, 133, 135, 137, 163, 164, 166, *170, 176*
Aprison, M. H., 87, *190*
Arion, W. J., 88, *191*
Arlinghouse, R., 120, *170*
Arnold, W. N., 259, *294*
Aronson, N. N., Jr., 164, 167, *170*
Arvidson, E. O., 241, *248*
Arvy, L., 58, *69*
Asai, J., 107, 127, 133, 145, *180, 181, 188, 193*
Asakura, T., 90, *200*
Ashmore, J., 101, 137, *171, 199*
Ashwell, G., 82, 89, 111, 131, 136, *184, 189*
Ashworth, C. T., 146, *170*
Ashworth, L. A. E., 7, 8, *30*
Atkinson, A., 90, *170*
Augenstein, L. G., 92, *184*, 206, 209, *250*
Avi-Dor, Y., 85, *170*
Avignon, J., 83, *170*
Axelrod, J., 82, 84, 85, 111, *170, 199*
Azzi, A., 230, *248*

297

Champagne, M., 158, *171*
Chan, H., 218, 220, 228, *248*
Chance, B., 26, *31*, 51, *71*, 100, 129, 133, 138, 139, 140, 144, *174*, *192*, 230, 235, *248*, *251*
Chang, T. M. S., 257, 263, *295*
Chang, Y.-Y., 103, *174*
Chantrenne, H., 87, 89, 136, *174*
Chanutin, A., 86, 87, 89, 135, *187*
Chapman, D., 214, 217, 239, 240, 241, 244, *249*, *250*
Chappel, J. B., 131, *174*
Chassy, B. M., 100, *174*
Chatterjee, G. C., 82, 111, 131, *174*
Chatterjee, I. B., 82, 83, 111, 131, *174*, *179*, *185*
Chattopadhyay, S. K., 79, 81, 82, 83, 84, 86, 90, 91, 92, 98, 106, 107, 109, 110, 111, 112, 113, 130, 135, *173*, *174*, *192*, *194*, 258, *295*
Chauveau, J., 83, *175*
Chaykin, S., 89, *185*
Cheatum, S. G., 82, *175*
Cheeseman, D. F., 262, *295*
Chen, R. F., 148, *175*
Chen, Y. T., 82, 83, *183*
Cherry, R., 217, *249*
Cheruy, A., 135, *175*
Chi, Y. M., 85, 101, *185*, *190*
Chibata, I., 258, 285, *296*
Chignell, C. F., 90, *175*
Chrisman, C. R., 111, *192*
Christensen, H. N., 118, *175*
Christie, G. S., 89, 130, 131, 132, 136, *175*
Christodoulou, C., 35, *70*
Chung, S.-T., 258, *295*
Chytil, F., 100, *175*
Cirillo, V. J., 100, *175*
Clark, D., 116, *183*
Claude, A., 74, *175*
Clausen, J., 82, 111, 130, *181*
Cleland, W. W., 80, 85, 131, 133, 135, 136, *170*, *175*, *180*, *195*, *203*
Clifton, K. H., 136, *189*
Cline, G. B., 40, 41, *68*
Cochran, D. G., 137, *189*
Coffey, J. W., 167, *175*
Cohen, B., 75, 80, 83, 93, 94, *177*
Cohen, E., 136, *185*
Cohen, G., 106, *175*

Cohen, P. P., 134, 135, *180*
Cohen, Y., 132, *200*
Colacicco, G., 208, 210, *249*
Cole, K. S., 236, *249*
Coleman, J. E., 109, *200*
Collins, C. G. S., 132, *175*
Collins, D. C., 85, *175*
Collins, F. D., 235, *249*
Colomb, M. G., 135, *175*
Colombo, B., 62, 63, *68*
Colowick, S. P., 133, 144, *184*
Conchie, J., 164, *175*, *187*
Conger, N., 134, *195*
Conney, A. H., 84, 93, *175*, *179*
Conover, T. E., 51, *71*
Conway, T., 120, *175*
Cook, W. H., 213, *249*
Coon, M. J., 84, *187*
Cooper, C., 128, 138, 139, *182*
Cooper, D. Y., 85, 93, 143, *175*, *190*, *197*
Cooper, J. R., 84, *173*
Cooper, T. G., 52, 54, *69*
Copenhaver, J. H., Jr., 90, *200*
Corchiello, E., 137, *186*
Cordes, E. H., 229, *250*
Corey, E. J., 80, 84, 115, *176*
Corman, L., 142, *175*
Cornatzer, W. E., 88, 135, *170*
Cornforth, J. W., 80, 115, *177*
Cornwell, D. G., 236, *248*
Cosmides, G. J., 93, *179*
Cotzias, G. C., 132, *175*
Coutsogeorgopoulos, C., 158, *186*
Cox, G. F., 130, *175*
Crane, F. L., 26, *30*, 246, *249*, *252*
Crane, R. K., 86, 135, *175*, *182*
Craston, A., 89, *193*
Craven, P. A., 135, 139, *175*
Creange, J. E., 100, *175*
Creasey, W. A., 91, *175*
Cremer, J. E., 84, *175*
Cremese, R., 136, *176*
Criddle, R. S., 36, 52, 56, *69*, 235, *249*
Crist, E. J., 150, *175*
Crook, E. M., 258, 260, 261, 262, 267, 284, 285, 287, 290, 291, 293, *295*, *296*
Cryer, P. E., 85, *175*
Cumar, F. A., 111, *170*
Curti, B., 161, *188*
Curtis, P. J., 154, *175*

J

Jackson, C., 163, *183*
Jackson, R. T., 81, *173*
Jackson, S., 130, 132, *187*
Jacobs, E. E., 133, *199*
Jacobs, R. J., 132, *195*
Jacobson, K. B., 89, 91, *183*
Jacobson, M., 84, 98, *187*
Jacques, P., 83, 88, 132, 133, 161, 163, 164, 166, *176, 184, 196*
Jaffe, H., 84, *184*
Jain, M. K., 229, *250*
Jakobsson, S. V., 88, *184*
James, A. T., 159, *191*
James, L. K., 92, *184*, 206, 209, 210, *250*
Jamieson, J. D., 52, 60, 61, 62, 63, *69, 70*
Jarett, L., 85, *175*
Jarnefelt, J., 90, 110, *184*
Jarrott, B., 83, 132, 154, *184*
Jayaraman, J., 86, 134, *186, 196*
Jean, D. H., 109, *184*
Jeanloz, R. W., 164, *184*
Jefcoate, C. R. E., 98, *184*
Jefferson, L. S., 101, *178*
Jensen, T. E., 55, *70*
Jerina, D. M., 84, *175, 184*
Ji, T. H., 241, *250*
Jirku, H., 85, *175*
John, K., 241, *252*
Johns, P. T., 88, *191*
Johnson, D., 147, *186*
Johnson, M. K., 86, 87, 132, 135, *169, 184*
Johnston, J. M., 87, 115, 116, *173, 184, 194*
Johnston, R. E., 88, *172*
Jòlles, P., 164, *184*
Jones, D. C., 132, *187*
Jones, L. C., 90, *177*
Jones, M. S., 127, 187, *184*
Jones, O. T. G., 127, 187, *184*
Jones, P. D., 83, 92, *184*
Jones, R. T., 134, *178*
Jonxis, J. H. P., 262, *295*
Jordan, W. K., 89, 136, *184*
Jorgensen, P. L., 107, *184*
Josefsson, L. I., 163, *184*
Joshi, V. C., 84, *184*
Jost, J. P., 100, *184*
Judah, J. D., 89, 130, 131, 132, 136. *175*
Jun, K. W., 84, *187*

Jung, C. Y., 220, *250*
Jung, W., 87, *198*
Junger, E., 131, *194*
Jurtshuk, P., Jr., 144, *184. 196*

K

Kaback, H. R., 110, *184*
Kabat, E. A., 87, *184*
Kadis, B., 85, *170*
Kafer, E., 134, *184*
Kagawa, Y., 51, *70*, 129, 137, 146, *184*
Kager, J. S., 133, 142, *192*
Kahlenberg, A., 90, 110, 112, *184, 200*
Kahn, A., 34, 52, 53, *68*
Kahn, J. S., 90, *174*
Kalant, H., 90, *183*
Kalckar, H. M., 82, 85, 111, *199*
Kalina, M., 144, *184*
Kalinsky, H. J., 89, *193*
Kallman, F. G., 87, *184*
Kalnitsky, G., 131, 136, 137, *196, 197*
Kamei, M., 90, 91, *202*
Kamin, H., 83, 98, 133, *184, 202*
Kamm, J. J., 83, *178*
Kanazawa, T., 90, *190*
Kandutsch, A. A., 80, 82, *184*
Kanfer, J., 82, 111, *184*
Kaplan, G., 209, *249*
Kaplan, J. G., 257, 258, 260, 262, 285, *295*
Kaplan, N. O., 81, 89, 91, 130, 133, 142, 144, *175, 183, 184, 185*
Kaputo, R., 111, *170*
Kar, N. C., 82, *179, 185*
Karaboyas, G. C., 100, *185*
Kaplan, N. M., 100, *184*
Karlsson, U., 51, *71*
Karnowsky, M. J., 130, *178*
Karovnikore, K. A., 130, *193*
Kasbekar, D. K., 87, 90, 136, *185*
Katchalski, E., 109, *196*, 257, 258, 266, 277, 280, 281, 282, 284, 285, 287, *294, 295*
Katchalsky, A., 288, *295*
Kato, M., 90, *186*
Kato, R., 83, 84, 98, *185*
Katz, A. M., 90, *195*
Kavanau, J. L., 238, *250*
Kawada, M., 89, *191*
Kawai, K., 132, *185*
Kay, E. R. M., 133, *177*
Kay, G., 257, 258, 263, *295, 296*

Kume, S., 106, 107, *193*
Kun, E., 86, 89, 130, 135, 136, 143, 148, *172, 176, 177, 178, 186*
Kundig, W., 110, *186*
Kuntzman, R., 84, 98, *187*
Kuo, J. F., 100, *186, 189*
Kurnick, N. B., 163, *186*
Kurokawa, M., 90, *186*
Kurup, C. K. R., 84, *184*
Kushnir, L. D., 228, *250*
Kuylenstierna, B., 127, 133, 135, 138, 139, *198*
Kwant, W. O., 244, *252*

L

Labbe, R. F., 137, *191*
LaBella, F. S., 87, 131, 136, *186*
Ladbrooke, B. D., 240, 241, *250*
La Du, B. N., 80, 84, 93, 98, *173, 179*
Lagerstedt, S., 163, *184*
Laidler, K. J., 266, 267, *294, 295*
Lamb, A. J., 35, *70*
Lamirande, G., *169*
Lands, W. E., 85, *186, 192*
Lang, H., 132, *186*
Lang, K., 132, *186*
Langan, T. A., 100, *186*
Langdon, R. G., 83, 137, *186, 193*
Langemann, H., 131, *185*
Langford, P., 118, *186*
Langmuir, I., 3, *31*, 92, *186*, 236, *250*
Lapresle, C., 165, *186*
Lardy, H. A., 128, 130, 132, 146, 147, 154, *186, 195, 203*
Larner, J., 100, *172, 186*
Laseter, A. H., 90, *196*
Laskowski, M., 163, *186*
Laufer, I., 90, *183*
Lauger, P., 222, 223, *250*
Lavate, W. V., 87, 90, 136, *185*
Lawford, G. R., 118, 119, *186*
Lawrence, A. S. C., 236, *250*
Layani, M., 161, *197*
Layne, D. S., 85, *175*
Lazarus, S. S., 146, *186*
Leaback, D. H., 164, *193*
Lee, C. P., 129, *174, 177*, 230, *248*
Lee, S. H., 134, *186*
Lee, Y.-P., 131, *186*
Leech, R. M., 159, *186*

Legler, B., 88, *198*
Lehninger, A. L., 17, 26, *31*, 56, *70*, 87, 128, 129, 130, 131, 135, 137, 143, *180, 185, 186, 187, 202*, 222, *248*
Lejeune, N., 164, 167, *186*
LeJohn, H. B., 130, 132, *186, 187*
Leloir, L. F., 88, 111, *187, 190, 192*
Lemberg, R., 98, *182*
Lemonde, A., 130, *190*
Lenard, J., 238, 240, 243, 245, *250*
Lenaz, G., 128, 129, *187*
Lengyel, P., 80, 119, *187*
Lenz, W. R., Jr., 132, 154, *187*
Leong, V., 164, 167, *183*
Leslie, I., 157, *187, 188*
Leslie, R. B., 90, *170*, 214, 240, *250*
Lesslauer, W., 222, 223, *250*
Letelier, M. E., 82, 84, *169*
Leuthardt, F., 134, 135, 137, 154, *187, 190, 191*
Lev, A. A., 223, *250*
Leventer, L. L., 85, *187*
Levey, G. S., 85, *177*
Levin, W., 84, 98, *187*
Levin, Y., 258, 266, 280, 281, 282, 285, 287, *295*
Levvy, G. A., 164, *175, 187*
Levy, M., 128, 133, *187*
Levy, R. I., 213, *249*
Lewin, I., 88, 89, 136, *194*
Li, Y. W., 137, *197*
Liberman, E. A., 222, 228, *251, 253*
Lichtenthaler, H. K., 235, *251*
Liddle, G. W., 100, *169*
Liddle, R. A., 100, *169*
Lieberman, I., 80, 121, *200*
Lieberman, M., 159, *187*
Light, P. A., 144, *174*
Ligla, K. T., 133, *202*
Likar, I. N., 125, *187*
Lilly, M. D., 258, 260, 262, 263, 267, 284, 285, 287, 290, 291, 293, 294, *295*
Lin, J.-K., 84, *187*
Lin, K., 84, *202*
Linberg, O., 147, *197*
Lindahl, T., 120, *182, 187*
Lindberg, M., 115, *192*
Lindenmayer, A., 52, 56, *70*
Linker, A., 164, *189*
Linn, T. C., 82, *187*

Linnane, A. W., 35, 52, 56, *70*, *71*
Lipmann, F., 120, *169*, *175*
Lipner, H. J., 131, *187*
Lipscomb, M. D., 144, *199*
Lipton, S. H., 140, *171*, 235, *250*
Listowsky, I., 130, 143, *180*
Littauer, U. Z., 86, *185*
Littleton, G. K., 100, *200*
Litwack, G., 85, 100, 134, *189*, *200*
Lloyd, J. B., 164, *187*
Lockshin, A., 235, *251*
Loeb, J. N., 118, 120, *183*
Loening, U. E., 35, 58, *70*
Lorand, L., 135, *193*
Lotlikar, P. D., 84, *187*
Lotspeich, W. D., 136, *191*
Loud, A. V., 37, 49, *70*
Love, L. L., 100, *174*
Lovrien, R., 241, *251*
Low, H., 147, *197*
Lowe, A. G., 90, *170*
Lowe, C. U., 131, *187*
Lowe, E., 80, *174*
Lowe, P. A., 87, 115, *184*
Lowenstein, J. M., 129, 130, 144, *187*
Lowinger, J., 90, *194*
Lu, A. Y. H., 84, *187*
Lubin, M., 62, *69*
Lucy, J. A., 16, *31*, 209, 210, 236, 238, *251*
Ludewig, S., 86, 87, 89, 135, *187*
Luibel, F. J., 146, *170*
Lukes, B., 120, 121, *174*
Lukins, H. B., 35, *70*
Lundquist, A., 163, 164, *187*
Luzzati, V., 211, 220, *251*
Lygre, D. G., 88, *187*, *191*
Lynen, F., 82, 137, 150, 151, 152, 153, *174*, *187*
Lynn, D., 144, *200*

M

Maass, H., 159, *182*
McArdle, B., 134, *187*
McBain, J. W., 211, 215, *251*
McCaman, R. E., 138, 139, *169*
McCann, W. P., 144, *187*
McClure, R. H., 235, *249*
McClure, W. O., 230, *249*
McComb, R. B., 86, *187*
McConnell, D. G., 235, 244, *250*

McConnell, H. M., 240, *250*
McCrea, B. E., 130, 132, *187*
MacDonald, R. A., 35, *70*, 88, *174*
McEwen, C. M., Jr., 132, 154, *187*
Macey, R. I., 90, *185*
McGarrahn, K., 82, 115, *174*
Machinist, J. M., 84, *187*, *188*
Machino, A., 85, *183*
McIlwain, H., 81, 90, *198*, *199*
MacIntosh, F. C., 257, 263, *295*
McKay, R., 137, *188*
Mackenzie, C. G., 133, *179*
Mackler, B., 133, 144, *185*
McKnight, R. C., 147, *188*
McLagan, N. F., 84, *188*
McLaren, A. D., 258, 260, 262, 285, *295*
McLean, P., 88, *179*
MacLennan, D. H., 11, 23, 26, *31*, 127, 129, 137, 145, *180*, *188*, 235, *250*
McLeod, R. M., 84, *188*
McMurray, W. C., 147, *186*
McSham, W. H., 131, *188*
McWhorter, W. P., 80, *178*
Maddy, A. H., 4, 6, 11, 14, 19, *31*, 229, 234, 240, 243, *251*
Madison, J. D., 120, *182*
Mahadevan, S., 163, 165, *188*
Mahesh, V. B., 82, *188*
Mahler, H. R., 35, *69*, 130, 131, 133, 137, 149, 161, *172*, *188*
Malamud, D., 100, *188*
Malcolm, B. R., 240, 243, *251*
Mandel, P., 157, 158, 159, *174*, *188*
Mandl, I., 164, *188*
Mann, K. G., 130, 143, *188*
Mannering, G. J., 93, *179*
Mano, Y., 82, *199*
Manoilov, S. E., 258, *295*
Mansour, T. E., 100, *188*, *199*
Mapson, L. W., 82, 83, *183*
March, R., 89, 136, *184*
Mardashev, S. R., 258, *295*
Margolis, S. A., 91, *188*
Margreth, A., 86, *188*
Marinetti, G. V., 85, *188*
Mariyama, T., 80, *200*
Marker, C. L., 86, *188*
Marguisee, M., 120, *182*
Marsh, C. A., 164, *187*
Marsh, J. B., 233, *253*

Moretz, R. C., 233, 244, *251*
Morgan, H. E., 220, *253*
Morgan, T. E., 234, *251*
Mori, T., 113, *188*, 258, 285, *296*
Morino, Y., 134, 155, *190*
Morré, D., 159, *190*
Morris, A. J., 60, *70*
Morris, H. P., 84, 98, 160, *173, 194*
Morris, M. L., 84, *198*
Morrison, M., 82, 111, *171*
Morton, R. K., 83, 87, 131, 133, *170, 190*
Moses, J. L., 146, *190*
Moss, D. W., 87, 89, *177, 190*
Moulé, Y., 83, *175*
Moustacchi, E., 35, *70*
Moyle, J., 142, 145, *189*
Mudd, J. B., 117, *190*
Müller, A. F., 134, 135, 154, *187, 190*
Mueller, G. C., 83, *190*
Mueller, P., 215, 216, 217, 218, 219, 221, 223, 224, 228, *251*
Mühlethaler, K., 58, *70*, 233, 238, 247, *251*
Mukherjee, S., 80, *190*
Muller, M., 133, 160, *170, 186*
Munday, K. A., 80, 115, *169*
Munk, K., 89, 131, *188*
Munkres, K. D., 11, *32*
Munoz, E., 107, *190*
Murad, F., 85, *190, 201*
Myers, D. K., 147, *190*
Myers, L. T., 83, 121, *190*

N

Nachbar, M. S., 107, *190*
Nachbaur, J., 135, *190*
Nagai, K., 90, 91, *202*
Nagano, K., 90, 109, 110, 112, *179, 190*
Nakagawa, T., 211, *252*
Nakagawa, Y., 90, 91, *202*
Nakamaru, Y., 90, 113, *190*
Nakamura, A., 90, *183*
Nakano, Y., 134, *190*
Nakao, M., 90, 109, 110, 112, *179, 190*
Nakao, T., 90, 109, 110, 112, *179, 190*
Narasimhulu, S., 85, 93, *175, 190*
Nath, M. C., 146, *194*
Nathan, P., 87, *190*
Navazio, F., 130, 132, *177*
Nayler, W. G., 90, *185*
Nebert, D. W., 84, *190, 191*
Needham, D. M., 136, *201*

Neet, K. E., 42, *70*
Nelson, G. J., 234, *252*
Nelson, J., 138, 150, 153, *202*
Nelson, L., 89, 131, 133, 136, *191*
Neuberger, A., 164, *176*
Neucere, N. J., 90, 106, 107, 109, 110, *173*
Neufeld, E. F., 80, 133, 144, *179, 184*
Neves, A. G., 164, 167, *189*
Neville, A. M., 82, 91, *191*
Newsholme, E. A., 86, 135, *191*
Ney, R. L., 101, *180*
Nichoalds, G. E., 104, *191*
Nicholls, D. G., 130, 139, *179, 191*
Nichols, D. W., 159, *191*
Nichols, G., 164, *202*
Nicholson, W. E., 100, *169*
Nielsen, H., 134, 135, 137, *187, 191*
Nikolaev, A. Ya., 258, *295*
Nirenberg, M. W., 118, *191*
Nishibayashi, H., 80, 98, *191, 195*
Nishida, G., 137, *191*
Nishimura, E. T., 161, *192*
Nisselbaum, J. S., 88, 155, *191, 198*
Nissley, S. P., 144, *177*
Nocito, V., 160, *172*
Noe, E., 86, 131, 136, *191*
Noguchi, S., 222, *251*
Noll, H., 118, *202*
Nordlie, R. C., 88, 89, *177, 181, 187, 191, 198*
Noronha, J. M., 132, 134, *191*
North, J. C., 84, 93, *188*
North, R. J., 56, *71*
Northcote, D. H., 233, *251*
Norum, K. R., 127, 130, 137, 139, 154, *178, 191*
Novack, B. G., 87, *195*
Noveles, R. R., 100, *191*
Novikoff, A. B., 33, 34, 55, 56, *70*, 82, 83, 86, 87, 88, 89, 90, 123, 125, 131, 132, 133, 135, 145, 146, 163, 164, *177, 181, 191, 195*
Nowinski, W. W., 26, *30*
Nozaki, M., 133, 154, *192*
Nunley, C. E., 40, 41, *68*

O

O'Brien, J. S., 6, 7, 10, 17, 18, 23, 24, 25, *31*, 233, 234, 235, *251*
Ochoa, S., 154, *182*
Ockerman, P. A., 163, *187*
O'Connell, D. J., 133, *193*

Schweet, R., 80, 120, *170, 196*
Scott, S. J., 160, *189*
Seal, U. S., 84, 89, *196*
Seamond, B., 144, *200*
Searcy, R. L., 212, *252*
Sebald, W., 133, *171*
Sedgwick, B., 80, 87, 115, 116, 136, 163, 167, *196, 198*
Seeman, P., 244, *252*
Segal, H. L., 88, *196*
Sekuzu, I., 144, *184, 196*
Sellinger, O. Z., 83, 86, 88, 132, 133, 135, 161, 164, 166, *176, 196, 199*
Selwyn, M. J., 146, 147, *196*
Semandeni, E., 169, *189*
Sen, A. K., 90, 91, 106, 107, 109, *170, 193, 196, 200*
Sengupta, M., 84, *184*
Senior, J. R., 91, 116, *196*
Serck-Hanssen, G., 80, *174*
Seshadri Sastri, P., 86, 87, 88, 131, 133, 134, *186, 196*
Sessa, G., 227, *252, 253*
Sgaragli, G., 87, *189*
Shah, D. O., 209, *252*
Shamberger, R. J., 163, 164, *196*
Shannon, L. M., 42, *71*
Shanthaveerappa, T. R., 125, *196*
Shapiro, B., 80, 85, 91, 116, *170, 196, 198*
Shapiro, H., 236, *250*
Sharon, N., 163, 164, *184, 196*
Sharp, A. K., 258, 290, 293, 294, *295*
Shashimoto, S., 161, *189*
Sheffield, H. G., 126, *201*
Shepard, T. H., 132, *176*
Sheperd, D., 137, 139, 149, *179, 196*
Shepherd, J. A., 131, 136, 137, *196, 197*
Shepherd, W. C., 218, 220, 228, *248*
Sherratt, H. S. A., 116, *183*
Shibko, S., 86, 87, 163, *197*
Shiga, T., 161, *197*
Shikita, M., 85, *197*
Shimakawa, H., 84, *201*
Shimazono, N., 82, *199*
Shinoda, K., 211, *252*
Siebert, G., 155, 156, 157, 158, *178, 185, 197*
Siegel, G. J., 90, 109, *169, 197*
Siegel, L., 130, 150, *197, 201*
Siegel, M. R., 62, *71*
Siekevitz, P., 34, 35, 50, 52, 60, 62, *69, 70,*

71, 80, 83, 88, 93, 134 135, 145, 147, 177, 193, 194, 197, 201
Siggins, G. R., 100, *197*
Silbert, C. K., 131, *197*
Silman, I., 51, 56, *69*, 140, *171*, 257, 258, 284, 285, *295*
Silverman, L., 37, *71*
Silverstein, E., 130, *197, 199*
Simmons, P., 134, 155, *172*
Simon, G., 234, *252*
Simon, M. R., 80, *188*
Simonson, H. C., 145, *193*
Simpson, E. R., 143, *197*
Simpson, M. V., 135, 154, *189*
Singer, S. J., 238, 240, 243, 245, *250*
Singer, T. P., 131, 133, 149, *183, 194, 197*
Sisler, H. D., 62, *71*
Sitte, P., 232, *252*
Sizer, I. W., 162, *194*
Sjöstrand, F. S., 15, 16, *31*, 45, 46, 47, 51, *71*, 123, *197, 198*, 233, 235, 238, 244, *252*
Skou, J. C., 106, 107, 108, 109, *184, 197*, 230, *252*
Skrivanova, J., 100, *175*
Skulachev, V. P., 222, *251*
Slater, E. C., 108, 128, 129, 131, 133, 135, 136, 142, 145, 146, 147, *170, 175, 190, 197, 199, 201*
Slater, T. F., 83, 87, 132, 133, 136, 149, *180, 197*
Slautterback, D. B., 26, 28, *31*, 81, *178*, 241, 245, *249*
Slenczka, W., 142, *185*
Slifer, E. D., 235, *249*
Slingerland, D. W., 84, *202*
Slotboom, A. J., 118, *200, 201*
Small, D. M., 207, *252*
Smallman, B. N., 87, *197*
Smekal, E., 229, 230, *252*
Smellie, R. M. S., 154, *175*
Smith, A. D., 163, *197*
Smith, C. A., 235, *252*
Smith, D. S., 52, *71*, 127, *198*
Smith, E. L., 123, *198*
Smith, J. C., 87, *198*
Smith, J. K., 89, *190*
Smith, L., 56, *70*
Smith, M. E., 80, 115, 116, *183, 198*
Smith, S., 132, 137, *178*
Smith, S. R., 130, 144, *187*

Tyler, A., 85, *174*
Tyler, D. D., 51, *71*
Tzagoloff, A., 19, 28, *31*, 127, 137, *180, 188*, 235, *250*

U

Udenfriend, S., 84, 132, *172, 173, 175, 184*, *189, 193*
Uemura, T., 258, *295*
Uesugi, S., 90, 110, *112*, *200*
Ugazio, G., 163, *200*
Uhr, M. L., 90, 113, *194*
Ulrich, F., 82, 147, *188*, *200*
Umberger, F. T., 84, *199*
Underhay, E., 86, 87, 88, *200*
Unemoto, T., 88, *200*
Updike, S. J., 263, *296*
Urry, D. W., 240, 245, *253*
Ussing, H. H., 230, *253*
Utsumi, K., 126, *176*
Utter, M. F., 137, *183, 185*
Uziel, M., 86, *170*

V

Vaes, G., 164, *200*
Valdovinos, J. G., 55, *70*
Valette, G., 132, *200*
Vallee, B. L., 109, *200*
Van, N. T., 224, *253*
Vanbellinghen, P. J., 134, *178*
Van Bruggen, J. T., 17, *31*
Van Deenen, L. L. M., 7, 25, 26, *30, 31*, 86, 117, 135, 163, *190, 193, 200, 201*, 233, 234, 239, *253*
van den Berg, H. J., 229, *253*
Van Den Bergh, S. G., 137, 146, *170*
Van den Bosch, H., 118, *200, 201*
Vandenheuval, F. A., 17, 23, *32*
Vaneste, M., 84, 93, *188*
Van Golde, L. M. G., 117, 118, *193, 200*
Van Lancker, J. L., 88, 133, *201*
Van Reen, R., 84, 133, *171*
Van Tol, A., 91, *201*
Van Tubergen, R. P., 48, 60, *68, 71*
Van Weenen, M. F., 120, *172*
Van Zutphen, H., 223, *253*
Varmus, H. E., 100, *176*
Vars, H. M., 87, 89, 135, *194, 195*
Vary, M. J., 130, 137, *201*

Vatter, A. E., 225, *253*
Vaughan, G. D., 100, *199*
Vaughan, M., 85, *201*
Veldsema-Currie, R. D., 137, 146, *201*
Velick, S. F., 83, *199*
Vergani, C., 82, *169*
Verma, S. P., 224, *253*
Vesell, E. S., 79, *201*
Vestling, C. S., 130, 143, 155, *182, 188*
Viala, R., 88, 163, 166, *179, 201*
Vigers, G. A., 137, 146, *201*
Vignais, P. V., 135, *175, 190*
Vilallonga, F., 210, *253*
Villa, L., 82, *169*
Villee, C. A., 82, 134, *179, 201*
Vitols, E., 56, *71*
Vogel, H. J., 80, *201*
Vogel, R. H., 80, *201*
Vogell, W., 130, *201*
Voynick, I. M., 164, 167, *201*
Vreeman, H. J., 220, *253*

W

Wada, F., 84, *201*
Wada, H., 134, 155, *190*
Waggoner, A. S., 231, 240, *250*
Waite, M., 86, 135, *201*
Waitzman, M. B., 81, *173*
Wakid, N. W., 87, 88, 132, 136, *201*
Wakil, S. J., 83, 92, 150, *181, 184*
Waksman, A., 134, 135, 154, 155, *194, 201*
Wald, G., 235, 236, *252, 253*
Waldner, H., 58, *70*
Waldschmidt, M., 132, 145, *201*
Walek-Szernecka, A., 168, *201*
Walker, P. G., 87, 89, 164, *178, 193, 201*
Wallace, A., 149, *172*
Wallace, P. G., 52, 56, *71*
Wallach, D. F. H., 5, 19, 20, 21, 25, 27, *32*, 240, 243, 245, 246, *253*
Wallenfels, K., 164, *201*
Walsh, D. A., 100, *201*
Wang, D. Y., 136, *180*
Wang, R. I. H., 83, 84, 92, 133, *173*
Warren, J. C., 82, 88, *175, 179*
Washko, M. E., 88, *196*
Wassarman, P. M., 130, *185*
Watanake, T., 134, *190*
Waters, D. A., 40, 41, *68*

Subject Index

Extracorporeal shunts, 264
Extraction, differential, 4

F

Fat-synthesizing complex, 153
Fatty acid(s), 2, 7, 11, 57, 117, 150, 206, 212, 234, 241, 242
  activated, 115
  biosynthesis, 9
  CoA esters, 116
  elongation, 27
  long-chain, 18
  oxidation, 27, 52
  saturated, 117
  syntheses, 117, 150, 151, 152, 153
  unsaturated, 18
Fatty acid synthetase, 152, 153
Fatty acyl chains, 15, 24
Fatty acyl-CoA dehydrogenases, 145
Ferrochelatase, 127
Fick's law, 270
Ficin, 287, 292, 293
Filipin, 223, 227
Flavin, 149, 161
Flavin adenine dinucleotide, 149
Flavoprotein, 144, 149, 160, 161
Flow
  bulk solvent, 17
  vs. reaction rate, 292
  systems, 256, 290
Fluid-fluid interface, 205
Fluorescence, 231
  probe, 230
  spectroscopy, 243
Fluorooxalacetate, 143
Forces
  organizing, 19
  pairable, 20
Freeze-etching, 58
Fumarase, 53, 148, 150
Functional groups, 255

G

Galactose, 13
β-Galactose-xylose-serine-peptide, 167
Galactosidase, 106, 166, 167
β-Galactoside, 45
Gel filtration, 241
Gel-included enzymes, 263

Gels
  charged, 282
  embedded, 282
Genetic
  control, 155
Genetic information, 33
Genetic material, 33
Genetic mutants, 36
Genetic systems, 128
Genome, 4, 33, 35, 65
Gland cells, 58
Glandular cells, 58
Gliadin, 208
Globoid, 78
Globosides, 14
Globulin, 208, 260
γ-Globulin, 229
Glucose, 226
  permeability, 220
Glucose oxidase, 263
α-Glucosidase, 167
Glucosylacylsphingosine, 14
β-Glucuronidase, 125, 166
Glutamate dehydrogenase, 121, 142, 143
Glutamic acid dehydrogenase, 45
Glutaraldehyde, 45
Glutathionase, 121
Glycerate pathway, 116
Glycerides, 235
Glycerol, 117, 226
Glycerol phosphate, 145
Glycerol 3-phosphate, 115
  pathway, 115
Glycerolphosphatides, 5, 6, 24
Glycerol 3-phosphoric acid, 116
Glycerophosphate dehydrogenase, 145
Glycine acyltransferase, 155
Glycogen, 41
  stores, 77
Glycogen synthetase, 111
Glycolic oxidase, 53
Glycolipids, 14
Glycolysis, nuclear, 156, 157
Glycoproteins, 30
Glycosidase, 167, 168
Glyoxylate cycle, 52, 54
  enzymes in, 53
Glyoxysomes, 33, 34, 52, 53, 54
  castor bean endosperm, 53
  corn, 53

Phosphatidylethanolamine, 6, 116, 117, 211, 234, 247
Phosphatidyl glycerol, 235
Phosphatidylinositol, 116, 211
Phosphatidylserine, 5, 92, 116, 231, 234
Phosphodiesterase, 101, 121
Phosphofructokinase, 157
Phosphoglyceric kinase, 157
Phospholipase, 17, 116, 128, 144, 208, 209, 243
  dispersion, 129
Phospholipid(s),
  6, 15, 28, 57, 75, 115, 117, 212, 214, 226, 231, 234, 240, 241, 245, 247
  bilayers, 244
  dispersions, 211
  microvesicles, 231
  mitochondrial, 92
  monolayers, 209
  vesicles, 244
  -water interface, 209
Phosphoprotein phosphatase, 75, 167
Phosphorylation, oxidative, 129, 140, 157, 222
Phosphothionate oxidation, 97
Phosphotransferase, 287
Photoaction spectrum, 224
Photochemical reactions, 233
Photoelectric action spectrum, 224
Photoelectromotive force, 224
Photolytic transfer, 247
Photophosphorylation, 34
Photophysical reactions, 233
Photosynthesis, 34, 159
Phytanic acid, 24
Pigments, 221
  photoactive, 224
Pinocytosis, 29, 160
Plant cells, 65
Plant embryo, 65
Plant particles, 169
Plasma albumin, bovine, 231
Plasma membrane, 6, 15, 16, 29, 74, 75, 121, 122, 232–235, 243, 244, 248
  enzymes, 12
  structure, 15
Plasmalemma, 233
Plasmodesmata, 66, 67, 78

Plastids, 233
Plateu–Gibbs border, 219
Pleated sheet, 246
  antiparallel, 246
Polyacrylamide gels, 263
Polyadenylate-synthesizing enzyme, 158
Polyelectrolyte, 19, 289
Polyelectrolyte gel, 282
Polylysine, 211
Poly-L-lysine, 240
Polymerase, 158
Polypeptide hormones, 99
Polypeptides, 207, 221
  helical, 245
Polyribosomes, 35
Polysomes, 34, 41
  bound, 118
Pore, 17, 243
  nuclear, 64
  radii, 17
  size, 17
  structures, 19
Postfixative, 46
Potassium permanganate fixation, 244
Potential energy, 207
Primer specificity, 153
Progesterone, 227
Promitochondria, 52, 55–57, 128
Pronase, 208
Propionamide, 226
Proplastids, 53, 54, 78
Protamine, 19, 211
Proteases, 167, 169
Protein(s), 8, 34, 206, 207, 212, 221, 227
  α-helical, 20
  arrangement, 238
  ascites cells, 20
  bacterial membranes, 20
  basic, 211
  β-conformation, 20
  chain synthesis, 119
  classification, 8
  complexes, 12
  configurations, 19, 244
  conformation, 81
  cross-links, 45
  denaturation, 259
  globular, 19, 243
  helical, 19